Robot Rules

Jacob Turner

Robot Rules

Regulating Artificial Intelligence

Jacob Turner
Fountain Court Chambers
London, UK

ISBN 978-3-319-96234-4   ISBN 978-3-319-96235-1   (eBook)
https://doi.org/10.1007/978-3-319-96235-1

Library of Congress Control Number: 2018951042

This book is written by the author in a personal capacity and none of the opinions herein should be attributed to any other entity or organisation with which the author is associated.

This Palgrave Macmillan imprint is published by the registered company Springer Nature Switzerland AG
The registered company address is: Gewerbestrasse 11, 6330 Cham, Switzerland

To Joanne, Caroline and Jonathan

FOREWORD

This is a very timely, thought-provoking and significant book.

These days, even half-serious newspapers contain at least one article every week, sometimes an article almost every day, on some aspect of the imminent and fundamental changes which are (mostly justifiably) said to be about to be wrought to our private, social and working lives by artificial intelligence, or robots. As with many prospective developments, both the precise nature and extent of the changes which AI will cause and the timing of any changes are to a significant extent a matter of conjecture, and so, there is room for a range of respectable predictions. More troublingly, the more extreme, imminent and confident (almost, one can say, the more unrespectable) any prediction about the future, the greater the prominence of the coverage it receives in the popular media. However, there is force in the point that virtually any discussion of the likely effect of a significant prospective development is to be welcomed, as it plays an essential part in the vital exercise of encouraging us to think about and prepare for that development when it comes to pass. Because the potential changes resulting from artificial intelligence will almost certainly be more revolutionary and more widespread than any development since homo sapiens evolved, these factors are all particularly in point when it comes to AI.

Having said that, in a somewhat paradoxical way, the current sensationalist coverage of the likely effects of artificial intelligence seems almost more to mask, rather than to get people ready for, the extraordinary changes which will result from AI. I think that this is due partly to a sort of novelty inoculation or exhaustion—in other words, the popular media crying wolf too often, too thoughtlessly and too loudly. But an at least equally important factor is, I believe, that the potential effects of AI are so far-reaching in relation to all aspects of our physical, mental, social and moral lives that most people find these changes too challenging to think about in any constructive or practical way. And yet it is both very important and very urgent that we prepare ourselves, both mentally as individuals and structurally as a society, for the AI revolution.

Amidst all the sensationalist, generalised noise, there are a number of much more considered and expert treatments of artificial intelligence, in the form of books and reports. Because the effect of AI is almost certain to be so very far-reaching, there is a need in particular for a considered and informed study of the legal, ethical, and regulatory implications of AI, bearing in mind the many individual areas which are liable to be seriously disrupted, challenged, marginalised or revolutionised as it is rolled out. As Jacob Turner says in this book, the world needs to be as well prepared as it can be for what has been, sensationally if not inaccurately, described as the unstoppable march of the robots—and the sooner we start seriously preparing the better.

A thoughtful and informed book which analyses the implications of current and future developments in AI and how we should plan to deal with them is therefore to be unreservedly welcomed. To write such a book requires a combination of many abilities—including a proper appreciation of the capabilities, functioning, and limits of computer science and technology, a combination of common sense and imagination, an understanding of society, human nature, and economics, and a real appreciation of morality, law and ethics. Not many people have this combination of talents, but any reader of *Robot Rules: Regulating Artificial Intelligence* will, I think, agree that Jacob Turner has demonstrated that he has.

The earlier chapters in this book set the scene and then discuss a number of important and challenging issues of principle and practice which will be thrown up by AI. These chapters include some facts about AI which are not only little known and interesting, but help to explain where we have got to so far. For instance, AI has been with us for well over half a century, in ways which Jacob Turner describes, and this means that we have experience as well as imagination to guide us to the future.

He also explains that AI involves different concepts; indeed, its very definition is a matter of contention, and he provides his own, to my mind rather satisfying, definition.

In addition, when discussing concepts, Jacob Turner brings what could be a dry topic to life by briefly, but illuminatingly, tracing their history and by raising very profound questions. Thus, when considering the question whether robots should have rights, he traces the development of animal rights. And his discussion of the debate as to whether robots can be said to have feelings raises deep metaphysical and moral questions as to the nature of consciousness and compassion, not to mention sex, and even the existence of the human soul. And in the chapter discussing whether robots should have a legal personality, a number of vivid examples are given, including robots in the boardroom and the Random Darknet Shopper.

In two chapters of particular fascination for lawyers, but also for interested non-lawyers, Jacob Turner explains why AI is already starting to require changes to some fundamental legal concepts, such as agency and causation, and he considers how certain principles of liability could be adjusted to incorporate AI—in criminal law, and in negligence, product liability, vicarious liability, contract, insurance and IP in civil law.

There is also an explanation as to how and why AI is an unprecedented technological development, particularly with the advent of unsupervised machine learning—i.e. machines learning without human input (as famously recently occurred with AlphaGo Zero) and no doubt in due course learning from other machines. In a nutshell, it is not only because AI will be so far-reaching in its effects, but also because it will be able to consider issues and resolve them both independently of, and unpredictably to, humans. This gives rise to a host of specific problems, which this book identifies and illuminatingly discusses. In effect, as Jacob Turner suggests, these problems can be divided into three categories, albeit that, at least when it comes to solutions, the three categories will, I think, be interconnected.

First, the issue of rights: should we be granting robots legal personality, like we treat companies, for example, as having a legal personality? To me, the argument that we should do so has real logical attraction. A company cannot act off its own bat: it can only act through humans. AI by contrast, although formed by humans (albeit maybe only indirectly), will be able to act off its own initiative. But the very fact that companies can only act through humans renders the notion of their having legal personality and liability less threatening to our ideas of normality. Giving robots legal personality brings home to us that, at least in some important respects, they are really like artificial people.

Second, the issue of responsibility: who is liable if AI causes any sort of damage, and who owns the intellectual property which is created by AI? If robots are granted a legal personality, the answer may be simple: the robots themselves. If they are not, then these questions become very thorny, but the answer may lie with their creator or vendor, or, if they are altered or not properly maintained, their operator (if there is such a person). As this book explains, it is probable that issues familiar to lawyers, such as foreseeability and remoteness, will come into play in rather new forms.

Third, ethics: how should AI make choices, and are there any categories of decision which AI should not take? This may well be the most difficult and challenging of the question, particularly if one considers the political and military implications. As Jacob Turner says, the biggest question is how humanity should live alongside AI; some experts believe that the survival of the human race could depend on solving this sort of issue. Further, it is an aspect of AI whose resolution particularly requires worldwide agreement and consistency, and worldwide enforcement that is seen to be effective.

Having raised these questions, Jacob Turner discusses them in a readable and thought-provoking way, which demonstrates that he has studied and thought about the technicalities, principles and practicalities in depth. However, he does not blind or bore the reader with too much or too detail or technicality. He focusses, quite rightly, on both principle and practicality. And, while, very sensibly, he does not suggest that there are any quick and easy answers, he raises and discusses the various options and clearly examines their respective pros and cons.

Having discussed these issues, the book, in an important chapter, containing an interesting review of the current state of play in a number of leading countries, discusses regulation and emphasises the need for global, rather than merely national, rules. Rather than leaving the issue to private groups or companies or to judges, Jacob Turner convincingly opts for legislation and accordingly recommends the development of new public institutions in order to formulate or suggest rules and principles on a cross-discipline and cross-border basis, citing domain names and space law as examples. This book is thus aimed at multidisciplinary audience—from lawyers and politicians to engineers and philosophers, not only because every thoughtful and responsible person should be interested in this topic, but also because people with all sorts of different expertise and experience will need to contribute to resolving the issues thrown up by AI.

The book then goes on to examine in two chapters the extent and ways in which both the creators of robots and the robots themselves might be controlled, characteristically giving examples of both the provisions of established rules in other fields and how they were actually agreed. And, as Jacob Turner explains, there has already been much work done on these topics in the field of AI itself, and the effect of that work is clearly and trenchantly summarised and assessed. These two chapters, whose contents may sound rather dry, in fact provide a different, and interesting, perspective on the fundamental issues discussed in the earlier chapters.

The book concludes with an Epilogue which in turn ends with the three sentences "In order to write rules for robots, the challenge is clear. The tools are at our disposal. The question is not whether we can, but whether we will". Thanks to Jacob Turner's book, the tools are now more readily at everyone's disposal, and the likelihood of writing the rules and doing so successfully has been substantially increased.

Temple, London EC4   David Neuberger
August 2018   The Rt. Hon. Lord Neuberger
of Abbotsbury, PC, President of
the UK Supreme Court 2012–2017

Acknowledgements

I have been fortunate to have had the help of many people in writing this book.

The idea for *Robot Rules* came from a lecture which I helped Lord Mance to prepare in 2016, former Deputy President of the UK Supreme Court, when I was his Judicial Assistant (Law Clerk). He was asked to address a conference on the "Future of Law" and the topic we chose was how AI should be regulated. What began as a ten-minute speech has ended two years later as a book spanning several hundred pages.

From the outset, Rob Pilgrim and I shared many stimulating and enjoyable discussions in developing the concept for the book. I owe much to all those who read and provided helpful comments on my drafts, including Oliver Nash, James Tobias, Sean Legassick, Dr. Shahar Avin, Professor José Hernández-Orallo, Matthew Fisher, Nicholas Paisner, Jakob Gleim, Camilla Turner and Gabriel Turner, and several anonymous reviewers.

I have also learned a great deal from exchanges with experts on AI policy including Hal Hodson, Georgia Frances King, Ed Leon Klinger, Dr. Tanya Filer, Guy Cohen, Haydn Bellfield, Rob McCargow, Professor Toby Walsh and Professor Stuart Russell. Jeffrey Ding's insights into China's AI policy were instrumental to the sections on that topic, and Professor Fumio Shimpo provided very helpful information on Japan.

I received valuable feedback on my ideas from the students and researchers on the Masters in Business Technology Course at the Judge Business School, Cambridge University; the Leverhulme Centre for the Future of Intelligence, Cambridge University; Queen's University,

Canada; the Oxford University Computer Science and Law Faculties; and the Future of Humanity Institute, Oxford University. I am very grateful to the conveners of the various courses, seminars and lectures: Assistant Professor Samuel Dahan, Aaron Libbey, Professor Jaideep Prabhu, Neil Gough, Professor Rebecca Williams, Assistant Professor Konatsu Nishigai, Dr. Steven Cave and Dr. Kanta Dihal. In addition, I have found the UK's All Party Parliamentary Group on AI, organized brilliantly by Professor Birgitte Andersen and Niki Iliadis, to be a great source of ideas and debate. I am very grateful to Lord Neuberger, former President of the UK Supreme Court, for kindly agreeing to write the Foreword.

I am indebted to my excellent editorial team at Palgrave Macmillan/ Springer, in particular Kyra Saniewski and Rachel Krause Daniel, who have been extremely helpful and informative throughout the commissioning, writing and editing process.

My parents Jonathan and Caroline continue to inspire me, and I am grateful to them for reading and commenting on drafts throughout the process. I would also like to thank my parents in law, Martin and Susan Paisner for all their support and hospitality; it was under their roof that much of this book was written. Finally, I could not have written this book without the love, patience and proof-reading of my wife Joanne.

London, England   Jacob Turner
August 2018

CONTENTS

CHAPTER 1

Introduction

He had not a minute more to lose. He pulled the axe quite out, swung it with both arms, scarcely conscious of himself, and almost without effort, almost mechanically, brought the blunt side down on her head. He seemed not to use his own strength in this. But as soon as he had once brought the axe down, his strength returned to him.... Then he dealt her another and another blow with the blunt side and on the same spot. The blood gushed as from an overturned glass, the body fell back. He stepped back, let it fall, and at once bent over her face; she was dead.[1]

Fyodor Dostoyevsky, *Crime and Punishment*

Our immediate reaction is emotional: anger, horror, disgust. And then reason sets in. A crime has been committed. A punishment must follow.

Now imagine the perpetrator is not a human, but a robot. Does your response change? What if the victim is another robot? How should society, and the legal system, react?

For millennia, laws have ordered society, kept people safe and promoted commerce and prosperity. But until now, laws have only had one subject: humans. The rise of artificial intelligence (AI) presents novel issues for which current legal systems are only partially equipped. Who or what should be liable if an intelligent machine harms a person or property? Is it ever wrong to damage or destroy a robot? Can AI be made to follow any moral rules?

---

[1] Fyodor Dostoyevsky, *Crime and Punishment*, translated by Constance Garnett (Urbana, IL: Project Gutenberg, 2006), Chapter VII.

© The Author(s) 2019
J. Turner, *Robot Rules*, https://doi.org/10.1007/978-3-319-96235-1_1

1

The best-known answers to any of these questions are Isaac Asimov's Laws of Robotics, from 1942:

First: A robot may not injure a human being or, through inaction, allow a human being to come to harm.

Second: A robot must obey orders given it by human beings except where such orders would conflict with the First Law.

Third: A robot must protect its own existence as long as such protection does not conflict with the First or Second Law.

Fourth: A robot may not harm humanity or, by inaction, allow humanity to come to harm.[2]

But Asimov's rules were never meant to serve as a blueprint for humanity's actual interaction with AI. Far from it, they were written as science fiction and were always intended to lead to problems. Asimov himself said: "These laws are sufficiently ambiguous so that I can write story after story in which something strange happens, in which the robots don't behave properly, in which the robots become positively dangerous".[3] Although they are simple and superficially attractive, it is easy to conceive of situations in which Asimov's Laws are inadequate. They do not say what a robot should do if it is given contradictory orders by different humans. Nor do they account for orders which are iniquitous but fall short of requiring a robot to harm humans, such as commanding a

---

[2]Isaac Asimov, "Runaround", in *I, Robot* (London: HarperVoyager, 2013), 31. Runaround was originally published in *Astounding Science Fiction* (New York: Street & Smith, March 1942). Owing to the potential weaknesses in his first three laws, Asimov later added the Fourth or Zeroth law. See Isaac Asimov, "The Evitable Conflict", *Astounding Science Fiction* (New York: Street & Smith, 1950).

[3]Isaac Asimov, "Interview with Isaac Asimov", interview on Horizon, BBC, 1965, http://www.bbc.co.uk/sn/tvradio/programmes/horizon/broadband/archive/asimov/, accessed 1 June 2018. Asimov made a similar statement in the introduction to his collection *The Rest of Robots*: "[t]here was just enough ambiguity in the Three Laws to provide the conflicts and uncertainties required for new stories, and, to my great relief, it seemed always to be possible to think up a new angle out of the sixty-one words of the Three Laws". Isaac Asimov, *The Rest of Robots* (New York: Doubleday, 1964), 43.

robot to steal. They are hardly a complete code for managing our relationship with AI.

This book provides a roadmap for a new set of regulations, asking not just what the rules should be but—more importantly—who should shape them and how can they be upheld.

There is much fear and confusion surrounding AI and other developments in computing. A lot has already been written on near-term problems including data privacy and technological unemployment.[4] Many writers have also speculated about events in the distant future, such as an AI apocalypse at one extreme,[5] or a time when AI will bring a new age of peace and prosperity, at the other.[6] All these matters are important, but they are not the focus of this book. The discussion here is not about robots taking our jobs, or taking over the world. Our aim is to set out how humanity and AI can coexist.

1   Origins of AI

Modern AI research began on a summer programme at Dartmouth College, New Hampshire, in 1956, when a group of academics and students set out to explore how machines could intelligently think.[7]

---

[4] As to data, see "Data Management and Use: Governance in the 21st Century a Joint Report by the British Academy and the Royal Society", *British Academy and the Royal Society*, June 2017, https://royalsociety.org/~/media/policy/projects/data-governance/data-management-governance.pdf, accessed 1 June 2018. As to unemployment, see Carl Benedikt Frey and Michael A. Osborne, "The Future of Employment: How Susceptible Are Jobs to Computerisation?", *Oxford Martin Programme on the Impacts of Future Technology Working Paper*, September 2013, http://www.oxfordmartin.ox.ac.uk/downloads/academic/future-of-employment.pdf, accessed 1 June 2018. See also Daniel Susskind and Richard Susskind, *The Future of the Professions: How Technology Will Transform the Work of Human Experts* (Oxford: Oxford University Press, 2015).

[5] See Nick Bostrom, *Superintelligence* (Oxford: Oxford University Press, 2014).

[6] See Ray Kurzweil, *The Singularity Is Near: When Humans Transcend Biology* (New York: Viking Press, 2005).

[7] Several nineteenth-century thinkers including Charles Babbage and Ada Lovelace arguably predicted the advent of AI and even prepared designs for machines capable of carrying out intelligent tasks. There is some debate as to whether Babbage actually believed that such a machine was capable of cognition. See, for example, Christopher D. Green, "Charles Babbage, the Analytical Engine, and the Possibility of a 19th-Century Cognitive Science", in *The Transformation of Psychology*, edited by Christopher D. Green, Thomas Teo, and Marlene Shore (Washington, DC: American Psychological Association Press, 2001), 133–152. See also Ada Lovelace, "Notes by the Translator", Reprinted in R.A. Hyman, ed. *Science and Reform: Selected Works of Charles Babbage* (Cambridge: Cambridge University Press, 1989), 267–311.

However, the idea of AI goes back much further.[8] The creation of intelligent beings from inanimate materials can be traced to the very earliest stories known to humanity. Ancient Sumerian creation myths speak of a servant for the Gods being created from clay and blood.[9] In Chinese mythology, the Goddess Nüwa made mankind from the yellow earth.[10] The Judeo-Christian Bible and the Quran have words to similar effect: "And the Lord God formed man of the dust of the ground, and breathed into his nostrils the breath of life; and man became a living soul".[11] In one sense, humans were really the first AI.

In literature and the arts, the idea of technology being used to create sentient assistants for humans or Gods has been around for thousands of years. In Homer's *Iliad*, which dates to around the eighth century BC, Hephaestus the blacksmith is "assisted by servant maids that he had made from gold to look like women".[12] In Eastern European Jewish folklore, there are tales of a rabbi in sixteenth century Prague who created the Golem, a giant human-like figure made from clay, in order to defend his ghetto from anti-Semitic pogroms.[13] In the nineteenth century, Frankenstein's monster brought to the popular imagination the dangers of humans attempting to create or recreate, intelligence through science and technology. In the twentieth century, ever since the term "robot" was popularised by Karel Čapek's screenplay *Rossum's Universal Robots*,[14] there have been many examples of AI in films, television and

[8] What follows is by no means intended to be exhaustive. For a far more comprehensive survey of AI and robotics in popular culture, religion and science, see George Zarkadakis, *In Our Image: Will Artificial Intelligence Save or Destroy Us?* (London: Rider, 2015).

[9] T. Abusch, "Blood in Israel and Mesopotamia", in *Emanuel: Studies in the Hebrew Bible, the Septuagint, and the Dead Sea Scrolls in Honor of Emanuel Tov*, edited by Shalom M. Paul, Robert A. Kraft, Eva Ben-David, Lawrence H. Schiffman, and Weston W. Fields (Leiden, The Netherlands: Brill, 2003), 675–684, especially at 682.

[10] New World Encyclopedia, Entry on Nuwa (quoting Qu Yuan (屈原), book: "Elegies of Chu" (楚辞, or Chuci), Chapter 3: "Asking Heaven" (天問)), http://www.newworldencyclopedia.org/entry/Nuwa, accessed 1 June 2018.

[11] Genesis 2:7, King James Bible.

[12] Homer, *The Iliad*, translated by Herbert Jordan (Oklahoma: University of Oklahoma Press: Norman, 2008), 352.

[13] Eden Dekel and David G. Gurley, "How the Golem Came to Prague", *The Jewish Quarterly Review*, Vol. 103, No. 2 (Spring 2013), 241–258.

[14] The original Czech is "Rossumovi Univerzální Roboti". Roboti translates roughly to "slaves". We will return to this feature in Chapter 4.

other media forms. But now for the first time in human history, these concepts are no longer limited to the pages of books or the imagination of storytellers.

Today, many of our impressions of AI come from science fiction and involve anthropomorphic manifestations that are either friendly or, more usually, unfriendly. These might include the bumbling C-3PO from *Star Wars*, Arnold Schwarzenegger's noble Terminator or the demonic HAL from *2001: A Space Odyssey*.

On the one hand, these humanoid representations of AI constitute a simplified caricature—something to which people can easily relate, but which bears little resemblance to AI technology as it stands. On the other hand, they represent a paradigm which has influenced and shaped AI as successive generations of programmers are inspired to attempt to recreate versions of entities from books, films and other media. In the field of AI, first science then life imitates art. In 2017, Neuralink, a company backed by serial technology entrepreneur Elon Musk, announced that it was developing a "neural lace" interface between human brain tissue and artificial processors.[15] Neural lace is—by Musk's own admission—heavily influenced by the writings of science fiction authors including in particular the *Culture* novels of Iain M. Banks.[16] Technologists have taken inspiration from stories found in faith as well as popular culture: Robert M. Geraci argues that, "[t]o understand robots, we must understand how the history of religion and the history of science have twined around each other, quite often working towards the same ends and quite often influencing another's methods and objectives".[17]

---

[15] "Homepage", Neuralink Website, https://www.neuralink.com/, accessed 1 June 2018; Chantal Da Silva, "Elon Musk Startup 'to Spend £100m' Linking Human Brains to Computers", *The Independent*, 29 August 2017, http://www.independent.co.uk/news/world/americas/elon-musk-neuralink-brain-computer-startup-a7916891.html, accessed 1 June 2018. For commentary on Neuralink, see Tim Urban's provocative blog post "Neuralink and the Brain's Magical Future", *Wait But Why*, 20 April 2017, https://waitbutwhy.com/2017/04/neuralink.html, accessed 1 June 2018.

[16] Tim Cross, "The Novelist Who Inspired Elon Musk", *1843 Magazine*, 31 March 2017, https://www.1843magazine.com/culture/the-daily/the-novelist-who-inspired-elon-musk, accessed 1 June 2018.

[17] Robert M. Geraci, *Apocalyptic AI: Visions of Heaven in Robotics, Artificial Intelligence, and Virtual Reality* (New York: Oxford University Press, 2010), 147.

Although popular culture and religion have helped to shape the development of AI, these portrayals have also given rise to a misleading impression of AI in the minds of many people. The idea of AI as only meaning humanoid robots which look, sound and think like us, is mistaken. Such conceptions of AI make its advent appear to be distant, given that no technology at present comes remotely close to resembling the type of human-level functionality made familiar by science fiction.

The lack of a universal definition for AI means that those attempting to discuss it may end up speaking at cross-purposes. Therefore, before it is possible to demonstrate the spreading influence of AI or the need for legal controls, we must first set out what we mean by this term.

2   NARROW AND GENERAL AI

It is helpful at the outset to distinguish two classifications for AI: narrow and general.[18] Narrow (sometimes referred to as "weak") AI denotes the ability of a system to achieve a certain stipulated goal or set of goals, in a manner or using techniques which qualify as intelligent (the meaning of "intelligence" is addressed below). These limited goals might include natural language processing functions like translation, or navigating through an unfamiliar physical environment. A narrow AI system is suited only to the task for which it is designed. The great majority of AI systems in the world today are closer to this narrow and limited type.

General (or "strong") AI is the ability to achieve an unlimited range of goals, and even to set new goals independently, including in situations of uncertainty or vagueness. This encompasses many of the attributes we think of as intelligence in humans. Indeed, general AI is what we see portrayed in the robots and AI of popular culture discussed above. As yet, general AI approaching the level of human capabilities does not exist and some have even cast doubt on whether it is possible.[19]

---

[18] For the distinction, see David Weinbaum and Viktoras Veitas, "Open Ended Intelligence: The Individuation of Intelligent Agents", *Journal of Experimental & Theoretical Artificial Intelligence*, Vol. 29, No. 2 (2017), 371–396.

[19] See Roger Penrose, *The Emperor's New Mind: Concerning Computers, Minds, and the Laws of Physics* (Oxford: Oxford University Press, 1989). The number of sceptics may be shrinking. As Wallach and Allen comment: "pessimists tend to get weeded out of the profession", Wendell Wallach and Colin Allen, *Moral Machines: Teaching Robots Right from Wrong* (Oxford: Oxford University Press, 2009), 68. For instance, Margaret Boden was one of the most well-known proponents of the sceptical view, although in her latest work, Margaret Boden, *AI: Its nature and Future* (Oxford: Oxford University Press, 2016), 119

Narrow and general AI are not hermetically sealed from each other. They represent different points on a continuum. As AI becomes more advanced, it will move further away from the narrow paradigm and closer to the general one.[20] This trend may be hastened as AI systems learn to upgrade themselves[21] and acquire greater capabilities than those with which they were originally programmed.[22]

3   Defining AI

The word "artificial" is relatively uncontroversial. It means something synthetic and which does not occur in nature. The key difficulty is with the word "intelligence", which can describe a range of attributes or abilities. As computer science expert and futurist Jerry Kaplan says, the question "what is artificial intelligence?" is an "easy question to ask and a hard one to answer" because "there's little agreement about what intelligence is".[23]

Some have suggested that the lack of general agreement on a definition of AI is beneficial. The authors of Stanford University's *One Hundred Year Study on Artificial Intelligence* state:

> Curiously, the lack of a precise, universally accepted definition of AI probably has helped the field to grow, blossom, and advance at an ever-accelerating pace. Practitioners, researchers, and developers of AI are instead guided by a rough sense of direction and an imperative to "get on with it".[24]

---

et seq she acknowledges the potential for "real" artificial intelligence, but maintains that "...*no one knows* for sure, whether [technology described as Artificial General Intelligence] could really be intelligent".

[20] See further Chapter 3 at s. 2.1.2.

[21] As to AI systems developing the capacity to self-improve, see further FN 114 below and more generally Chapter 2 at s. 3.2.

[22] Our prediction for the process of narrow AI gradually coming closer to general AI is similar to evolution. *Homo sapiens* did not appear overnight as if by magic. Instead, we developed iteratively through a series of gradual upgrades to our hardware (bodies) and software (minds) on the basis of trial and error experiments, otherwise known as natural selection.

[23] Jerry Kaplan, *Artificial Intelligence: What Everyone Needs to Know* (New York: Oxford University Press, 2016), 1.

[24] Peter Stone et al., "Defining AI", in *"Artificial Intelligence and Life in 2030". One Hundred Year Study on Artificial Intelligence: Report of the 2015–2016 Study Panel* (Stanford, CA: Stanford University, September 2016), http://ai100.stanford.edu/2016-report, accessed 1 June 2018.

Defining AI can resemble chasing the horizon: as soon as you get to where it was, it has moved somewhere into the distance. In the same way, many have observed that AI is the name we give to technological processes which we do not understand.[25] When we have familiarised ourselves with a process, it stops being called AI and becomes just another clever computer programme. This phenomenon is known as the "AI effect".[26]

Rather than asking "what is AI?" it is better to start with the question: "why do we need to define AI at all?" Many books are written on energy, medicine and other general concepts which do not start with a chapter on the definition of these terms.[27] In fact, we go through life with a functional understanding of many abstract notions and ideas without necessarily being able to describe them perfectly. Time, irony and happiness are just a few examples of concepts that most people understand but would find difficult to define. Justice Potter Stewart of the US Supreme Court once said that he could not define hardcore pornography "But I know it when I see it".[28]

However, when considering how to regulate AI, it is not sufficient to follow Justice Stewart. In order for a legal system to function effectively, its subjects must be able to understand the ambit and application of its rules. To this end, legal theorist Lon L. Fuller set out eight formal requirements for a system of law to satisfy certain basic moral norms—principally that humans have an opportunity to engage with them and shape their behaviour accordingly. Fuller's desiderata include requirements that law should be promulgated so that citizens know the standards to which they are being held, and that laws should be understandable.[29] To pass Fuller's tests, legal systems must use specific and workable definitions

[25] Pamela McCorduck, *Machines Who Think: A Personal Inquiry into the History and Prospects of Artificial Intelligence* (Natick, MA: A.K. Peters, 2004), 133.

[26] Peter Stone et al., "Defining AI", in *"Artificial Intelligence and Life in 2030". One Hundred Year Study on Artificial Intelligence: Report of the 2015–2016 Study Panel* (Stanford, CA: Stanford University, September 2016), http://ai100.stanford. edu/2016-report, accessed 1 June 2018. See also Pamela McCorduck, *Machines Who Think: A Personal Inquiry into the History and Prospects of Artificial Intelligence* (Natick, MA: A.K. Peters, 2004), 204.

[27] The same observation might be made of law itself. See H.L.A. Hart, *The Concept of Law* (2nd edn. Oxford: Clarendon, 1997).

[28] *Jacobellis v. Ohio*, 378 U.S. 184 (1964), 197.

[29] Lon L. Fuller, *The Morality of Law* (New Haven, CT: Yale University Press, 1969).

when describing the conduct and phenomena which are subject to regulation. As Fuller says: "We need to share the anguish of the weary legislative draftsman who at 2:00 a.m. says to himself 'I know this has got to be right, and if it isn't people may be hauled into Court for things we don't mean to cover at all. But for how long must I go on rewriting it?'".[30]

In short, people cannot choose to comply with rules they do not understand. If the law is impossible to know in advance, then its role in guiding action is diminished if not destroyed. Unknown laws become little more than tools of the powerful. They can lead ultimately to the absurd and frightening scenario imagined in Kafka's *The Trial*, where the protagonist is accused, condemned and ultimately executed for a crime which is never explained to him.[31]

Most of the universal definitions of AI that have been suggested to date fall into one of two categories: human-centric and rationalist.[32]

3.1 Human-Centric Definitions

Humanity has named itself *homo sapiens*: "wise man". It is therefore perhaps unsurprising that some of the first attempts at defining intelligence in other entities referred to human characteristics. The most famous

---

[30] Ibid., 107.

[31] Franz Kafka, *The Trial*, translated by Idris Parry (London: Penguin Modern Classics, 2000).

[32] Stuart Russell and Peter Norvig divide definitions into four categories: (i) thinking like a human: AI systems adopt similar thought processes to human beings; (ii) acting like a human: AI systems are behaviourally equivalent to human beings; (iii) thinking rationally: AI systems have goals and reason their way towards achieving those goals; (iv) acting rationally: AI systems act in a manner that can be described as goal-directed and goal-achieving. Stuart Russell and Peter Norvig, *Artificial Intelligence: International Version: A Modern Approach* (Englewood Cliffs, NJ: Prentice Hall, 2010), para. 1.1 (hereafter "Russell and Norvig, *Artificial Intelligence*"). However, John Searle's "Chinese Room" thought experiment demonstrates the difficulty of distinguishing between acts and thoughts. In short, the Chinese Room experiment suggests that we cannot distinguish between intelligence of Russell and Norvig's types (i) and (ii), or types (iii) and (iv) John R. Searle, "Minds, Brains, and Programs", *Behavioral and Brain Sciences*, Vol. 3, No. 3 (1980), 417–457. Searle's experiment has been met with various numbers of replies and criticisms, which are set out in the entry on The Chinese Room Argument, Stanford Encyclopedia of Philosophy, First published 19 March 2004; substantive revision 9 April 2014, https://plato.stanford.edu/entries/chinese-room/, accessed 1 June 2018.

example of a human-centric definition of AI is known popularly as the "Turing Test".

In a seminal 1950 paper, Alan Turing asked whether machines could think. He suggested an experiment called the "Imitation Game".[33] In the exercise, a human invigilator must try to identify which of the two players is a man pretending to be a woman, using only written questions and answers. Turing proposed a version of the game in which the AI machine takes the place of the man. If the machine is able to succeed in persuading the invigilator not only that it is human but also that it is the female player, then it has demonstrated intelligence.[34] Modern versions of the Imitation Game simplify the task by asking a computer program as well as several human blind control subjects to each hold a five-minute typed conversation with a panel of human judges in a different room. The judges have to decide whether or not the entity with which they are corresponding is a human; if the computer can fool a sufficient proportion of them (a popular competition sets this at just 30%), then it has won.[35]

A major problem with Turing's Imitation Game is that it tests only the ability to mimic a human in typed conversation, and that skilful impersonation does not equate to intelligence.[36] Indeed, in some of the more "successful" tests of programmes designed to succeed in the Imitation Game, the programmers prevailed by creating a computer which exhibited frailties which we tend to associate with humans, such as spelling errors.[37] Another tactic favoured by programmers in modern Turing tests is to use stock humorous responses so as to deflect attention

---

[33] Alan M. Turing, "Computing Machinery and Intelligence", *Mind: A Quarterly Review of Psychology and Philosophy*, Vol. 59, No. 236 (October 1950), 433–460, 460.

[34] Yuval Harari has offered the interesting explanation that the form of Turing's Imitation Game resulted in part from Turing's own need to suppress his homosexuality, to fool society and the authorities into thinking he was something that he was not. The focus on gender and subterfuge in the first iteration of the test is, perhaps, not accidental. Yuval Harari, *Homo Deus* (London: Harvill Secker, 2016), 120.

[35] See, for example, the website of The Loebner Prize in Artificial Intelligence, http://www.loebner.net/Prizef/loebner-prize.html, accessed 1 June 2018.

[36] José Hernández-Orallo, "Beyond the Turing Test", *Journal of Logic, Language and Information*, Vol. 9, No. 4 (2000), 447–466.

[37] "Turing Test Transcripts Reveal How Chatbot 'Eugene' Duped the Judges", Coventry University, 30 June 2015, http://www.coventry.ac.uk/primary-news/turing-test-transcripts-reveal-how-chatbot-eugene-duped-the-judges/, accessed 1 June 2018.

away from their program's lack of substantive answers to the judges' questions.[38]

To avoid the deficiencies in Turing's test, others have suggested definitions of intelligence which do not rely on the replication of *one* aspect of human behaviour or thought and are instead parasitic on society's vague and shifting notion of what makes humans intelligent. Definitions of this type are often variants of the following: "AI is technology with the ability to perform tasks that would otherwise require human intelligence".[39]

The inventor of the term AI, John McCarthy, has said that there is not yet "a solid definition of intelligence that doesn't depend on relating it to human intelligence".[40] Similarly, futurist Ray Kurzweil wrote in 1992 that the most durable definition of AI is "[t]he art of creating machines that perform functions that require intelligence when performed by people".[41] The main problem with parasitic tests is that they

---

[38]Various competitions are now held around the world in an attempt to find a 'chatbot', as conversational programs are known, which is able to pass the Imitation Game. In 2014, a chatbot called 'Eugene Goostman', which claimed to be a 13-year-old Ukrainian boy, convinced 33% of the judging panel that he was a human, in a competition held by the University of Reading. Factors which assisted Goostman included that English (the language in which the test was held) was not his first language, his apparent immaturity and answers which were designed to use humour to deflect the attention of the questioner from the accuracy of the response. Unsurprisingly, the world did not herald a new age in AI design. For criticism of the Goostman 'success', see Celeste Biever, "No Skynet: Turing Test 'Success' Isn't All It Seems", *The New Scientist*, 9 June 2014, http://www.newscientist.com/article/dn25692-no-skynet-turing-test-success-isnt-all-it-seems.html, accessed 1 June 2018. The author Ian McDonald offers another objection: "Any AI smart enough to pass a Turing test is smart enough to know to fail it". Ian McDonald, *River of Gods* (London: Simon & Schuster, 2004), 42.

[39]This definition is adapted from that used by the UK Department for Business, Energy and Industrial Strategy, *Industrial Strategy: Building a Britain Fit for the Future* (November 2017), 37, https://www.gov.uk/government/uploads/system/uploads/attachment_data/file/664563/industrial-strategy-white-paper-web-ready-version.pdf, accessed 1 June 2018.

[40]"What Is Artificial Intelligence?", Website of John McCarthy, last modified 12 November 2007, http://www-formal.stanford.edu/jmc/whatisai/node1.html, accessed 1 June 2018.

[41]Ray Kurzweil, *The Age of Intelligent Machines* (Cambridge, MA: MIT Press, 1992), Chapter 1.

are circular. Kurzweil admitted that his own definition, "... does not say a great deal beyond the words 'artificial intelligence'".[42]

In 2011, Nevada adopted the following human-centric definition for the purpose of legislation regulating self-driving cars: "the use of computers and related equipment to enable a machine to duplicate or mimic the behavior of human beings".[43] The definition was repealed in 2013 and replaced with a more detailed definition of "autonomous vehicle", which was not tied to human actions at all.[44]

Although it is no longer on the statute books, Nevada's 2011 law remains an instructive example of why human-centric definitions of intelligence are flawed. Like many human-centric approaches, this was both over- and under-inclusive. It was over-inclusive because humans do many things which are not "intelligent". These include getting bored, tired or frustrated, as well as making mistakes such as forgetting to indicate when changing lanes. Furthermore, many cars already have non-AI features which could fall within this definition. For instance, automatic headlights which turn on at night would be mimicking the behaviour of a human being turning the lights on manually, but the behaviour would have been triggered by nothing more complex or mysterious than a light sensor coupled to simple logic gate.[45]

[42] Ibid.

[43] NV Rev Stat § 482A.020 (2011), https://law.justia.com/codes/nevada/2011/chapter-482a/statute-482a.020/, accessed 1 June 2018.

[44] For the new law, see NRS 482A.030. "Autonomous vehicle" now means a motor vehicle that is equipped with autonomous technology (Added to NRS by 2011, 2876; A 2013, 2010). NRS 482A.025 "Autonomous technology" means technology which is installed on a motor vehicle and which has the capability to drive the motor vehicle without the active control or monitoring of a human operator. The term does not include an active safety system or a system for driver assistance, including without limitation, a system to provide electronic blind spot detection, crash avoidance, emergency braking, parking assistance, adaptive cruise control, lane keeping assistance, lane departure warning, or traffic jam and queuing assistance, unless any such system, alone or in combination with any other system, enables the vehicle on which the system is installed to be driven without the active control or monitoring of a human operator (Added to NRS by 2013, 2009). Chapter 482A—Autonomous Vehicles, https://www.leg.state.nv.us/NRS/NRS-482A.html, accessed 1 June 2018.

[45] Ryan Calo, "Nevada Bill Would Pave the Road to Autonomous Cars", *Centre for Internet and Society Blog*, 27 April 2011, http://cyberlaw.stanford.edu/blog/2011/04/nevada-bill-would-pave-road-autonomous-cars, accessed 1 June 2018.

The 2011 Nevada definition was also under-inclusive because there are various emergent qualities that computer programs can display which go well beyond human capabilities. The manner in which humans solve problems is limited by the hardware available to us: our brains. AI has no such limits. DeepMind's AlphaGo program achieved superhuman capabilities in Chess, Go, and other board games. DeepMind CEO Demis Hassabis explained: "It doesn't play like a human, and it doesn't play like a program, it plays in a third, almost alien, way".[46] At a sufficient point of advancement, it will no longer be accurate to describe AI as duplicating or mimicking the behaviour of humans—it will have surpassed us.

3.2 Rationalist Definitions

More recent AI definitions avoid the link to humanity by focussing on thinking or acting *rationally*. To think rationally means that an AI system has goals and reasons towards these goals. To act rationally is for the AI systems to perform in a manner that can be described as goal-directed.[47] In this vein, Nils J. Nilsson says intelligence is "that quality that enables an entity to function appropriately and with foresight in its environment".[48]

---

[46] Will Knight, "Alpha Zero's "Alien" Chess Shows the Power, and the Peculiarity, of AI", *MIT Technology Review*, https://www.technologyreview.com/s/609736/alpha-zeros-alien-chess-shows-the-power-and-the-peculiarity-of-ai/, accessed 1 June 2018. See for the academic paper: David Silver, Thomas Hubert, Julian Schrittwieser, Ioannis Antonoglou, Matthew Lai, Arthur Guez, Marc Lanctot, Laurent Sifre, Dharshan Kumaran, Thore Graepel, Timothy Lillicrap, Karen Simonyan, and Demis Hassabis, "Mastering Chess and Shogi by Self-Play with a General Reinforcement Learning Algorithm", *Cornell University Library Research Paper*, 5 December 2017, https://arxiv.org/abs/1712.01815, accessed 1 June 2018. See also Cade Metz, What the AI Behind AlphaGo Can Teach Us About Being Human", *Wired*, 19 May 2016, https://www.wired.com/2016/05/google-alpha-go-ai/, accessed 1 June 2018.

[47] Russell and Norvig, *Artificial Intelligence*, para. 1.1.

[48] Nils J. Nilsson, *The Quest for Artificial Intelligence: A History of Ideas and Achievements* (Cambridge, UK: Cambridge University Press, 2010), Preface. Similarly, Shane Legg (one of the co-founders of the leading AI company DeepMind), writing with his doctoral supervisor Professor Marcus Hutter, also supports a rationalist definition of intelligence: "Intelligence measures an agent's ability to achieve goals in a wide range of environments". Shane Legg, "Machine Super Intelligence" (Doctoral Dissertation submitted to the Faculty of Informatics of the University of Lugano in partial fulfillment of the requirements for the degree of Doctor of Philosophy, June 2008).

Although rationalist definitions are suitable to describe narrow AI systems which have a known set of functions or aims, later developments may come to pose problems.[49] This is because rationalist definitions of AI are often premised, whether implicitly or explicitly, on the existence of external goals for the AI. The difficulty which may arise when applying such definitions to more advanced, general AI is that it is unlikely to have static goals by which its behaviour or computational processes can be assessed. Indeed, the existence of static goals is arguably anathema to the idea of all-purpose AI. Unsupervised machine learning by its nature does not have a set goal, except perhaps at a high level of abstraction—for instance to "sort data and recognise patterns".[50] The same can be said of AI systems which are capable of rewriting their own source code. Thus, whilst rationalist definitions of intelligence are adopted now by many in the AI community, they may not be appropriate to tomorrow's technology.

Another type of rationalist definition for AI focusses on "doing the right thing at the right time".[51] This too is flawed. Having the quality of *intelligence* is not the same as selecting the option which is deemed the most *intelligent* in any given situation. *First*, it is likely to be impossible to know what the "right thing" is without (a) possessing an infallible moral system, which does not exist, and (b) having a perfect knowledge of the outcomes of a given action. Just as humans can be intelligent but also fallible, an entity which possesses the quality of AI may not always select the best outcome (whatever "best" might mean). Indeed, if AI was automatically imbued with an ability to always to the right thing, then there would be little need to regulate it.

Secondly, a test which relies on an entity doing the right thing at the right time tends to anthropomorphise the program or entity in question, by imposing human volitions and motivations on to it. This leads to the results of that test being over-inclusive. As the leading AI text book authors Stuart Russell and Peter Norvig point out, a clock which is designed to update its time when its wearer changes time zone would be displaying "successful behaviour" (or doing the right thing), but

---

[49] Another way of putting this is that rationalist definitions are appropriate for narrow AI but are less well suited to general AI.

[50] For a discussion of unsupervised machine learning, see Chapter 2 at s. 3.2.1.

[51] See, for example, Stuart Russell and Eric Wefald, *Do the Right Thing: Studies in Limited Rationality* (Cambridge, MA: MIT Press, 1991).

nonetheless it seems to fall somewhat short of true intelligence. Russell and Norvig explain: "...the intelligence in question belongs to the clock's designer, rather than to the clock itself".[52]

3.3 The Sceptics

Sceptics doubt the possibility of a universal definition for intelligence. Robert Sternberg, a psychologist, is reported to have said "there seem[ed] to be almost as many definitions of intelligence as there were experts asked to define it".[53] Edwin G. Boring, another psychologist, wrote "[i]ntelligence is what is measured by intelligence tests".[54] At first glance, Sternberg and Boring's points may seem glib. In fact, they contain important insights. Boring shows that the quality of intelligence can differ depending on what the person seeking to define it is, or setting the test, is looking for. Sternberg made a similar observation: different experts look for different things, meaning that it is of little use comparing their tests side by side.

3.4 Our Definition

Unlike most of the examples above, this book does not seek to lay down a universal, all-purpose definition of AI which can be applied in any context. Its aim is much less ambitious: to arrive at a definition which is suited to the legal regulation of AI. One of the main principles of legal interpretation is to find out the purpose of the speaker.[55] Our purpose is to regulate AI. In order to regulate AI, we must therefore ask: what is the unique factor of AI that *needs* regulation?

In this book, intelligence is used to refer to the ability to make choices. It is the nature of these choices—and their effect on the world—which is our key concern.

---

[52] Russell and Norvig, *Artificial Intelligence*, paras. 2.3, 35.

[53] Robert Sternberg, quoted in Richard Langton Gregory, *The Oxford Companion to the Mind* (Oxford: Oxford University Press, 2004), 472.

[54] Ernest G. Boring, "Intelligence As the Tests Test It", *New Republic*, Vol. 35 (1923), 35–37.

[55] See, for example, Aharon Barak, *Purposive Interpretation in Law*, translated by Sari Bashi (Princeton, NJ: Princeton University Press, 2007).

Our definition of AI is therefore as follows:
Artificial Intelligence Is the Ability of a Non-natural Entity to Make Choices by an Evaluative Process

We will use the term "robot" to refer to a physical entity or system which uses AI. Although the word robot is frequently used to describe *any* type of automation of a process by a machine, here we add an extra requirement that the action is carried out by an entity using AI.[56]

As to the "artificial" part of the definition, "non-natural" is preferable to "man-made" because of the propensity of AI to design and create other AI. At some point, mankind may drop out of the picture. This is one of the emergent features of AI which means that it requires novel legal treatment, in that the chain of causation between AI and its original human "creator" can no longer be sustained.[57]

It is implicit in the definition's reference to making choices that such decisions be autonomous: self-governing.[58] *Autonomy* (from the Greek

---

[56]Elsewhere, the terms "robots" and "robotics" are sometimes used to describe any type of automation, whether involving AI or not (see, for example, definition of "robot" in the Merriam-Webster Dictionary, https://www.merriam-webster.com/dictionary/robot, accessed 1 June 2018. This book's definition is closer to the original meaning of the term "robot"—as intelligent servants—as used by Capek (see FN 14 above). Others have taken a contrary view: that artificial intelligence cannot exist without physical embodiment. See Ryan Calo, "Robotics and the Lessons of Cyberlaw", *California Law Review*, Vol. 103 (2015), 513–563, 529: "A robot in the strongest, fullest sense of the term exists in the world as a corporeal object with the capacity to exert itself physically". See also Jean-Christophe Baillie, "Why AlphaGo Is Not AI", *IEEE Spectrum*, 17 March 2016, https://spectrum.ieee.org/automaton/robotics/artificial-intelligence/why-alphago-is-not-ai, accessed 1 June 2018.

[57]As to the unique nature of this aspect of AI, see further Chapter 2.

[58]The Society of Automobile Engineers has provided a useful primer of five levels of autonomy for autonomous vehicles. These are as follows:

Level 0—No Automation: The full-time performance by the human driver of all aspects of the dynamic driving task, even when enhanced by warning or intervention systems; Level 1—Driver Assistance: The driving mode-specific execution by a driver assistance system of either steering or acceleration/deceleration using information about the driving environment and with the expectation that the human driver performs all remaining aspects of the dynamic driving task; Level 2—Partial Automation: The driving mode-specific execution by one or more driver assistance systems of both steering and acceleration/deceleration using information about the driving environment and with the expectation that the human driver performs all remaining aspects of the dynamic driving task; Level 3—Conditional Automation: The driving mode-specific performance by an Automated Driving System of all aspects of the dynamic driving task with the expectation that the human driver will respond appropriately to a request to intervene; Level 4—

auto: self, and *nomos*: law) is different from *automation*, where a process is repeated by a machine. Autonomy does not require that AI instigates its own functioning; it can make an autonomous choice even if has interacted with a human in taking that decision. For instance, if a human types a query into a search engine, she has clearly had a causal impact on the AI functioning, and indeed, the AI might take into account her preferences in returning search engine results (based on her past searches, as well as many other variables such as her age or location). But ultimately the *choice* of what results are displayed remains that of the search engine.[59]

Turning to the final aspect of this book's definition, an "evaluative process" is one where principles are weighed against each other before a conclusion is reached. Principles can be contrasted with rules. Rules are applicable in an "are-all-or-nothing" fashion.[60] When a valid rule applies in a given case, it is conclusive. If two rules conflict, then one of

High Automation: The driving mode-specific performance by an Automated Driving System of all aspects of the dynamic driving task, even if a human driver does not respond appropriately to a request to intervene; Level 5—Full Automation: The full-time performance by an Automated Driving System of all aspects of the dynamic driving task under all roadway and environmental conditions that can be managed by a human driver.

To put this book's definition into context, AI might be displayed even at Level 1—provided the system is making choices based on evaluative principles, even if this is only within a narrow sphere, and even if it is only providing advice to the human driver. Of course, the more potential for human oversight of the process, the less need there will be for a separate legal regime, but the same principles apply nonetheless. More difficult questions apply at level 2 onwards, where power is actually delegated to the AI system. See SAE International, J3016, https://www.sae.org/misc/pdfs/automated_driving.pdf, accessed 1 June 2018.

This classification was adopted by the US Department of Transport in September 2016. SAE, "U.S. Department of Transportation's New Policy on Automated Vehicles Adopts SAE International's Levels of Automation for Defining Driving Automation in On-Road Motor Vehicles", *SAE Website*, https://www.sae.org/news/3544/, accessed 1 June 2018.

[59] In his discussion of how robots are to be regulated, Bertolini has eschewed a definition of "robots", calling this a pointless exercise, and instead focussed on autonomy as the relevant criterion justifying special legal treatment. However, in seeking to describe autonomy, Bertolini relies on undefined and highly debated concepts, including "self-awareness or self-consciousness, leading to free will and thus identifying a moral agent", and "the ability to intelligently interact in the operating environment". In so doing, Bertolini avoids the key question of what it is that should be regulated. Andrea Bertolini, "Robots as Products: The Case for a Realistic Analysis of Robotic Applications and Liability Rules", *Law Innovation and Technology*, Vol. 5, No. 2 (2013), 214–247, 217–221.

[60] Ronald Dworkin, "The Model of Rules", *The University of Chicago Law Review*, Vol. 35 (1967), 14, 14–46, 25.

them cannot be a valid rule. Principles give justificatory support to various courses of actions, but they are not necessarily conclusive. Unlike rules, principles have "weight". When valid principles conflict, the proper method for resolving the conflict is to select the position that is supported by the principles that have the greatest aggregate weight.[61]

To illustrate the difference between systems involving principles (requiring evaluation) and rules (which do not), it is necessary to describe in very brief terms two types of technologies which have traditionally been described as intelligent.

In "symbolic AI", sometimes known as "Good Old Fashioned AI",[62] programs consist of logical decision trees (in the format: if X, then Y).[63] The decision trees are a set of rules or instructions as to what to do with a given input. Complex examples are known as "expert systems". When programmed with a set of rules, expert systems use deductive reasoning to follow the decision tree through a series of yes or no answers so as to arrive at a predetermined final output.[64] The decision-making process is deterministic, meaning that each step can in theory be traced back to decisions made by a programmer no matter how numerous the stages.

Artificial neural networks are computer systems made up of large number of interconnected units, each of which can usually compute only one thing.[65] Whereas conventional networks fix the architecture before training starts, artificial neural networks use "weights" in order to

---

[61] Ibid., and see also Scott Shapiro, "The Hart-Dworkin Debate: A Short Guide for the Perplexed", *Working Paper No. 77*, University of Michigan Law School, 9, https://law.yale.edu/system/files/documents/pdf/Faculty/Shapiro_Hart_Dworkin_Debate.pdf, accessed 1 June 2018.

[62] Another term for this technology is "classical AI".

[63] Though not an exact match, programs described as classical or symbolic AI (sometimes referred to as "Good Old Fashioned AI", see Margaret Boden, *AI: Its Nature and Future* (Oxford: Oxford University Press, 2016), 6–7, bear somewhat more resemblance to the decision tree format than do programs based on neural networks—the other main branch of AI technology.

[64] For a discussion of the distinction between systems based on "Good Old Fashioned AI" versus neural networks, see Lefteri H. Tsoukalas and Robert E. Uhrig, *Fuzzy and Neural Approaches in Engineering* (New York, NY: Wiley, 1996).

[65] Originally, they were inspired by the functioning of brains.

determine the connectivity between inputs and outputs.[66] Artificial neural networks can be designed to alter themselves by changing the weights on the connections which makes activity in one unit more or less likely to excite activity in another unit.[67] In "machine learning" systems, the weights can be re-calibrated by the system over time—often using a process called backpropagation—in order to optimise outcomes.[68]

Broadly, symbolic programs are not AI under this book's functional definition, whereas neural networks and machine learning systems are AI.[69] Like Russell and Norvig's clock, any intelligence reflected in a symbolic system is that of the programmer and not the system itself.[70] By contrast, the independent ability of neural networks to determine weights between connections is an evaluative function characteristic of intelligence.

Neural networks and machine learning are techniques which fall within this book's definition of AI, but they are not the only technologies capable of doing so. This book's definition of AI is intended to cover neural networks but also to be sufficiently flexible to encompass also other technologies which may become more prevalent in the future—one example being whole brain emulation (the science of attempting to map and then reproduce the entire structure of an animal brain).

[66] Song Han, Jeff Pool, John Tran, and William Dall, "Learning Both Weights and Connections for Efficient Neural Network", *Advances in Neural Information Processing Systems* (2015), 1135–1143, http://papers.nips.cc/paper/5784-learning-both-weights-and-connections-for-efficient-neural-network.pdf, accessed 1 June 2018.

[67] Margaret Boden, "On Deep Learning, Artificial neural Networks, Artificial Life, and Good Old-Fashioned AI", Oxford University Press Website, 16 June 2016, https://blog.oup.com/2016/06/artificial-neural-networks-ai/, accessed 1 June 2018.

[68] David E. Rumelhart, Geoffrey E. Hinton, and Ronald J. Williams, "Learning Representations by Back-Propagating Errors", *Nature*, Vol. 323 (9 October 1986), 533–536.

[69] Admittedly, setting up a hard distinction between symbolic AI and neural networks may be a false dichotomy, as there are systems which utilise both elements. In those situations, provided that the neural network, or other evaluative process, has a determinative effect on the choice made, then the entity as a whole will pass the test for intelligence under this book's definition.

[70] Karnow adopts a similar distinction, describing "expert" versus "fluid" systems. The latter, he says, necessitate different legal treatment, based on their unpredictability. Curtis E.A. Karnow, "Liability for Distributed Artificial Intelligences", *Berkeley Technology Law Journal*, Vol. 147 (1996), 11, http://scholarship.law.berkeley.edu/btlj/vol11/iss1/3, accessed 1 June 2018.

This functional definition may be under-inclusive from the perspective of those seeking a universal measure of intelligence. Unlike most other definitions, it does not attempt to encompass all the technologies which have traditionally been described as "intelligent". However, as noted above, the intention is only to cover those aspects of technology which are salient from a legal perspective. Chapter 2 will discuss features of AI as defined in this book which make it unique as a phenomenon; expert systems would not meet this threshold.

In addition, the functional definition could also be seen as over-inclusive. Although there are debates as to whether general intelligence must include features such as imagination, emotions or consciousness, these capabilities are not relevant to the majority of aspects of AI which need to be regulated.[71] Regulation is needed where AI has an impact on the world, and it can do so even without these additional features.[72]

The functional definition does not offer a simple "yes or no" answer as to whether any given piece of technology has AI or not. However, it is common for there to be some uncertainty at the outer boundaries of any legislative ordinance. This is the result of the inherently imprecise nature of language.[73] For instance, a sign might stipulate "no vehicles

[71] The situation is slightly different with regard to "rights" for AI, which we discuss in Chapter 4. As we explain there, certain rights might best be reserved to AI which is indeed conscious and can suffer. However, the better way to account for this issue is not to say that an entity is not AI *unless* it can suffer, but rather to say that AI *which can also suffer* ought to be accorded an enhanced set of rights or legal status. See further Chapter 4 at s. 1.

[72] Indeed, an absence of features such as imagination, emotions or consciousness may contribute to situations in which an AI system is liable to act differently from humans. For instance, an AI system which lacks the ability to empathise with human suffering might present more danger to people than a human carrying out the same task. This phenomenon, of itself, is one reason why new rules are desirable to guide and constrain choices made by AI.

[73] For a famous application of this principle, see Lewis Carroll's *Through the Looking Glass*: "When *I* use a word", Humpty Dumpty said, in rather a scornful tone, "it means just what I choose it to mean—neither more nor less". "The question is", said Alice, "whether you *can* make words mean so many different things". "The question is", said Humpty Dumpty, "which is to be master—that's all". Lewis Carroll, *Through the Looking-Glass* (Plain Label Books, 2007), 112 (originally published 1872). See also the UK House of Lords case *Liversidge v Anderson* [1942] A.C. 206, 245.

are allowed in the park".[74] Most would agree that this prohibits cars and motorbikes, but it is unclear from the wording alone whether skateboards, bicycles or wheelchairs are also banned.[75] Legislators can seek to avoid uncertainty by setting out a list of what is and is not allowed. The difficulty with using lists is that they ossify the law, and may be difficult to update or to apply to situations which were not contemplated at the time the list was drafted. The highly technical and fast-developing nature of AI renders the list-based approach unsuitable as a workable mechanism.

An alternative approach (and the one suggested here) is to set a core definition which captures the essence of a term, without delimiting its precise boundaries.[76] Often the task of applying ambiguous legislation falls in the first instance to regulatory agencies, for example a park warden, and then in the second instance to a judge (if the decision of a warden to issue a fine is challenged).

As AI advances, questions as to its boundaries may well—at least using this book's definition—become less difficult to draw. AI experts might point out that even deep learning systems, which involve multiple layers of neural networks, are far from being independent of human input and are instead constantly monitored and nudged by humans. However, it is suggested here that the further AI improves in terms of capability and the more it is deployed for use by non-experts, such human input is liable to decrease. The more remote that the actual decision-making procedure becomes from the original designer, the clearer it will be that the entity is making choices.

José Hernández-Orallo has proposed a universal test of intelligence, capable of covering the entire "machine world", which includes not just artificial entities but also animals, humans and any hybrids of these

---

[74]H.L.A Hart, "Positivism and the Separation of Law and Morals", *Harvard Law Review*, Vol. 71 (1958), 593, 607.

[75]See, for example, Ann Seidman, Robert B. Seidman, and Nalin Abeyesekere, *Legislative Drafting for Democratic Social Change* (London: Kluwer Law International, 2001), 307.

[76]Those interpreting the core definition can use various tools so as to ascertain the proper scope of application of the provision in question. These might include the legislative history of the provision, the mischief to which it was directed or even shifting social norms. See Ronald Dworkin, "Law as Interpretation", *University of Texas Law Review*, Vol. 529 (1982), 60.

groups.[77] Hernández-Orallo focusses on computational principles for the measurement of intelligence, which are capable of scoring an entity as to the degree of its intelligence. Relevant features include "compositionality", namely the capability of a system of building new concepts and skills over previous ones.[78] If AI does need to be regulated separately merely automated machines and programs, then tests such as that proposed by Hernández-Orallo could become very significant in assisting authorities delineating questions at the boundary of what is and is not intelligent, as well as to track the progress of the field through advances in AI powers.

4   AI, AI Everywhere

Armed with a definition of AI, it is now possible to identify its current uses and growing prevalence.

It might be objected that some of the examples of AI suggested below do not fulfil our functional definition. It is indeed true that certain of the outcomes could be achieved without using AI, either because the entities use deterministic rules or because humans are actually making the choices. This could be called the "Mechanical Turk" objection, after the chess-playing machine which astounded audiences in the late eighteenth and early nineteenth centuries. As the name suggests, it resembled a Turban-wearing "Turkish" man, sitting at a desk. The Turk's designer, Baron von Kempelen, claimed that it was able to use a mysterious form of mechanical intelligence to defeat opponents at chess. In fact, the Turk was merely a complex illusion. The Turk's desk concealed a chamber in which a human chess player sat, directing the mechanical arms to move pieces.[79] As with the Turk, in order to determine whether a process or a

---

[77] José Hernández-Orallo, *The Measure of All Minds: Evaluating Natural and Artificial Intelligence* (Cambridge: Cambridge University Press, 2017). See also José Hernández-Orallo and David L. Dowe, "Measuring Universal Intelligence: Towards an Anytime Intelligence Test", *Artificial Intelligence*, Vol. 174 (2010), 1508–1539. For an important early examination of algorithmic information theory and universal distributions, see Ray Solomonoff, "A Formal Theory of Inductive Inference: Part I", *Information and Control*, Vol. 7, No. 1 (1964), 1–22.

[78] See further Chapter 6 at s. 2.1 in which it is argued that there is a spectrum of intelligences between narrow and general artificial intelligence, using the increasing ability of programs to display compositionality as an example.

[79] Discussed in Gerald M. Levitt, *The Turk, Chess Automaton* (Jefferson, NC: McFarland & Co., 2007).

program uses real AI according to our definition, it is necessary to check under the bonnet and ascertain exactly how a decision is taken. More important than the outcome is how that outcome was reached.

The founding members of the Dartmouth College summer school expressed a desire to "find how to make machines use language, form abstractions and concepts, solve kinds of problems now reserved for humans, and improve themselves".[80] Over 60 years later, we interact with such machines on a daily basis. The smartphone is an instructive example. The Pew Research Centre calculated in 2016 that 68% of adults in the world's 11 most advanced economies owned a smartphone, a device which provides instant access to the power of both the Internet and machine learning.[81] Smartphone applications (or "apps") including music library recommendations based on past listening history, as well as predictive text suggestions for messaging, are all potentially examples of AI. The complex algorithms behind search engines improve themselves based on our searches and reaction to the results. Every time we use a search engine, that search engine is using us.[82]

Virtual Personal Assistants including Apple's Siri, the Google Assistant, Amazon's Alexa and Microsoft's Cortana are now commonplace. This trend is connected to the growth of the "Internet of Things", where household devices are connected to the Internet.[83] Whether it is

---

[80] John McCarthy, Marvin L. Minsky, Nathaniel Rochester, and Claude E. Shannon, "A Proposal for the Dartmouth Summer Research Project on Artificial Intelligence", 31 August 1955, full text available at: http://www-formal.stanford.edu/jmc/history/dartmouth/dartmouth.html, accessed 1 June 2018.

[81] Jacob Poushter, "Smartphone Ownership and Internet Usage Continues to Climb in Emerging Economies", *Pew Research Centre*, 22 February 2016, http://www.pewglobal.org/2016/02/22/smartphone-ownership-and-internet-usage-continues-to-climb-in-emerging-economies/, accessed 1 June 2018. The global median smartphone ownership was at the time of the poll 43% but this rate is climbing fastest in developing countries.

[82] As Ariel Ezrachi and Maurice E. Stucke chart in their book, *Virtual Competition* (Oxford: Oxford University Press, 2016), Internet sites can use an increasingly sophisticated set of data including the time users spend hovering their mouse over a particular part of a page in order to predict, and shape, their preferences.

[83] Perhaps surprisingly, the idea of household appliances connected to the Internet has a fairly long history. In 1990, a toaster was reportedly connected to the then fledgling Internet via TCP/IP networking. The power could be remote controlled, allowing a user to determine how darkened the toast should be, http://www.livinginternet.com/i/ia_myths_toast.htm, accessed 1 June 2018.

a fridge which learns when you need eggs and orders them for you, or a hoover which can tell which parts of your floor need the most cleaning, AI is coming to fulfil the roles once played by domestic servants.[84]

The uses of AI as an aid to or even as a replacement for human judgement and decision-making can go from the immaterial—selection of which song to play next—to the highly consequential. For instance, in early 2017, a UK police force announced it was piloting a program called the Harm Assessment Risk Tool to determine whether a suspect should be kept in custody or released on bail, based on various data.[85]

Self-driving cars are among the most well-known examples of AI. Advanced prototypes are now being tested on our roads by both technology companies like Google and Uber, but also traditional car makers such as Tesla and Toyota.[86] AI has also caused its first fatalities: in 2017, a Tesla Model S driving on autopilot crashed into a truck, killing

---

[84]David Schatsky, Navya Kumar, and Sourabh Bumb, "Intelligent IoT: Bringing the Power of AI to the Internet of Things", Deloitte, 12 December 2017, https://www2.deloitte.com/insights/us/en/focus/signals-for-strategists/intelligent-iot-internet-of-things-artificial-intelligence.html, accessed 1 June 2018.

[85]Aatif Sulleyman, "Durham Police to Use AI to Predict Future Crimes of Suspects, Despite Racial Bias Concerns", *Independent*, 12 May 2017, http://www.independent.co.uk/life-style/gadgets-and-tech/news/durham-police-ai-predict-crimes-artificial-intelligence-future-suspects-racial-bias-minority-report-a7732641.html, accessed 1 June 2018. For criticism of such technology and its tendency to adopt racial biases, see Julia Angwin, Jeff Larson, Surya Mattu and Lauren Kirchner, "Machine Bias: There's Software Used Across the Country to Predict Future Criminals: And It's Biased Against Blacks", *ProPublica*, May 2016, https://www.propublica.org/article/machine-bias-risk-assessments-in-criminal-sentencing, accessed 1 June 2018. We will return in Chapter 8 at s. 3 to the propensity for such decision-making AI to adopt human biases and the ways in which regulation might stop it.

[86]See, for example, the U.S. Department of Transportation, "Federal Automated Vehicles Policy", September 2016, https://www.transportation.gov/AV, accessed 1 June 2018, as well as the UK House of Lords Science and Technology Select Committee, 2nd Report of Session 2016–2017, "Connected and Autonomous Vehicles: The Future?", 15 March 2017, https://www.publications.parliament.uk/pa/ld201617/ldselect/ldsctech/115/115.pdf, accessed 1 June 2018.

its passenger[87]; and in 2018, an Uber test car in autonomous mode hit and killed a woman in Arizona.[88] They will not be the last.

From AI which kills accidentally to AI which kills deliberately: several militaries are developing semi and even fully autonomous weapons systems. In the skies, AI drones are able to identify, track and potentially kill targets without the need for human input. A 2016 report of the US Department of Defense research division explored the potential for AI to become a cornerstone of US defense policy.[89] A 2017 Chatham House Report concluded that militaries around the world were developing AI weapons capabilities "that could make them capable of undertaking tasks and missions on their own".[90] Allowing AI to kill targets without human intervention remains one of its most controversial potential uses. At the time of writing the most lethal known use of autonomous ground-based weapons was in a friendly fire incident, when a South African artillery cannon malfunctioned and killed nine soldiers.[91] It is unlikely to be long before enemies too are in the crosshairs.

[87] Gareth Corfield, "Tesla Death Smash Probe: Neither Driver nor Autopilot Saw the Truck", *The Register*, 20 July 2017, https://www.theregister.co.uk/2017/06/20/tesla_death_crash_accident_report_ntsb/, accessed 1 June 2018.

[88] Sam Levin and Julia Carrie Wong, "Self-driving Uber Kills Arizona Woman in First Fatal Crash Involving Pedestrian", *The Guardian*, 19 March 2018, https://www.theguardian.com/technology/2018/mar/19/uber-self-driving-car-kills-woman-arizona-tempe, accessed 1 June 2018.

[89] Department of Defense, "Defense Science Board, Office of the Under Secretary of Defense for Acquisition, Technology and Logistics, Summer Study on Autonomy", June 2016, http://web.archive.org/web/20170113220254/http://www.acq.osd.mil/dsb/reports/DSBSS15.pdf, accessed 1 June 2018.

[90] Mary L. Cummings, Artificial Intelligence and the Future of Warfare, *Chatham House*, 26 January 2017, https://www.chathamhouse.org/publication/artificial-intelligence-and-future-warfare, accessed 1 June 2018.

[91] Some reports cast doubt on whether the malfunction was as a result of software or human error. See, for example, Tom Simonite, "'Robotic Rampage' Unlikely Reason for Deaths", *New Scientist*, 19 October 2007, available at: https://www.newscientist.com/article/dn12812-robotic-rampage-unlikely-reason-for-deaths/, accessed 1 June 2018.

Robots can care as well as kill. Increasingly sophisticated AI systems are being used to provide physical and emotional support to older people in Israel and Japan,[92] a trend which is surely likely to grow, both in those countries and elsewhere as the richer world continues to adapt to ageing populations. AI is also being used in medicine as an aid to clinical decision-making. Other systems under development and in operation allow for diagnosis and treatment to be fully automated.[93]

In commerce, the US Congressional Research Service estimates that algorithmic programs account for roughly 55% of trading volume in the US equities market and around 40% of European equities markets.[94] Under our definition, most algorithmic trading does not involve the use of AI as yet. However, its capability of taking complex strategic decisions in a manner which surpasses human reasoning seems likely to make AI particularly well suited to this task.[95]

Even the creative industries are taking advantage of AI. Music composition programs were among the first examples of this development.[96] In 1997, the *New Scientist* reported that a computer in California had

---

[92] An example of this is Elli.Q, a social care robot which has been designed to convey emotion through speech tones, light and movement or body language. See Darcie Thompson-Fields, "AI Companion Aims to Improve Life for the Elderly", *Access AI*, 12 January 2017, http://www.access-ai.com/news/511/ai-companion-aims-to-improve-life-for-the-elderly/, accessed 1 June 2018.

[93] Daniela Hernandez, "Artificial Intelligence Is Now Telling Doctors How to Treat You", *Wired Business/Kaiser Health News*, 2 June 2014, https://www.wired.com/2014/06/ai-healthcare/. Alphabet's DeepMind has been partnering with healthcare providers, including the NHS, on a variety of initiatives, including an app called Streams, which has the capability to analyse medical history and test results to alert doctors and nurses of potential dangers which might not have otherwise been spotted, see "DeepMind—Health", https://deepmind.com/applied/deepmind-health/, accessed 1 June 2018.

[94] Rena S. Miller and Gary Shoerter, "High Frequency Trading: Overview of Recent Developments", *US Congressional Research Service*, 4 April 2016, 1, https://fas.org/sgp/crs/misc/R44443.pdf, accessed 1 June 2018.

[95] Laura Noonan, "ING Launches Artificial Intelligence Bond Trading Tool Katana", *Financial Times*, 12 December 2017, https://www.ft.com/content/1c63c498-de79-11e7-a8a4-0a1e63a52f9c, accessed 1 June 2018.

[96] Alex Marshall, "From Jingles to Pop Hits, A.I. Is Music to Some Ears", *New York Times*, 22 January 2017, https://www.nytimes.com/2017/01/22/arts/music/juke-deck-artificial-intelligence-songwriting.html, accessed 1 June 2018.

written Mozart's 42nd Symphony, a feat not even Mozart himself could manage.[97] A program called Mubert is able to compose entirely new tracks which, its creators say, are "based on the laws of musical theory, mathematics and creative experience".[98] In 2016, a director and a New York University AI researcher collaborated to create an AI system which created a new horror film script, after being "fed" dozens of successful scripts. The neural network highlighted the recurrent themes and created a new work: *Sunspring*. *The Guardian* described it as "a weirdly entertaining, strangely moving dark sci-fi story of love and despair".[99]

AI is now creating works of semi-abstract art. One of the most famous examples is Google's DeepDream, a neural net which scans millions of images and can generate hybrid creations on demand.[100] In early 2017,

[97] Bob Holmes, "Requiem for the Soul", *New Scientist*, 9 August 1997, https://www.newscientist.com/article/mg15520945-100-requiem-for-the-soul/, accessed 1 June 2018. For criticism, see Bayan Northcott, "But Is It Mozart?", *Independent*, 4 September 1997, http://www.independent.co.uk/arts-entertainment/music/but-is-it-mozart-1237509.html, accessed 1 June 2018.

[98] "Homepage", Mubert Website, http://mubert.com/en/, accessed 1 June 2018.

[99] Hal 90210, "This Is What Happens When an AI-Written Screenplay Is Made into a Film", *The Guardian*, 10 June 2016, https://www.theguardian.com/technology/2016/jun/10/artificial-intelligence-screenplay-sunspring-silicon-valley-thomas-middleditch-ai, accessed 1 June 2018.

[100] The process used to create such visualisations was revealed first on two blog posts of 17 June 2015 and 1 July 2015 by Alexander Mordvintsev, Christopher Olah, and Mike Tyka, See "Inceptionism: Going Deeper into Neural Networks", Google Research Blog, 17 June 2015, https://research.googleblog.com/2015/06/inceptionism-going-deeper-into-neural.html, accessed 1 June 2018. The name DeepDream was first used in the latter, https://web.archive.org/web/20150708233542/http://googleresearch.blogspot.co.uk/2015/07/deepdream-code-example-for-visualizing.html, accessed 1 June 2018. Like many scientific breakthroughs, and innovations, the DeepDream generator was discovered as a by-product of other research, into the use of neural networks. Its designers explained: "Two weeks ago we blogged about a visualization tool designed to help us understand how neural networks work and what each layer has learned. In addition to gaining some insight on how these networks carry out classification tasks, we found that this process also generated some beautiful art". The program used to create the visualisations is now available online at: https://deepdreamgenerator.com/, accessed 1 June 2018. See also Cade Metz, "Google's Artificial Brain Is Pumping Our Trippy—And Pricey—Art", *Wired*, 29 February 2016, https://www.wired.com/2016/02/googles-artificial-intelligence-gets-first-art-show/, accessed 1 June 2018.

the Chinese company Tencent reported that it had successfully used deep learning techniques to identify fashion trends among millennials. Apparently, China's post-1995 generation is particularly fond of "light black".[101]

Even more ethically challenging uses of AI are in development or use. These include robots designed to satisfy human sexual desires (sex-bots),[102] as well as the potential for humans to physically augment themselves with AI capabilities, giving rise to hybrids or cyborgs.[103]

From this brief and by no means exhaustive survey of its impact, it is clear that AI is already in our homes, workplaces, hospitals, roads, cities and skies. The Dartmouth College group's original funding proposal suggested that AI could "solve the kinds of problems now reserved for humans...if a carefully selected group of scientists work on it together for a summer".[104] The initial estimate may have been somewhat optimistic, but the scale of humanity's achievements in AI in the past 60 years compared to the previous 200,000 of *homo sapiens'* existence suggests that the Dartmouth group's guess was not as wild as it may have seemed.

5   SUPERINTELLIGENCE

In 1965, mathematician and former Second World War code-breaker I.J. Good predicted that "...an ultraintelligent machine could design even better machines; there would then unquestionably be an 'intelligence explosion,' and the intelligence of man would be left far behind".[105] This remains the operating assumption of some AI experts today. In his influential book, *Superintelligence* Nick Bostrom describes the consequences

---

[101] Tencent "Not Your Father's AI: Artificial Intelligence Hits the Catwalk at NYFW 2017", *PR Newswire*, http://www.prnewswire.com/news-releases/not-your-fathers-ai-artificial-intelligence-hits-the-catwalk-at-nyfw-2017-300407584.html, accessed 1 June 2018.

[102] For an in-depth treatment of love between humans and robots, see D. Levy, *Love and Sex with Robots* (New York: Harper Perennial, 2004).

[103] See Chapter 4 at s. 4.4.

[104] John McCarthy, Marvin L. Minsky, Nathaniel Rochester, and Claude E. Shannon, "A Proposal for the Dartmouth Summer Research Project on Artificial Intelligence", 31 August 1955, full text available at: http://www-formal.stanford.edu/jmc/history/dartmouth/dartmouth.html, accessed 1 June 2018.

[105] Ian J. Good, "Speculations Concerning the First Ultraintelligent Machine", in *Advances in Computers*, edited by F. Alt and M. Ruminoff, Vol. 6 (New York: Academic Press, 1965).

of the AI explosion in dramatic terms, explaining that in some models it could be a matter of days between the development of the initial
"seed" superintelligence and its spawn becoming so powerful that no
human-controlled force is able to reassert control: "Once artificial intelligence reaches human level, there will be a positive feedback loop that will
give the development a further boost. AIs would help constructing better
AIs, which in turn would help building better AIs, and so forth".[106]

The advent of fully general AI is associated by many writers with a
phenomenon some have predicted, known as "the singularity".[107] This
term is usually used to describe the point at which AI matches and
then surpasses human intelligence. However, the conception of the singularity as a single discernible moment is unlikely to be accurate. Like
the move from weak AI to general AI, the singularity is best seen as a
process rather than a single event. There is no reason to think AI will
match every human capability at once. Indeed, in many fields (such as
the ability to undertake complex calculations), AI is already well ahead of
humans, whereas in others such as the ability to recognise human emotions, it lags behind.

Proponents of superintelligence argue that AI has repeatedly surpassed
expectations in recent years. In the mid to late twentieth century, many
thought that a computer could never defeat a human Grandmaster at
chess.[108] Then, in 1997, IBM's Deep Blue defeated former world champion[109] Garry Kasparov in a best of six match. In the early 2000s, many

---

[106] Nick Bostrom, "How Long Before Superintelligence?", *International Journal of Future Studies*, 1998, vol. 2..

[107] The singularity was conceived of shortly after the advent of modern AI studies, having been introduced by John von Neumann in 1958 and then popularised by Vernor Vinge, in "The Coming Technological Singularity: How to Survive in the Post-human Era" (1993), available at: https://edoras.sdsu.edu/~vinge/misc/singularity.html, accessed 22 June 2018 and subsequently by Ray Kurzweil, *The Singularity Is Near: When Humans Transcend Biology* (New York: Viking Press, 2005).

[108] In 1968, a Scottish chess champion bet AI pioneer John McCarthy £500 that a com-
puter would not be able to beat him by 1979. Levy won that wager (though was eventually beaten by a computer in 1989). For an account, see Chris Baraniuk, "The Cyborg Chess Player Who Can't Be Beaten", BBC Website, 4 December 2015, http://www.bbc.com/future/story/20151201-the-cyborg-chess-players-that-cant-be-beaten, accessed 1 June 2018.

[109] The situation is somewhat complicated in that Kasparov had held the Fédération Internationale des Échecs (FIDE) world title until 1993, when a dispute with FIDE led him to set up a rival organization, the Professional Chess Association.

thought that a computer could never defeat a human champion at Go, a vastly more complex board game popular in Asia. In fact, as late as 2013 Bostrom wrote "Go-playing programs have been improving at a rate of about 1 dan [a level of accomplishment]/year in recent years. If this rate of improvement continues, they might beat the human world champion in about a decade".[110] Just three years later, in March 2016, DeepMind's AlphaGo defeated champion player Lee Sedol by four games to one— with the human champion even resigning in the final game, having been tactically and emotionally crushed.[111] The killer move by AlphaGo was successful precisely because it used tactics which went against all traditional human schools of thought.[112] Of course, winning board games is one thing but taking over the world is quite another.

[110] Nick Bostrom, *Superintelligence: Paths, Dangers and Strategies* (Oxford: Oxford University Press, 2014), 16.

[111] In May 2017, a subsequent version of the program, "AlphaGo Master", defeated the world champion Go player, Ke Jie by three games to nil. See "AlphaGo at The Future of Go Summit, 23–27 May 2017", DeepMind Website, https://deepmind.com/research/alphago/alphago-china/, accessed 16 August 2018. Perhaps as a control against accusations that top players were being beaten psychologically by the prospect of playing an AI system rather than on the basis of skill, DeepMind had initially deployed AlphaGo Master in secret, during which period it beat 50 of the world's top players online, playing under the pseudonym "Master". See "Explore the AlphaGo Master series", DeepMind Website, https://deepmind.com/research/alphago/match-archive/master/, accessed 16 August 2018. DeepMind, promptly announced AlphaGo's retirement from the game to pursue other interests. See Jon Russell, "After Beating the World's Elite Go Players, Google's AlphaGo AI Is Retiring", *Tech Crunch*, 27 May 2017, https://techcrunch.com/2017/05/27/googles-alphago-ai-is-retiring/ accessed 1 June 2018. Rather like a champion boxer tempted out of retirement for one more fight, AlphaGo (or at least a new program bearing a similar name, AlphaGo Zero) returned a year later to face a new challenge: AlphaGo Zero. This is discussed in Chapter 2 at s. 3.2.1, and FN 130 and 131.

[112] Cade Metz, "In Two Moves, AlphaGo and Lee Sedol Redefined the Future", *Wired*, 16 March 2016, https://www.wired.com/2016/03/two-moves-alphago-lee-sedol-redefined-future/, accessed 1 June 2018. In October 2017, DeepMind announced yet another breakthrough involving Go: a computer which was able to master the game without access to any data generated by human players. Instead, it was provided only with the rules and, within a number of hours, had mastered the game to such an extent that it was able to beat the previous version of AlphaGo by 100 games to 0. See "AlphaGo Zero: Learning from Scratch", DeepMind Website, 18 October 2017, https://deepmind.com/blog/alphago-zero-learning-scratch/, accessed 1 June 2018. See also Chapter 2 at s. 3.2.1.

The quality of intelligence to improve itself is separate from its capacity to solve other problems. Though humans have displayed general intelligence for hundreds of thousands of years, we have not yet managed to design programs with superior general intelligence to our own. We cannot be sure that AI technology will not meet a similar plateau, even after it achieves a form of general intelligence.[113]

Notwithstanding these limitations, in recent years there have been several significant developments in the capabilities of AI. In January 2017, Google Brain announced that technicians had created AI software which could itself develop further AI software.[114] Similar announcements were made around this time by the research group OpenAI,[115] MIT,[116] the University of California, Berkeley and DeepMind.[117] And these are only the ones we know about—companies, governments and even some independent individual AI engineers are likely to be working on processes which go far beyond what those have yet made public.

[113] For a helpful analysis of the barriers to the singularity, see Toby Walsh, *Android Dreams* (London: Hurst & Co., 2017), 89–136.

[114] Barret Zoph and Quoc V. Le, "Neural Architecture Search with Reinforcement Learning", *Cornell University Library Research Paper*, 15 February 2017, https://arxiv.org/abs/1611.01578, accessed 1 June 2018. See also Tom Simonite, "AI Software Learns to Make AI Software", *MIT Technology Review*, 17 January 2017, https://www.technologyreview.com/s/603381/ai-software-learns-to-make-ai-software/, accessed 1 June 2018.

[115] Yan Duan, John Schulman, Xi Chen, Peter L. Bartlett, Ilya Sutskever, and Pieter Abbeel, "RL2: Fast Reinforcement Learning via Slow Reinforcement Learning", *Cornell University Library Research Paper*, 10 November 2016, https://arxiv.org/abs/1611.02779, accessed 1 June 2018.

[116] Bowen Baker, Otkrist Gupta, Nikhil Naik, and Ramesh Raskar, "Designing Neural Network Architectures Using Reinforcement Learning", *Cornell University Library Research Paper*, 22 March 2017, https://arxiv.org/abs/1611.02167, accessed 1 June 2018.

[117] Jane X. Wang, Zeb Kurth-Nelson, Dhruva Tirumala, Hubert Soyer, Joel Z Leibo, Remi Munos, Charles Blundell, Dharshan Kumaran, and Matt Botvinick, "Learning to Reinforcement Learn", *Cornell University Library Research Paper*, 23 January 2017, https://arxiv.org/abs/1611.05763, accessed 1 June 2018.

6   OPTIMISTS, PESSIMISTS AND PRAGMATISTS

Commentators on the future of AI can be grouped into three camps: the optimists, the pessimists and the pragmatists.[118]

The optimists emphasise the benefits of AI and downplay any dangers. Ray Kurzweil has argued "… we have encountered comparable specters, like the possibility of a bioterrorist creating a new virus for which humankind has no defence. Technology has always been a double edged sword, since fire kept us warm but also burned down our villages".[119] Similarly, engineer and roboethicist Alan Winfield said in a 2014 article: "*If* we succeed in building human equivalent AI and *if* that AI acquires a full understanding of how it works, and *if* it then succeeds in improving itself to produce super-intelligent AI, and *if* that super-AI, accidentally or maliciously, starts to consume resources, and *if* we fail to pull the plug, then, yes, we may well have a problem. The risk, while not impossible, is improbable".[120] Fundamentally, optimists think humanity can and will overcome any challenges AI poses.

The pessimists include Nick Bostrom, whose "paperclip machine" thought experiment imagines an AI system asked to make paperclips which decides to seize and consume all resources in existence, in its blind aderence to that goal.[121] Bostrom contemplates a form of superintelligence which is so powerful that humanity has no chance of stopping it from destroying the entire universe. Likewise, Elon Musk has said we risk "summoning a demon" and called AI "our biggest existential threat".[122]

---

[118] It may be objected that this is a simplification, or even a caricature, and indeed many have expressed sentiments at different times which could be covered by each of these categories, and in reality, there are more points on a spectrum than strict alternatives. Nonetheless, we think these labels provide a helpful summary of current attitudes.

[119] Ray Kurzweil, "Don't Fear Artificial Intelligence", *Time*, 19 December 2014, http://time.com/3641921/dont-fear-artificial-intelligence/, accessed 1 June 2018.

[120] Alan Winfield, "Artificial Intelligence Will Not Turn into a Frankenstein's Monster", *The Guardian*, 10 August 2014, https://www.theguardian.com/technology/2014/aug/10/artificial-intelligence-will-not-become-a-frankensteins-monster-ian-winfield, accessed 1 June 2018.

[121] Nick Bostrom, *Superintelligence*, (Oxford: Oxford University Press, 2014), 124–125.

[122] Elon Musk, as quoted in S. Gibbs, "Elon Musk: Artificial Intelligence Is Our Biggest Existential Threat", *The Guardian*, 27 October 2014, https://www.theguardian.com/technology/2014/oct/27/elon-musk-artificial-intelligence-ai-biggest-existential-threat, accessed 1 June 2018.

The pragmatists acknowledge the benefits predicted by the optimists as well as the potential disasters forecast by the pessimists. Pragmatists argue for caution and control. This view was endorsed by the thousands of eminent signatories of the Open Letter on AI, organised by the Future of Life Institute in 2015.[123] The letter states:

> There is now a broad consensus that AI research is progressing steadily, and that its impact on society is likely to increase. The potential benefits are huge, since everything that civilisation has to offer is a product of human intelligence; we cannot predict what we might achieve when this intelligence is magnified by the tools AI may provide, but the eradication of disease and poverty are not unfathomable. Because of the great potential of AI, it is important to research how to reap its benefits while avoiding potential pitfalls.

Combining optimism and pessimism, Stephen Hawking said that AI will be: "either the best, or the worst thing, ever to happen to humanity".[124]

The most prominent futurists tend to concentrate on the long-term impact of potential superintelligence, which may still be decades away. By contrast, many legislators concentrate on the extreme short term, or even the past. Often the time lag between the development of a new technology and its regulation means that the law has several years to catch up. Overzealous regulation of technology can seem absurd in retrospect. We do not want to be in the position of the first automobile drivers in the nineteenth century, who were required to drive at no greater than two miles per hour in cities and to employ someone to walk in front of their vehicle waving a red flag.[125]

Technology is not always adopted uncritically: progress for the majority can often conflict with vested interests. In the early nineteenth century, the "Luddites"—aggrieved agricultural workers supposedly led by Ned Ludd—rioted for several years, destroying mechanised power looms

---

[123] "Open Letter", Future of Life Institute, https://futureoflife.org/ai-open-letter/, accessed 1 June 2018.

[124] Alex Hern, "Stephen Hawking: AI Will Be 'Either Best or Worst Thing' for Humanity", *The Guardian*, 19 October 2016, https://www.theguardian.com/science/2016/oct/19/stephen-hawking-ai-best-or-worst-thing-for-humanity-cambridge, accessed 1 June 2018.

[125] See The Locomotives on Highways Act 1861, The Locomotive Act 1865 and the Highways and Locomotives (Amendment) Act 1878 (all UK legislation).

which threatened their employment.[126] Today debates continue as to whether countries should harness nuclear technology to satisfy insatiable demands for energy.

We are in danger of oscillating between the complacency of the optimists and the craven scruples of the pessimists. AI presents incredible opportunities for the benefit of humanity and we do not wish to fetter or shackle this progress unnecessarily.

The problem with headline-grabbing predictions about the destructive or beneficial potential of superintelligence or the singularity is that they distract the public from the more mundane, but ultimately far more important issues of how humanity and AI should interact now. As Pedro Domingos put it in a 2015 book: "People worry that computers will get too smart and take over the world, but the real problem is that they're too stupid and they've already taken over the world".[127]

7   IF NOT NOW, WHEN?

Some will say this book is premature: although AI might one day require a change in our laws, for the moment it is unnecessary. General AI does not yet exist, and until then, we should spend our time more productively, rather than speculating or even legislating idly about a technology which might never arrive.

This attitude is overly complacent and relies on two incorrect assumptions: *first*, it underestimates the penetration of AI technology in the world today, and *secondly*, it rests on a hubristic belief that somehow human ingenuity will be able to address any issues without extra cost or difficulty at some unspecified later stage.

It is not surprising that most people have failed to notice AI's tightening grip. Incremental developments in technology mean that we often do not even register its improvement. The significant upgrade of Google Translate in 2016 using machine learning is a rare outlier in that it was actually picked up by media.[128] Companies carefully stagger the release

---

[126]See, for example, Steven E. Jones, *Against Technology: From the Luddites to Neo-Luddism* (London: Routledge, 2013).

[127]Pedro Domingos, *The Master Algorithm: How the Quest for the Ultimate Learning Machine Will Remake Our World* (New York: Allen Lane, 2015), 286.

[128]This was due in large part to the publication of: Gideon Lewis-Kraus, "The Great A.I. Awakening", *The New York Times Magazine*, 14 December 2016, https://www.nytimes.com/2016/12/14/magazine/the-great-ai-awakening.html, accessed 1 June 2018.

of new technologies through software patches and upgrades, gradually immersing their users. Though barely noticeable at the time, the cumulative differences can be huge.[129] Because of the natural psychological tendency not to notice a series of small changes, humans risk becoming like frogs in a restaurant. If you drop a live frog into a pot of boiling water it will try to escape. But if you place a frog in a pot of cold water and slowly bring it to the boil the frog will sit calmly, even as it is cooked alive.

What if 200 years ago, at the dawn of the Industrial Revolution, we had known the dangers of global warming? Perhaps we would have created institutions to study man's impact on the environment. Perhaps we would have enshrined national laws and international treaties, agreeing for the good of humanity to constrain harmful activities and to promote sound ones. The world today could have been very different. We might be free from the scourge of rising sea temperatures and melting ice caps. We might have avoided decades of increasingly unpredictable weather cycles, bringing misery and destruction to millions of people. We might have achieved a fair and equitable settlement between richer and poorer nations, respected and honoured by all.

Instead, we are scrambling to legislate backwards to curb climate change. Relatively new innovations such as emissions-trading[130] and self-imposed greenhouse gas limits[131] are both projected to have a limited effect on reducing global warming, but climate scientists generally agree that enormously damaging changes will occur to our atmosphere without far more drastic action.

Humanity is unlikely to have to wait two centuries to see the enormous consequences of AI. The consultancy McKinsey has estimated that compared with the Industrial Revolution "this change is happening ten times faster and at 300 times the scale, or roughly 3,000 times the impact".[132]

---

[129]The changing appearance of the Facebook interface over time is a good example of a technology company using small updates to make large changes over time. See Jenna Mullins, "This Is How Facebook Has Changed Over the Past 12 Years", *ENews*, 4 February 2016, http://www.eonline.com/uk/news/736977/this-is-how-facebook-has-changed-over-the-past-12-years, accessed 1 June 2018.

[130]See the Kyoto Protocol to the United Nations Framework Convention on Climate Change, 1997.

[131]See the Paris Climate Agreement, 2016.

[132]Richard Dobbs, James Manyika, and Jonathan Woetzel, "No Ordinary Disruption: The Four Global Forces Breaking All the Trends", *McKinsey Global Institute*, April 2015, https://www.mckinsey.com/mgi/no-ordinary-disruption, accessed 1 June 2018.

8   ROBOT RULES

It may not be immediately obvious why law is relevant to the various industries and aspects of society affected by AI. In fact, legal regulation is as crucial to their smooth operation as it is to every other element in our lives. Just because we do not have daily interactions with lawyers, judges, courts or the police does not mean that our legal system is not having an effect.

Laws "work" even when they are not being used in courtrooms to convict criminals or to award damages to claimants. Indeed, laws are most effective when they are a silent background condition allowing parties to deal with each other in a fair and predictable atmosphere. The legal system is like oxygen. Day to day we do not notice it; in fact, many readers will not have given any thought to their own breathing before coming to this paragraph. However, if the amount of oxygen in the air drops even by a small amount, life quickly becomes intolerable.

The law plays a vital role in solving "coordination problems" which arise where agents can choose from several options, none of which is obviously right or wrong, but where the system as a whole will only function correctly if everyone acts in a similar manner.[133] It would not make sense to say that it is better to drive on the right or the left as a general moral proposition, but the laws of traffic in England dictate that all must drive on the left, because if people were allowed to choose for themselves, there would be chaos.[134]

Although autonomous vehicles may lack some of the fallibilities of human drivers, if there were multiple different AI systems using the roads each with their own internal safety systems, this could lead to more fatalities rather than fewer. Two cars heading in opposite directions might crash head-on because one takes evasive action by steering to its right and the other takes evasive action by steering to its left.[135]

---

[133] The observation that law is not simply a command backed by a threat (such as "do not steal or you will be punished") was made originally by H.L.A. Hart in *The Concept of Law* (2nd edn. Oxford: Clarendon, 1997). Hart observed that such models of the law do not fully account for law's role in other social functions, such as making certain agreements legally binding. For the command theory of law, see John Austin, *The Province of Jurisprudence Determined and the Uses of the Study of Jurisprudence* (London: John Murray, 1832), vii.

[134] See Gerald Postema, "Coordination and Convention at the Foundations of Law", *Journal of Legal Studies*, Vol. 165 (1982), 11, 172 *et seq.*

[135] As explained further in Chapter 6, without a new universal system to ensure that all AI vehicles adhere to the same rules, many of their potential advantages over human drivers in terms of safety and efficiency will be lost.

Just as AI is disrupting markets and industries, it will also come to disrupt the legal rules and principles which have, until now, underpinned the way that those industries function. There are three main areas in which AI will give rise to new challenges:

1. Responsibility: If AI were to cause harm, or to create something beneficial, who should be held responsible?
2. Rights: Are there moral or pragmatic grounds for granting AI legal protections and responsibilities?
3. Ethics: How should AI make important choices, and are there any decisions it should not be allowed to take?

The following chapters will expand on these themes, demonstrating the types of problems that are likely to arise and how they might be addressed by current legal systems. The latter part of the book will move on to examining how novel institutions and then rules could be designed, in order to solve these problems in a coherent, stable and politically legitimate manner.

Chapter 2 elaborates on why AI is unique as a legal phenomenon and calls into question certain fundamental assumptions across most if not all systems of law. Chapter 3 analyses various mechanisms for establishing who or what is responsible when AI causes harm or creates something beneficial. Chapter 4 discusses whether AI should at some point be granted rights from a moral perspective. Chapter 5 considers the pragmatic arguments for and against granting AI legal personality.[136] Chapter 6 sets out how we can design international systems to create the types of new laws and regulations needed. Chapter 7 looks at controls on the human creators of AI, and finally, Chapter 8 discusses the possibility of building in or teaching rules to AI itself.

The biggest question in the next ten to twenty years is not going to be how to stop AI from destroying humanity, but how humanity should live alongside it. Today's regulation is likely to influence how technology develops. In building structures for effective everyday legal regulation in the medium term, we can prepare ourselves far better for any existential threat.

---

[136] In philosophical terms, the concept of according rights and obligations to an entity is sometimes referred to as "personhood", but the preferred term in law is "legal personality", and that will be used here. For discussion of what legal personality entails, see Chapter 5 at s. 2.1. For the avoidance of doubt, *legal* personality does not refer to the collection of psychological traits which characterise an individual.

CHAPTER 2

Unique Features of AI

Laws have been adapted over thousands of years to regulate many different phenomena.[1] Some say the advent of AI is no different to other social and technological developments and can be addressed through established legal frameworks. This chapter explains why AI presents a unique difficulty for legal regulation. We do not need to do away with all existing laws and start afresh. However, certain fundamental principles will need to be reconsidered.

The chapter will begin by setting out arguments which have been raised against making major legal changes to accommodate AI. Next, it will analyse the concepts of agency[2] and causation which underpin current legal systems. Finally, it will identify properties of AI which do not fit easily into established legal structures.

1   Sceptics of Novelty: Of Horses and HTTP

AI is not the first phenomenon to raise issues of legal responsibility for the acts of other intelligent beings. Some of the world's oldest systems of law addressed responsibility for a semi-autonomous vehicle with more computational power and complexity than even the most advanced

---

[1] See John H. Farrar and Anthony M. Dugdale, *Introduction to Legal Method* (2nd edn. London: Sweet & Maxwell, 1982).

[2] For the difference between agency and patiency, see FN 16 below.

© The Author(s) 2019
J. Turner, *Robot Rules*, https://doi.org/10.1007/978-3-319-96235-1_2

self-driving car. That vehicle was the horse.[3] One policy response was to hold the animals themselves liable for harm caused. Other legal systems constructed means of ascribing liability for such intelligent (or somewhat intelligent) entities to humans. It is the latter which have proved more enduring.

US Judge and author Frank Easterbrook said in the 1990s that the idea of "cyberlaw", a separate set of rules to govern the Internet, made no more sense than to say that there should be a separate "Law of the Horse".[4] Easterbrook argued that "the best way to learn the law applicable to specialized endeavours is to study general rules", because:

> Lots of cases deal with sales of horses; others deal with people kicked by horses; still more deal with the licensing and racing of horses, or with the care veterinarians give to horses, or with prizes at horse shows. Any effort to collect these strands into a course on 'The Law of the Horse' is doomed to be shallow and to miss unifying principles.[5]

When invited to give the keynote address by the organisers of a University of Chicago legal forum on "Property in Cyberspace", Judge Easterbrook may have surprised his hosts by concluding:

> Error in legislation is common, and never more so than when the technology is galloping forward. Let us not struggle to match an imperfect legal system to an evolving world that we understand poorly. Let us instead do what is essential to permit the participants in this evolving world to make their own decisions. That means three things: make rules clear; create property rights where now there are none; and facilitate the formation of bargaining institutions. Then let the world of cyberspace evolve as it will, and enjoy the benefits.[6]

---

[3] D.I.C. Ashton-Cross, "Liability in Roman Law for Damage Caused by Animals", *The Cambridge Law Journal*, Vol. 11, No. 3 (1953), 395–403.

[4] Judge Frank H. Easterbrook, quoting Gerhard Casper, former Dean of the University of Chicago, in Frank H. Easterbrook, "Cyberspace and the Law of the Horse", *University of Chicago Legal Forum* (1996), 207–215, 207.

[5] Ibid.

[6] Ibid., 215.

As Harvard Law Professor and cyberlaw enthusiast Lawrence Lessig later put it, "'Go home,' in effect, was Judge Easterbrook's welcome".[7] Following the same reasoning as Easterbrook, critics of this book's thesis may argue that current legal concepts can be adapted to address AI. In part, this view stems from disagreement and uncertainty as to what is actually meant by the terms AI and robots.[8]

It may well be that when some commentators say that AI does not require any change in the law, they are talking about technologies which do not meet the test set out earlier in this book, namely that AI is the ability of a non-natural entity to make choices by an evaluative process. To the extent that an entity does not meet this threshold test, it will not require new legal principles.

Unhelpfully, some of those who argue for distinct laws to govern AI have avoided defining what needs to be regulated. For instance, Matthew Scherer wrote in a 2016 article which advocated new regulatory agencies for AI, that "[t]his paper will effectively punt on the definitional issue and 'define' AI for the purposes of this paper in a blissfully circular fashion: 'artificial intelligence' refers to machines that are capable of performing tasks that, if performed by a human, would be said to require intelligence".[9] Sceptics are unlikely to be convinced by this approach.

Uncertainty as to the definition of AI is not the only reason why some feel it does not merit separate legal treatment. A more fundamental objection is that AI can be regulated by the gradual development of existing legal principles. In a 2015 BBC radio programme on the topic of "The Law and Artificial Intelligence", the host asked various experts: "do we need new law in this area?" Professor of Internet Law Lillian Edwards summed up the sceptical approach:

---

[7] Lawrence Lessig, "The Law of the Horse: What Cyberlaw Might Teach", *Harvard Law Review*, Vol. 113, 501.

[8] As to the difficulty of defining AI from a legal perspective, see Matthew Scherer, "Regulating Artificial Intelligence Systems: Risks, Challenges, Competencies and Strategies", *Harvard Journal of Law & Technology*, Vol. 29, No. 2 (Spring 2016), 354–398, 359. The authors' definition of AI is set out in Chapter 1 at s. 3.4.

[9] Matthew Scherer, "Regulating Artificial Intelligence Systems: Risks, Challenges, Competencies and Strategies", *Harvard Journal of Law & Technology*, Vol. 29, No. 2 (Spring 2016), 354–398, 362.

I don't think we need much new law [for AI]. I think the nature of the law which most people don't really get who aren't lawyers is that the law is informed by principles and... there are already a very large list of principles of liability regimes. So we have rules of negligence, we have rules of product liability, we have rules of allocating risks in insurance law... We have troubles every time a new technology comes along. We have troubles applying laws to ships... we have troubles applying laws to horses. So it's not obvious that the law doesn't need to be adapted and disputed and litigated but I don't think we need much fundamental new law.[10]

Appearing on the same programme, technology lawyer Mark Deem, agreed with Edwards and advocated incremental development as the solution, saying "...the law has this ability to fill the gaps, and we should embrace that".[11] The remainder of this chapter suggests why the incremental approach is problematic.

2   FUNDAMENTAL LEGAL CONCEPTS

2.1   Subjects and Agents

The first major legal concept to be challenged by AI is agency. The word agency can mean several things in law. In this context, we are not referring to a principal-agent relationship, where one entity (the principal) appoints another (the agent) to act on its behalf. Rather, and as explained below, we use "agency" in a wider philosophical sense.

Any system of law—whether common, civil, national or international, secular or religious[12]—tells humans what they should and should not do. In more formal terms, systems of law regulate behaviour by stipulating legal subjects: those whose behaviour is to be regulated. A legal subject is an entity which holds rights and obligations in a given system. The status of legal subject is something which is thrust upon a person, animal or thing.

---

[10]Lillian Edwards, "The Law and Artificial Intelligence", *Unreliable Evidence*, interview by Clive Anderson on BBC Radio 4, first broadcast 10 January 2015, http://www.bbc.co.uk/programmes/b04wwgz9, accessed 1 June 2018.

[11]Ibid.

[12]As to the differences between these systems and their relative ability to manage change, see Chapter 6. By a system of "law", we refer here to prescriptive laws in social science, rather than descriptive scientific "laws", such as Newton's laws of physics, or the laws of thermodynamics.

A legal agent is a subject which can control and change its behaviour and understand the legal consequences of its actions or omissions.[13] Legal agency requires knowledge of and engagement with the relevant norms. Agency is not simply imbued on passive recipients. Rather it is an interactive process.[14] All legal agents must be subjects but not all subjects will be agents. Whereas there are many types of legal subjects—both human and non-human—legal agency is at present reserved only to humans. Advances in AI may undermine this monopoly.

In order for something to exercise agency, there are several prerequisites. The laws in question must be sufficiently clear and publicly promulgated so as to allow humans to regulate their behaviour on the basis of such norms.[15] Not all humans are legal agents. Young children are

---

[13] This definition draws upon Bruno Latour's depiction of "actants": "any thing that [modifies] a state of affairs by making a difference....": Bruno Latour, *Reassembling the Social: An Introduction to Actor-Network Theory* (Oxford: Oxford University Press, 2005), 71. For a different view, see Jack M. Balkin, "Understanding Legal Understanding: The Legal Subject and the Problem of Legal Coherence", *The Yale Law Journal*, Vol. 103 (1993), 105, 106–166, 106, in which the definition of legal subject is expanded to include what is described herein as agency. See also Lassa Oppenheim, *International Law: A Treatise* (1st edn. London: Longmans, Green and Co), 18–19, in which Oppenheim explains that "Since the Law of Nations is based on the common consent of individual States, and not of individual human beings, states solely and exclusively are the subjects of International Law. This means that the Law of Nations is a law for the international conduct of states and not of their citizens. Subjects of the rights and duties arising from the Law of Nations are States solely and exclusively". Though this is no longer an accurate description of the position under public international law, nonetheless the distinction and terminology concerning what it is to be a legal subject are instructive.

[14] Legal agency is related to the "internal aspect" of law, whereby a participant in a legal system regards its laws as norms for her behaviour. See H.L.A. Hart in *The Concept of Law* (2nd edn. Oxford: Clarendon 1972); Scott J. Shapiro, "What Is the Internal Point of View?" *Yale Faculty Scholarship Series* (2006) Paper 1336, http://digitalcommons.law.yale.edu/fss_papers/1336, accessed 1 June 2018.

[15] Fuller's eight principles for legal systems are a reasonable guide here. See Lon L. Fuller, *The Morality of Law* (Yale University Press, 1969), discussed in Chapter 1 at s. 3. In summary, these are as follows: (1) laws should be general; (2) laws should be promulgated, such that subjects might know the standards to which they are being held; (3) retroactive rule-making and application should be minimised; (4) laws should be understandable; (5) laws should not be contradictory; (6) laws should not require conduct beyond the abilities of those affected; (7) laws should remain relatively constant through time; and (8) laws should be administered in a manner which is consistent with the manner in which they are announced and described.

not capable of understanding laws and modifying their behaviour accordingly. The notional agency of humans who are not themselves capable of exercising it is generally imparted to another true agent, for example that human's parents or their doctors.[16] The same applies to those with cognitive impairments, in comas or similar. Agency is not a binary matter—it can exist to a greater or lesser degree. As children develop and learn, they gradually become aware of more of their legal rights and obligations, and at a certain (usually arbitrary) point, the law treats that human as being legally responsible for their own actions.[17]

Many legal systems have a concept of "personhood" or "personality",[18] which can be held by humans (natural persons) and non-human

[16] For a recent example of a situation in which two parties (the parents and the doctors) of a child disagreed on the appropriate manner of exercising the relevant imputed legal agency, see *In the matter of Charlie Gard* [2017] EWHC 972 (Fam). In that case, the parents of a terminally ill-child disagreed with the decision of the doctors not to send the child abroad for experimental treatment. In legal terms, the child was joined to the legal action against the doctors by his "guardian *ad litem*", a third party whose role it was to act "in the best interests of the child". In a sense though, each of the relevant parties to the litigation purported to be so acting, in the absence of the child's ability to exercise legal agency in its own right.

[17] The fact that children mature at different rates does not stop many legal systems setting arbitrary ages "of majority" and of criminal responsibility, at which children are held legally responsible for their own actions. As discussed further in Chapter 3, we therefore should not be averse to the setting of arbitrary thresholds for the responsibility of AI. It is important also to distinguish at the outset between agency and "patiency". Patients are those to whom moral rights and duties are owed, whereas agency is the ability to owe such rights and duties. Not all moral patients are moral agents. As noted above, young children do not meet the criteria for agency. However, children meet the criteria for patiency because adult agents owe duties to them. This chapter concerns agency rather than patiency. Whether AI qualifies for the latter (at least in moral terms) is addressed in Chapter 4.

[18] Hobbes wrote that "A person, is he, whose words or actions are considered, either as his own, or as representing the words or actions of an other man, or of any other thing to whom they are attributed, whether Truly or by Fiction". Thomas Hobbes, *Leviathan: Or, The Matter, Forme, & Power of a Common-Wealth Ecclesiasticall and Civill* (London: Andrew Crooke, 1651), 80. In this section, we avoid setting out our taxonomy by referring the term "legal persons" because the philosophical status of "personality" could lead to confusion by eliding various of the characteristics which we identify as features of being a subject and an agent, respectively. See, for example, Rodney Brooks, *Robot: The Future of Flesh and Machines* (London: Allen Lane/Penguin Press, 2002), 194–195; Benjamin Allgrove, *Legal Personality for Artificial Intellects: Pragmatic Solution or Science Fiction* (DPhil Dissertation, University of Oxford, 2004).

entities (legal persons). Although legal personality takes different forms across legal systems,[19] it only entails the status of subject and not agent. The following subsections analyse various categories of non-human subjects and legal persons through history, in order to demonstrate why none of these meet the threshold for agency described above. Chapter 5 addresses the separate question of whether, in light of its legal agency (amongst other features), AI should be granted legal personality.

2.1.1 Corporations

Companies (also known as corporations) are some of the oldest and today most common examples of non-human legal persons.[20] Companies are entities owned by shareholders (which can themselves be companies), and they are controlled by directors (which too can be companies). They can sue and be sued and can even, in some systems, be subject to criminal liability in their own right.[21]

Though it is common to talk of a corporation acting in its own name, as the English Lord Chancellor, Viscount Haldane, put it in 1915, the reality is that "...a corporation is an abstraction. It has no mind of its own any more than it has a body of its own...".[22] Historian Yuval

---

[19] See, for example, Shawn Bayern, Thomas Burri, Thomas D. Grant, Daniel M. Häusermann, Florian Möslein, and Richard Williams, "Company Law and Autonomous Systems: A Blueprint for Lawyers, Entrepreneurs, and Regulators", *Hastings Science and Technology Law Journal*, Vol. 9, No. 2 (Summer 2017), 135–161, for a discussion of different legal forms through which AI might be recognised as a legal subject in various systems. Such proposals are discussed in Chapter 5.

[20] On the history of company law, see Lorraine Talbot, *Critical Company Law* (Abingdon, UK: Routledge-Cavendish, 2007). We will use the terms "companies" and "corporations" interchangeably to refer generally to all forms of legal entities made up of collections of people.

[21] See Lord Sumption in the UK Supreme Court case *Petrodel Resources Ltd v. Prest* [2013] UKSC 34 at 8: "The separate personality and property of a company is sometimes described as a fiction, and in a sense it is. But the fiction is the whole foundation of English company and insolvency law. As Robert Goff L.J. once observed, in this domain "we are concerned not with economics but with law. The distinction between the two is, in law, fundamental": *Bank of Tokyo Ltd v. Karoon (Note)* [1987] AC 45, 64. He could justly have added that it is not just legally but economically fundamental, since limited companies have been the principal unit of commercial life for more than a century. Their separate personality and property are the basis on which third parties are entitled to deal with them and commonly do deal with them".

[22] *Lennard's Carrying Co Ltd v. Asiatic Petroleum Co Ltd* [1915] AC 705, 713.

Harari explains that limited liability companies are among humanity's most ingenious inventions but only exist as a "figment of our collective imagination".[23]

Although we *act* as if corporations can do things independently of their owners, directors and employees, in reality they cannot. Corporations like General Motors, Royal Dutch Shell, Tencent, Google and Apple certainly wield enormous power and hold vast amounts of assets but if we strip away the human input, nothing is left. True, these companies exist on paper and in electronic form, as holders of bank accounts, tax liabilities and in entries on property registers. But without humans there would be no one to take decisions that are then ascribed to the company, on which basis the rights and obligations may be altered, created and destroyed.

We should not confuse a company with its physical expression. Collective fictions including companies can have a secondary effect on the physical world. We can build towering corporate headquarters, magnificent temples and august courthouses, but these would just be empty edifices without the collective belief in whichever fiction we have used to justify their construction. The disparity between the company as a fiction and as a physical reality is illustrated by Ugland House, a building in the Cayman Islands, where over 18,000 companies are registered.[24] President Obama once said of Ugland House: "That's either the biggest building in the world or the biggest tax scam in the world".[25]

Nineteenth-century legal scholar Otto von Gierke argued that corporations are not mere fictions but in fact real "group-persons".[26] This concept can account for the fact that companies often take decisions which do not result from the choice of one single person being imputed to the company but rather from some expression of collective will, such

[23]Yuval Harari, *Sapiens: A Brief History of Humankind* (London: Random House, 2015), 19 and 363.

[24]"Frequently Asked Questions", Website of Ugland House, https://www.uglandhouse. ky/faqs.html, accessed 1 June 2018.

[25]Nick Davis, "Tax Spotlight Worries Cayman Islands", *BBC News Website*, 31 March 2009, http://news.bbc.co.uk/1/hi/world/americas/7972695.stm, accessed 1 June 2018.

[26]See, for example, Otto von Gierke, *Political Theories of the Middle Age*, edited and translated by F.W. Maitland (Cambridge: Cambridge University Press, 1927); Otto von Gierke, *Natural Law and the Theory of Society*, edited and translated by Ernest Baker (Cambridge: Cambridge University Press, 1934).

as a vote of board members. In this context, von Gierke's arguments rely on metaphysical and social constructs which turn on the "reality" of such collective will. But shared belief is clearly not the same as objective reality. Even if many people in medieval times believed that the English King's touch could cure the unpleasant disease of scrofula, this did not mean it was true.[27] A full critique of von Gierke's thesis is outside the scope of the present work,[28] but for present purposes it is sufficient to note that group personality rests ultimately on the collection of individual *human* decisions. To this extent, von Gierke's thesis does not solve the problem of how legal systems can accommodate non-human decision-making.[29]

2.1.2 Countries

For even longer than legal systems have recognised corporations, countries have been able to create and change legal relations, despite not having any independent directing mind of their own.[30] As jurist F.A. Mann commented, prior to recognition "...the non-recognised State does not exist. It is, if one prefers so to put it, a nullity".[31] In much the same

---

[27] See, for example, David J. Sturdy, "The Royal Touch in England", in *European Monarchy: Its Evolution and Practice from Roman Antiquity to Modern Times*, edited by Heinz Duchhardt, Richard A. Jackson, and David J. Sturdy (Stuttgart: Franz Steiner Verlag, 1992), 171–184.

[28] For discussion of the debate between fiction theorists and corporate realists, see, for example, S.J. Stoljar, *Groups and Entities: An Inquiry into Corporate Theory* (Canberra: Australian National University Press, 1973), 182–186; Gunther Teubner, "Enterprise Corporatism: New Industrial Policy and the 'Essence' of the Legal Person", in *A Reader on the Law of Business Enterprise*, edited by Sally Wheeler (Oxford: Oxford University Press, 1994).

[29] As is explored further in Chapters 3 and 5, the "housing" of AI within legal structures similar to corporations is one solution to addressing the legal responsibility for and rights of AI. See also Shawn Bayern, Thomas Burri, Thomas D. Grant, Daniel M. Häusermann, Florian Möslein, and Richard Williams, "Company Law and Autonomous Systems: A Blueprint for Lawyers, Entrepreneurs, and Regulators", *Hastings Science and Technology Law Journal*, Vol. 9, No. 2 (Summer 2017), 135–161.

[30] For comparative perspectives on different forms of legal personality, see Katsuhito Iwai, "Persons, Things and Corporations: Corporate Personality Controversy and Comparative Corporate Governance", *The American Journal of Comparative Law*, Vol. 47 (1999), 583–632.

[31] F.A. Mann, "The Judicial Recognition of an Unrecognised State", *International and Comparative Law Quarterly*, Vol. 36, No. 2 (1987), 348–350.

way as corporations, countries are accorded the legal status of personality and are consequently subjects of international law as well as national law. In his book *Imagined Communities*, historian and sociologist Benedict Anderson explained how countries have no objective reality beyond being a social construct. The nation, Anderson said, is "an imagined political community... *imagined* because the members of even the smallest nation will never know most of their fellow-members, meet them, or even hear of them, yet in the minds of each lives the image of their communion".[32]

Like corporations, whenever a country is said to have taken a decision, it is not in reality the country which has acted but rather one or more humans who are deemed to have the appropriate authority—whether the King, Queen, President, Prime Minister, Ambassador and so on.[33] Nations may act legally and politically through institutions such as governments and ministries but beyond headed notepaper and grand buildings, they too ultimately rely on human decision-makers. The same principles apply both to subnational entities, such as regions or districts, as well as to supranational entities and groupings, such as the European Union, or the Organization of Petroleum Exporting Countries.[34]

---

[32] Benedict Anderson, *Imagined Communities: Reflections on the Origin and Spread of Nationalism* (London: Verso, 1991), 6. Yuval Harari adopts the same approaches, grouping nations alongside laws, companies and religion as necessary "myths". Yuval Harari, *Sapiens: A Brief History of Humankind* (London: Random House, 2015).

[33] For a discussion of the authority of individuals to conclude legal relations on behalf of countries, see *Donegal International Ltd v. Zambia* [2007] 1 Lloyd's Rep 397. At a more philosophical level, see Quentin Skinner, "Hobbes and the Purely Artificial Person of the State", *The Journal of Political Philosophy*, Vol. 7, No. 1 (1999), 1–29, and David Runciman, "What Kind of Person Is Hobbes's State? A Reply to Skinner", *The Journal of Political Philosophy*, Vol. 8, No 2 (2000), 268–278.

[34] Art. 47 of the Treaty on European Union (TEU) provides that the EU itself has legal personality, making it an independent entity in its own right (at least as a matter of EU law). This is perhaps best seen as a collective agreement by the Member States to pool their sovereignty, to this extent. At least according to the official legal website of the EU, the conferral of legal personality on the EU means that it has the ability to: conclude and negotiate international agreements in accordance with its external commitments; become a member of international organisations; join international conventions, such as the European Convention on Human Rights. Glossary of Summaries, Eur-Lex: Access to European Union Law, http://eur-lex.europa.eu/summary/glossary/union_legal_personality.html, accessed 1 June 2018.

2.1.3 Buildings, Objects, Deities and Concepts

Buildings, objects, deities and concepts have been granted some legal rights. In the UK case *Bumper Development Corporation Ltd v. Commissioner of Police of the Metropolis*,[35] the Court of Appeal held that an Indian temple having a legal persona recognised in India could assert rights and make claims under English law. Even though it would not be recognised as a litigant if based in England and Wales, the temple was nonetheless entitled, in accordance with the principle of comity of nations, to sue in England for the return of a statue allegedly looted from it. Similarly in the US case *Autocephalus Greek Orthodox Church of Cyprus v. Goldberg*,[36] the US 7th Circuit Court of Appeals held that mosaics should be returned to a church which was deemed to be their legal owner.

In *Bumper*, the UK Court of Appeal drew a parallel with the legal recognition of corporations in England.[37] Crucially though, the temple acted through its human representatives. The temple building did not instruct lawyers, nor did the temple itself prepare statements of case, evidence and all the other steps needed to pursue its rights.[38] In the Indian appeal *Pramatha Nath Mullick v. Pradyumnakumar Mullick*, also known as the *Hindu Idol* case, ruled upon by the Judicial Committee of the Privy Council (then India's highest court), Lord Shaw held:

> A Hindu idol is, according to long established authority, founded upon the religious customs of the Hindus, and the recognition thereof by courts of law of 'juristic entity'. It has a juridical status with the power of suing and being sued. Its interests are attended to by the person who has the deity in his charge and who is in law its manager with all the powers which would, in such circumstances, on analogy be given to the manager of the estate of an infant heir.[39]

---

[35] [1991] 1 WLR 1362.

[36] 917 F. 2d 278 (7th Cir. 1990).

[37] For discussion of the *Bumper* case, see Mira T. Sundara Rajan, *Moral Rights: Principles, Practice and New Technology* (New York: Oxford University Press, 2011), 468–476.

[38] In Shakespeare's Macbeth, it was prophesied that a forest would move: "Macbeth shall never vanquish'd be until/Great Birnam Wood to high Dunsinane Hill/Shall come against him" (Act 4, Scene 1). In the event, the trees in the forest did not "Unfix his earth-bound root" (Act 4, Scene 1) of their own accord, but rather the soldiers in Malcolm's army cut down the trees and carried them as camouflage when storming Macbeth's castle.

[39] (1925) 52 Ind. App. 245 at 250.

There are a few, possibly apocryphal, historical accounts of objects being punished. One tells of a statue erected by Athenians in honour of a famous athlete, Nikon of Thasos, which was pushed from its pedestal by his envious foes. As it fell, the statue crushed one of its assailants. Instead of laying blame on the other members of the mob, and perhaps on the unfortunate victim himself, the Athenians put the statue before a tribunal. The statue was found guilty and sentenced to be cast into the sea; history does not relate whether it was allowed to plead self-defence.[40]

It is also said that the eighteenth-century Japanese samurai and jurist Ōoka Tadasuke ruled that a *jizo* (statue) in a temple be bound with rope as punishment for having been the only witness to a crime (the theft of a piece of silk) and not doing anything to stop it.[41] To this day, a statue—allegedly the same as punished by Ōoka Tadasuke—remains tied by a large number of ropes in Tokyo's Narihira Temple.[42]

The Japanese example is instructive, given that in Shintō, a Japanese religion, or belief-structure (which remains the largest in that country),[43] all things are said to possess *kami*, which translates as "spirit", "soul" or "energy". This includes people and animals, as well as inanimate objects or natural features, such as rocks, rivers and places.[44] Consequently, to a

---

[40] Evans, *Animals*, 172. A similar story (except this time referring to the athlete Theagenes) appears in the writing of Pausanias, a second century AD Greek traveler and geographer. See Pausanias, *Description of Greece*, translated by William H.S. Jones, D. Litt, and Henry A. Ormerod (Cambridge, MA: Harvard University Press; London, William Heinemann Ltd, 1918), 6.6.9–11. See also John Chipman Gray, *The Nature and Sources of the Law*, edited by Roland Gray (London: MacMillan, 1921), 46. As to the psychological reasons which might justify "punishment" of an inanimate object, see Chapter 8 at s. 5.3.

[41] Pascal Fauliot, *Samurai Wisdom Stories: Tales from the Golden Age of Bushido* (Boulder, CO: Shambhala Publications, 2017), 119–120. Fauliot writes that in fact the "punishment" of the statue was an elaborate ruse by the wise judge: when the gathered people laughed at this ridiculous sentence, he fined each of them one piece of silk for contempt of court. Once all the silk was gathered, the victim of the silk theft was able to identify the stolen piece and thereby identify the perpetrator.

[42] Pictures of this can be seen at Muza-chan's Gate to Japan website, http://muza-chan.net/japan/index.php/blog/unique-tradition-rope-wrapped-jizo-statue, accessed 1 June 2018.

[43] According to the CIA World Fact Book, Shintō is practised by approximately 80% of the Japanese population, "Entry on Japan", CIA World Fact Book, https://www.cia.gov/library/publications/the-world-factbook/geos/ja.html, accessed 1 June 2018.

[44] "Shinto at a Glance", *BBC Religions*, last updated 10 July 2011, http://www.bbc.co.uk/religion/religions/shinto/ataglance/glance.shtml, accessed 1 June 2018; See also Encyclopedia of Shinto, http://eos.kokugakuin.ac.jp/modules/xwords/, accessed 1 June 2018.

Japanese audience the notion of an object holding rights and responsibilities is perhaps not as farfetched as it may appear to a Western observer.[45]

More recently, legal scholars and policy-makers have given serious consideration to the question of whether parts of the environment, such as plants, trees or coral reefs, might have legal standing.[46] For instance, in 2010 Bolivia passed the "Law of the Rights of Mother Earth", which included in Article 5 the following pronouncement: "For the purpose of protecting and enforcing its rights, Mother Earth takes on the character of collective public interest. Mother Earth and all its components, including human communities, are entitled to all the inherent rights recognized in this Law".[47] On the basis of a similar law, in 2011 a group of Ecuadorian citizens brought a successful legal action on behalf of the environment against the Provincial Government of Loja to halt expansion of a roadway which they claimed that was damaging an important watershed.[48]

The endowment of non-human entities with rights will be considered further in Chapter 4, but for present purposes it is sufficient to note that even if natural entities such as trees, rivers, mountains or even the environment as a whole are granted standing to sue, in reality it is a human which must decide to pursue the claim.[49] As Bryson, Diamantis and Grant say: "Nature cannot protect itself in a court of law".[50]

---

[45] We return to this theme in Chapter 4, discussing whether AI should be accorded rights.

[46] Christopher Stone, "Should Trees Have Standing?-Toward Legal Rights for Natural Objects", *Southern California Law Review*, Vol. 45 (1972), 450, 453–457.

[47] "Law of Mother Earth: The Rights of Our Planet. A Vision from Bolivia", World Future Fund, http://www.worldfuturefund.org/Projects/Indicators/motherearthbolivia.html, accessed 18 July 2017. See also John Vidal, "Bolivia Enshrines Natural World's Rights with Equal Status for Mother Earth" *The Guardian*, 10 April 2011, https://www.theguardian.com/environment/2011/apr/10/bolivia-enshrines-natural-worlds-rights, accessed 1 June 2018.

[48] Natalia Greene, "The First Successful Case of the Rights of Nature Implementation in Ecuador", *The Rights of Nature* (2011), http://therightsofnature.org/first-ron-case-ecuador/, accessed 1 June 2018.

[49] Lawrence B. Solum, "Legal Personhood for Artificial Intelligences", *North Carolina Law Review*, Vol. 70, 1231–1287, 1239–1240.

[50] Joanna J. Bryson, Mihailis E. Diamantis, and Thomas D. Grant, "Of, for, and by the People: The Legal Lacuna of Synthetic Persons", *Artificial Intelligence and Law*, Vol. 25, No. 3 (September 2017), 273–291.

2.1.4 Animals

This section considers the legal regimes applicable to animals, both through history and in the current day. It is suggested here that although some legal systems now recognise animal rights[51] and in the past animals were also thought to be subject to responsibilities, animals do not meet the threshold for legal agency.

Historic legal treatment of animals

Edward Payson Evans described various instances of animals being tried for crimes in his 1906 work *The Criminal Prosecution and Capital Punishment of Animals*.[52] The punishment of animals for "wrongs" can be traced back at least as far as the Old Testament: "If an ox gore a man or a woman that they die, then the ox shall be surely stoned, and his flesh shall not be eaten; but the owner of the ox shall be quit".[53] The ox is punished, and its owner is spared. However, in an early example of foreseeability of harm giving rise to vicarious liability, the next verse provides: "But if the ox were wont to push with his horn in time past, and it hath been testified to his owner, and he hath not kept him in, but that he hath killed a man or a woman; the ox shall be stoned, *and his owner also shall be put to death*".[54]

Evans lists an extraordinary array of animals against which judicial proceedings were instituted, described by one commentator as "a veritable Noah's Ark of creatures", including "horseflies, Spanish flies and gad-flies, beetles, grasshoppers, locusts, caterpillars, termites, weevils, blood-suckers, snails, worms, rats, mice, moles, cows, bitches and she-asses, horses, mules, bulls, pigs, oxen, goats, cocks, cockchafers, dogs, wolves, snakes, eels, dolphins and turtledoves".[55] Among the crimes for which animals were put to death, Evans notes that "[i]n 1394, a pig was

---

[51] See Chapter 4 at s. 2.

[52] Edward Payson Evans, *The Criminal Prosecution and Capital Punishment of Animals* (London: William Heinemann, 1906). Hereafter "Evans, *Animals*".

[53] Exodus 21:28, King James Bible. Roman law, by contrast, does not appear to allow for any liability on the part of, or punishment for the animal. See, for example, D.I.C. Ashton-Cross, "Liability in Roman Law for Damage Caused by Animals", *The Cambridge Law Journal*, Vol. 11, No. 3 (1953), 395–403.

[54] Ibid., 21:29 (emphasis added).

[55] Piers Beirnes, "The Law Is an Ass: Reading E.P. Evans' The Medieval Prosecution and Capital Punishment of Animals", *Society and Animals*, Vol 2. No. 1, 27–46, 31–32.

hanged at Mortaign for having sacrilegiously eaten a consecrated wafer".[56]

Evans suggests various reasons as to why different societies saw fit to hold animals legally responsible for their actions. One justification relied simply on the aforementioned section in Exodus and reasoned by analogy from there that all animals should be subject to punishment where they cause harm. It is unclear from the Old Testament itself as to what was the reasoning behind such punishments, but it seems they could either be rationalised on the basis of (a) protection of society from an animal which, having caused harm in the past, might do so again; or (b) retribution against the animal.

Another justification for the punishment of animals, suggested by Esther Cohen, is that medieval society considered that animals were inferior to humans in the cosmic hierarchy, having been created for the latter's utility. Thus, any animal which killed a human had upset the cosmic order and thereby offended God.[57] Piers Beirnes contends that "there is no solid evidence of a general belief that the volition and intent of animals was of the same order as those of humans".[58] Generally though, the justifications for holding animals liable will have varied from place to place and time to time.[59] Moreover, the ostensible justification offered for bestial trial and punishment may well have differed from the underlying one. Reviewing Evans' work, psychologist Nicholas Humphrey concluded: "Taken together, Evans' cases suggest that again and again, the

---

[56] Evans, *Animals*, 156.

[57] Esther Cohen, "Animals in Medieval Perceptions: The Image of the Ubiquitous Other", *Animals and Human Society: Changing Perspectives*, edited by Aubrey Manning and James Serpell (London and New York: Routledge, 2002), 59–80.

[58] Piers Beirnes, "The Law Is an Ass: Reading E.P. Evans' The Medieval Prosecution and Capital Punishment of Animals", *Society and Animals*, Vol 2. No. 1, 27–46, 29.

[59] See, for example, the unusual *Coustumes et stilles de Bourgoigne*, a legal text from between 1270 and 1360, which made a distinction between homicides by an ox or a horse (following which the animal was to be spared), and a homicide committed by another animal "or a Jew" (following which the perpetrator was to be "hung by their rear legs"). Cited in Esther Cohen, "Animals in Medieval Perceptions: The Image of the Ubiquitous Other", *Animals and Human Society: Changing Perspectives*, edited by Aubrey Manning and James Serpell (London and New York: Routledge, 2002), 59–80.

true purpose of the [animal] trials was psychological. People were living at times of deep uncertainty".[60]

As with companies, it can be seen that the decision to hold animals legally liable for crimes—and thereby to make them legal subjects—was generally speaking divorced from any view that the animals actually were aware of their obligations and could have acted as agents.

Modern legal treatment of animals

It might be argued that because animals exhibit many of the same capabilities and tendencies as AI, we ought to apply the same legal principles to both.[61] On the surface, there are certainly some similarities between animals and AI: both can be trained (at least up to a point), both can follow simple commands, both can learn new skills or techniques based on their environments and the thought processes of both can at times be somewhat inscrutable to a human observer.

Broadly speaking, a balance must be struck between liability assumed by an animal's owner which is based in part on the tendencies of the animal, and the countervailing principle that "everyone must take the risks associated with the ordinary characteristics of animals commonly kept in this country. These risks are part of the normal give and take of life".[62] In the UK, the liability for animals is governed partly by judge-made common law, including negligence, and partly by legislation under the Animals Act 1971.[63] The latter provides for strict liability for the "keeper"[64] of an animal in certain defined circumstances.

Though mechanisms for accommodating responsibility for animals may provide some assistance in designing systems for AI, there are several factors render it difficult to apply to AI all of the laws on liability for animals, at least in the long term.

---

[60]Nicholas Humphrey, "Bugs and Beasts Before the Law", *The Public Domain Review*, http://publicdomainreview.org/2011/03/27/bugs-and-beasts-before-the-law/, accessed 1 June 2018.

[61]See, for example, Matthew Scherer, "Digital Analogues (Intro): Artificial Intelligence Systems Should Be Treated Like...", *Law and AI Blog*, 8 June 2016, http://www.lawandai.com/2016/06/08/digital-analogues/, accessed 1 June 2018.

[62]*Mirvahedy v. Henley* [2003] UKHL 16; [2003] 2 AC 491 [6].

[63]For discussion, see Rachael Mulheron, *Principles of Tort Law* (Cambridge: Cambridge University Press, 2016).

[64]UK Animals Act 1971, s. 6(3).

First, many laws maintain some form of distinction between wild and domesticated animals. This distinction is inexact even when it comes to animals, as was demonstrated by *McQuaker v. Goddard*,[65] a case which concerned the responsibility for a camel in the Chessington Zoological Garden which bit the hand of a visitor who had been feeding it apples. The Court of Appeal of England and Wales held, after some debate, that the camel was to be treated as "domesticated", with the effect that the zoo's owner was not held to be liable for its violent actions. Lord Justice Scott explained: "Wild animals are assumed to be dangerous to human beings because they have not been domesticated. Domestic animals are assumed not to be dangerous". Domesticated animals, on the other hand, could be assumed to be safe unless it was shown that the owner or keeper had specific knowledge of dangerous tendencies. Unlike wild animals, AI does not (by definition) naturally exist in a state of freedom. This might perhaps be said to occur if an AI entity was somehow to "escape" from human control and develop independently. However, for the moment, this fundamental distinction in animal law remains difficult to apply across to AI.

Secondly, animals are limited by their natural faculties. Depending on the species, animals can be trained to perform a range of tasks, but there is a certain level of complexity at which further tuition becomes impossible.[66] A dog may be taught to retrieve a ball, but it cannot be taught to fly an aeroplane or perform brain surgery. The eminent psychologist David Premack wrote: "A good rule of thumb is this: Concepts acquired by children after 3 years of age are never acquired by chimpanzees".[67] AI is not so limited. As discussed in Chapter 1 at Section 5, in recent years there have been significant advances in capabilities of AI systems. Even if there are further peaks and troughs of activity, it is reasonable to predict that in the coming decades the technology will continue to improve, and consequently to be delegated yet more important tasks by humans. Consequently, the legal and moral issues raised by the actions of AI are of a different order of complexity than those of animals.

---

[65][1940] 1 KN 687.

[66]Dorothy L. Cheney, "Extent and Limits of Cooperation in Animals", *Proceedings of the National Academy of Sciences*, Vol. 108, No. Supplement 2 (2011), 10902–10909; David Premack, "Human and Animal Cognition: Continuity and Discontinuity", *Proceedings of the National Academy of Sciences*, Vol. 104, No. 35 (2007), 13861–13867.

[67]David Premack, "Human and Animal Cognition: Continuity and Discontinuity", *Proceedings of the National Academy of Sciences*, Vol. 104, No. 35 (2007), 13861–13867.

Like AI, animals will not always act as expected. As the plaintiff in *McQuaker v. Goddard* discovered, a previously docile animal may suddenly lash out and bite a passer-by, or a trained horse may run into the middle of a road.[68] But it is inconceivable that an animal might commit a securities fraud.[69] The predictability of animals' range of actions is connected to the next difference, namely the manner in which animals will go about achieving such actions.

Thirdly, the manner in which an animal will achieve a goal is broadly predictable and is more often attributable to evolution rather than individual decision-making. Examples of animals "solving" problems are limited to fairly rudimentary tasks within narrow cognitive boundaries, such as a monkey using a stick to poke a termite's nest, or a bird dropping a snail's shell from height in order to access the animal inside. These are hardly on a par with defeating human champions at poker.[70]

AI pioneer Marvin Minsky said, in a discussion of the extent to which intelligence as demonstrated by both humans and artificial entities is different from that which is called "intelligence" in animals:

> …it is only an illusion that animals can 'solve' … problems. No individual bird *discovers* a way to fly. Instead, each bird exploits a solution that evolved from countless reptile years of evolution. Similarly, although a person might find it very hard to design an oriole's nest or a beaver's dam, no oriole or beaver ever figured out such things at all. Those animals don't 'solve' such problems themselves; they only exploit procedures available within their complicated gene-built brains.[71]

By contrast, AI can function not just by virtue of what it has been programmed to do but learns and changes of its own accord. It might be objected that Minsky's quote above is over-simplistic, and that some

---

[68] As occurred in *Searle v. Wallbank* [1947] AC 341; [1947] 1 All ER 12.

[69] John Markoff, "As Artificial Intelligence Evolves, So Does Its Criminal Potential", 23 October 2016, https://www.nytimes.com/2016/10/24/technology/artificial-intelligence-evolves-with-its-criminal-potential.html, accessed 1 June 2018.

[70] As Professor Toby Walsh notes, "Poker offers some interesting challenges… One is that it is a game of imperfect information… Another challenge of poker is that it is a game of psychology, requiring you to understand the strategy of your opponents… Despite these challenges, computers are now very good at playing poker". Toby Walsh, *Android Dreams* (London: Hurst & Co, 2017), 85.

[71] Marvin Minsky, *The Society of Mind* (London: Picador/Heinemann, 1987), para. 7.1.

animals are capable of learning and developing skills by themselves. This realisation may in turn require humans to re-think their relationship with animals and the rights which they are accorded.[72] Such discussion is outside the remit of this book. Perhaps then, it is more correct to say that the difference between AI's decision-making and that of animals is one of degree rather than type.

2.1.5 Conclusions on Agency

Though in common parlance we often speak of a company or a country "deciding" to do something, in reality this is shorthand for saying that the humans in control of that entity made such a decision. Animals may in a limited sense choose to take one action rather than another, but they lack the crucial second part of legal agency, namely the ability to understand and interact with a legal system. The final section of this chapter suggests that AI may meet both these requirements, independent of human input. The question of what it is to be legally "independent" of humans is addressed further below, in relation to causation.

2.2   Causation

The second fundamental principle challenged by AI is causation: the apparent connection between one event and others which follow.

The traditional view of causation is that events may be characterised as linked through relationships of cause and effect. This is easy to express in simple terms. If a brick is thrown at a glass window which then shatters, the brick being thrown is the cause and the window shattering is the effect. Many philosophical[73] and scientific[74] objections have been raised to this account of events, but it nonetheless remains the basis for most legal systems.

---

[72] See, for instance, Yuval Harari, "Industrial Farming Is One of the Worst Crimes in History", *The Guardian*, 25 September 2015, https://www.theguardian.com/books/2015/sep/25/industrial-farming-one-worst-crimes-history-ethical-question, accessed 1 June 2018.

[73] For a summary of such philosophical discussions, see Jonathan Schaffer, "The Metaphysics of Causation", *The Stanford Encyclopaedia of Philosophy (Fall 2016 Edition)*, edited by Edward N. Zalta, https://plato.stanford.edu/archives/fall2016/entries/causation-metaphysics/, accessed 1 June 2018.

[74] One of the most influential critiques of causality is that of physicist Niels Bohr, who in 1948 identified that quantum theory was "irreconcilable with the very idea of causality". Niels Bohr, "On the Notions of Causality and Complementarity", *Dialectica*, Vol. 2, No. 3–4 (1948), 312–319.

Without the notion of cause and effect, legal agency would not function. Legal agency is the ability to understand the consequences of one's actions in legal terms and to adapt one's behaviour accordingly, so as to bring about or avoid certain events. Causation provides the connection between acts or omissions and their consequences.

In law, the deemed cause of an event is not simply a question of objective fact but rather of policy and value judgements. The key question for present purposes is whether the relationships which we have to date treated as being causal can withstand the intervention of AI.

2.2.1 Factual Causation

Causation in law, at least with regard to allocating liability for harm, encompasses two separate elements: factual and legal. Donal Nolan explains that factual causation is "the question of whether or not there is a historical connection between the wrongful conduct of the defendant and the damage suffered by the claimant", which is "analytically different" from legal causation, or "proximate cause", namely "whether the historical connection between the defendant's wrongful conduct and the damage suffered by the claimant is strong enough to justify the imposition of liability".[75]

The most common expression of factual causation is to construct a hypothetical counterfactual, by asking whether "but for" the relevant potentially causative event, the relevant effect would have occurred.[76] As Wex Malone noted, the "but for" test is an artificial construct: "...this very announcement is a statement of legal policy. It marks an effort to point out the bare minimum requirement for imposing liability".[77]

If a murderer stabs a victim to death, then at an extreme one might say that the murderer's parents are the cause of the murder, because without them then the murderer would never have been born to go on

---

[75] Donal Nolan, "Causation and the Goals of Tort Law", in *The Goals of Private Law*, edited by Andrew Robertson and Hang Wu Tang (Oxford: Hart Publishing, 2009), 165–190, 165.

[76] This test was criticised by Hume, who had himself first articulated it in *An Enquiry Concerning Human Understanding*, V, Pt. I; Loewenberg, "The Elasticity of the Idea of Causality", *University of California Publications in Philosophy*, Vol. 15, No. 3 (1932).

[77] Wex S. Malone, "Ruminations on Cause-in-Fact", *Stanford Law Review*, Vol. 9, No. 1 (December 1956), 60–99, 66.

and commit the murder. Indeed, applying this reasoning one could keep going back indefinitely, to say that the grandparents, great-grandparents and so on were the factual cause of the murder.

Factual causation may appear at first glance to simply be a question of scientific evidence ("did he jump or was he pushed?"), but two examples show that it is treated in practice by legal systems as a policy-based issue.

"Under-determination" refers to situations where there is insufficient evidence to know whether a given event was a "but for" cause.[78] Strictly speaking it a question of the adequacy of evidence rather than a principled objection to the but for test. That said, in the real world principles of causation must still be used even where (as is often the case) humans lack perfect knowledge of what happened.

From the late twentieth century onwards, there has been much litigation concerning liability for illnesses caused by certain tiny carcinogenic particles.[79] These carcinogens, principally asbestos, were present in mining and industrial processes for many years. Scientific evidence suggests that exposure to just one molecule of the relevant carcinogen can lead to the development of a fatal cancer. Victims have often worked for more than one employer over the course of their careers in the relevant industries. Many years later when an unfortunate victim becomes ill, it is unclear which employer's actions or omissions were the but for cause of the damage.[80] In such circumstances, judges have departed from

[78] Few, if any, legal systems demand perfect knowledge of what happened. Usually, the standard it set somewhere lower. In the UK, the burden of proof applied by the courts in civil cases amounts to being able to prove (in civil cases) that one event "on the balance of probabilities" caused another event, meaning that it was more than 50% likely that the first event caused the second. Under-determination occurs where even this relatively low standard cannot be met.

[79] For discussion, see Jane Stapleton, "Factual Causation, Mesothelioma and Statistical Validity", *Law Quarterly Review*, Vol. 128 (April 2012), 221–231; John G. Fleming, "Probabilistic Causation in Tort Law", *The Canadian Bar Review*, Vol. 68, No. 4 (December 1989), 661–681.

[80] This was essentially the fact pattern in the English cases of *McGhee v. National Coal Board* [1973] 1 WLR 1 (HL); *Fairchild v. Glenhaven Funeral Services Ltd* [2003] 1 AC 32; and *Barker v. Corus* (UK) Ltd [2006] UKHL 20; [2006] 2 AC 572. For discussion of the principles of causation in Canada, see *Cook v. Lewis* [1951] SCR 830; *Lawson v. Laferriere* (1991) 78 DLR (4th) 609. For those in Australia, see *Rufo v. Hosking* [2004] NSWCA 391.

the usual "but for" test in order to provide a remedy for victims who would—under the normal principles—not have one.[81]

"Over-determination" occurs where there are two or more causes, each of which individually would have been sufficient to cause the effect in question. For instance, take a situation where person A lights a fire on one side of a house and person B independently lights another fire on the other side of the house, and the house burns down. But for the actions of either A or B then the house still would have burned down. In such circumstances, courts have recoiled from a strict interpretation of the "but for" test, according to which both would escape liability.[82]

2.2.2 Legal Causation

Once factual causation is established, the next stage is to inquire whether a particular factual cause was also a proximate or legal cause. Factual causation is a necessary, but not sufficient factor for legal causation.[83] In

---

[81] In the influential US case, *Sindell v. Abbott Laboratories* 607 P. 2d 924 (Cal. 1980), the court dispensed with the usual balance of probabilities test that the defendant was the "but for" cause of harm and instead apportioned liability against multiple defendants on the basis of their market share in a fungible good which caused injury. *Sindell* shifted the burden of proof on to the defendants but it remained open to them to prove that their products were not responsible for the injury. See also *Vigioltou v. Johns-Manville Corp.*, 543 F. Supp. 1454, 1460–1461 (W.D. Pa. 1986).

[82] See, for example, the Canadian Supreme Court decision in the "double hunter" case, *Cook v. Lewis* [1951] SCR 830, in which two hunters simultaneously shot at a grouse, missed and hit another member of their party. The solution selected by Rand J in that case was to shift the burden of proof from the victim to the hunters, to prove that they had *not* caused the injury. A similar issue occurred in *Jobling v. Associated Dairies* [1982] AC 794, where one causative event occurred to cause harm, but was then followed by another caus-ative event that would have caused the same harm and occurred prior to the first coming to trial. In the UK House of Lords, Lord Wilberforce acknowledged that in the interests of justice he was compelled to abandon the normal "but for" test, without being able to expound an alternative, saying at 805: "The result of the present case may be lacking in precision and rational justification, but so long as we are content to live in a mansion of so many different architectures, this is inevitable".

[83] It should be noted that the person deemed to be the legal cause of the event is usu-ally held liable, there are some situations in which a subject is held liable for consequences which they did not cause, including in situations of strict or vicarious liability. We will return to these legal mechanisms in Chapter 3.

legal causation, the question is not so much what was the cause of an event, but rather: what was the *relevant* cause?

To avoid the question of legal causation simply becoming a circular exercise (akin to saying "that which is legally relevant is selected because it is legally relevant"), there are certain meta-norms which inform many legal systems when a person will be held responsible for a given consequence.[84] These vary across legal system and between different contexts—such as criminal and private law.

The overarching ingredients of legal causation for harm include: (a) the free, deliberate and informed action or omission of a legal agent; (b) that the agent either knew or ought to have known of the potential consequences of such action or omission; and (c) that there has been no intervening act (sometimes referred to in Latin as a *novus actus interveniens*) splitting factors (a) and (b) from the eventual consequences.[85]

Part (b) is sometimes referred to as the "foreseeability" or "remoteness" of certain consequences. Applying this doctrine, where an unfortunate but unpredictable chain of events leads from one action to damage, the person who originally caused the damage may be excused liability. In a leading US case from 1928, *Palsgraf v. Long Island Railroad Co*,[86] a railway employee dropped a package on to the platform, which exploded, and caused a coin-operated scale standing at the other end of the platform to fall over, hitting Mrs. Palsgraf, and causing her psychological injury. The railway company was found not to be liable because the chain of events was deemed too unlikely as to have been foreseeable. Mrs. Palsgraf had not come within the group of people for whom "hazard was apparent to the eye of ordinary vigilance", and therefore no actionable wrong had been committed towards her.

---

[84] For a discussion of such meta-norms in the context of tort, see Allen Linden, *Canadian Tort Law* (5th edn. Toronto: Butterworths, 1993), Chapter 1.

[85] The US Restatement (Third) of Torts: Physical and Emotional Harm provides that a superseding cause is "an intervening force or act that is deemed sufficient to prevent liability for an actor whose tortious conduct was a factual cause of harm", para 34 (American Law Institute, 2010). For an extensive discussion of causation and its underlying moral and legal principles, see H.L.A. Hart and Anthony M. Honoré, *Causation in the Law* (2nd edn. Oxford: Clarendon Press, 1988). For a helpful summary and criticism, see Jane Stapleton, "Law, Causation and Common Sense", *Oxford Journal of Legal Studies*, Vol. 8, No. 1 (1988), 111–131.

[86] 248 N.Y. 339, 162 N.E. 99 (1928).

The three ingredients of legal causation in turn support the underlying tenets of legal agency, namely the ability of humans to understand the consequences of their actions and adapt their behaviour accordingly. If a person's action is compelled, for example by force, then that person's agency has been compromised. Likewise, if a person's action was free but the consequences of it were unforeseeable, then the person cannot be said to have exercised full agency with regard to the result, because agency requires that the result at least could have been reasonably predicted. The emphasis on the free will of agents extends not just to the causation of damage but also the ability to create legal agreements. Where a person's freedom of choice is vitiated by duress, or even misrepresentation, then a contract to which they have apparently agreed may be void.[87] Finally, the emphasis on intervening acts upholds agency in general because it gives legal effect to the third party's free and deliberate action.[88]

The above analysis has focussed primarily on situations where injury and damage have been caused, but causation can also play a significant role in establishing who is responsible for beneficial events, such as the creation of intellectual property rights in inventions and designs—which often have several factual sources.[89] For instance, where an AI system

---

[87] See, for example, *Pau On v. Lau Yiu Long* [1980] AC 614.

[88] Though the three ingredients of legal causation usually produce an answer which accords to common sense, in some circumstances they can lead to striking results, when different legal fictions are applied. One of the most striking such result occurred in South Africa where, in 2012, several hundred miners were charged with the murder of other miners who had been shot by police during a riot. South African prosecutors reasoned that because the police shot at the crowd of miners in self-defence, the actions of the police were not treated as being sufficiently free, willing and informed as to break the chain of causation. The police shooting was a foreseeable consequence of the riot, and therefore, all of the miners who participated were charged with having caused the murders of their fellow miners. Some British newspapers called this the result of an "apartheid" law, but in fact it was at least in part the result of an application of standard principles of legal causation equally applicable in the UK. See Jacob Turner, "Do the English and South African Criminal Justice Systems Share a 'Common Purpose'?" *African Journal of International and Comparative Law*, Vol. 21, No. 2 (2013), 295–300. For a US case in which a similar result was reached, see *People v. Caldwell*, 681 P. 2d 274 (Cal. 1984).

[89] See, for example, the leading UK House of Lords case, *Designers Guild Ltd v. Russell Williams (Textiles) Ltd (t/a Washington DC)* [2000] 1 WLR 2416, concerning the question of what constituted copying a "substantial part" of another's design.

writes a best-selling book, or creates a valuable work of art, questions arise as to who owns the relevant property.[90] The creation and ascription of intellectual property rights is partly a question of fact ("Is this painting by Matisse?"), but also to a significant degree a question of policy ("How far should the design of a fabric be protected if it has been heavily influenced by a Matisse painting?").[91]

The legal system in question might promote a range of aims, from fostering creativity for its own sake to increasing economic output.[92]

2.2.3 Conclusions on Causation

Whether it is a question of determining liability for harm or responsibility for beneficial events, causation is not simply a question of objective fact but rather one of economic, social and legal policy.[93] The analysis encompasses, whether overtly or covertly, judgements about what types of behaviour we want to promote or discourage, as well as issues of justice and distribution. Seen in this light, it should become clear that seeking a human or even a fictional corporate agent behind every AI act is just one of many policy responses that could be chosen.[94]

---

[90] This issue is discussed in greater detail in Chapter 3 at s. 4.

[91] For an enlightening discussion of some of these policies, see the judgment of Lord Hoffmann in the House of Lords in the *Designers Guild* case (FN 88 above), where he commented that "Copyright law protects foxes better than hedgehogs". By this, he meant that copyright in the UK protects those with many small ideas better than it protects those with one big idea. The phrase derives from the Greek philosopher Archilochus. The key point for present purposes is that another system of law might provide greater support to the hedgehogs.

[92] See, for example, Jane C. Ginsburg, "The Concept of Authorship in Comparative Copyright Law", *DePaul Law Review*, Vol. 52 (2003), 1063. We address potential ways of determining the ownership of such designs and products in Chapter 4.

[93] Curtis E.A. Karnow, "Liability for Distributed Artificial Intelligences", *Berkeley Technology Law Journal*, Vol. 11 (1996), 147, 191–192, http://scholarship.law.berkeley.edu/btlj/vol11/iss1/3, accessed 1 June 2018.

[94] In Chapter 3, we elaborate further on other options concerning the responsibility for what AI does.

3   FEATURES OF AI WHICH CHALLENGE FUNDAMENTAL LEGAL CONCEPTS

AI law expert Ryan Calo says, in a paper which argues for "a moderate conception of legal exceptionalism for purposes of assessing robotics",[95] a technology is exceptional "when its introduction into the mainstream requires a systematic change to the law or legal institutions".[96]

This section provides two reasons as to why AI is exceptional: it makes moral choices; and it can develop independently. As a result of these features, the fundamental legal concepts of agency and causation—at least in their present human-centric form—are likely to be stretched to breaking point.[97]

3.1   AI Makes Moral Choices

Given that the essence of AI is its autonomous choice-making function, AI is qualitatively unlike existing technologies in that it must sometimes take independent "moral" decisions. This is challenging to established legal systems because for the first time a piece of technology is interposing itself between humans and an eventual outcome. Rather than

---

[95] Ryan Calo, "Robotics and the Lessons of Cyberlaw", *California Law Review*, Vol.103, 513–563. Balkin criticises Calo, correctly, for placing an undue emphasis on robots, as opposed to AI more generally. Jack B. Balkin, "The Path of Robotics Law", *The Circuit* (2015), Paper 72, Berkeley Law Scholarship Repository, http://scholarship.law.berkeley. edu/clrcircuit/72, accessed 1 June 2018: "We may be misled if we insist on too sharp a distinction between robotics and AI systems, because we do not yet know all the ways that technology will be developed and deployed".

[96] Calo also argues that "robotics" have a distinct set of legal properties. However, his analysis focuses primarily on embodied technologies. Calo states: "robots are best thought of as artificial objects or systems that sense, process, and act upon the world to at least some degree". In our analysis, AI is the starting point and robots are a subset thereof. Consequently, this book's treatment of the unique features of AI differs somewhat from that of Calo. See also Jack B. Balkin "The Path of Robotics Law", *The Circuit* (2015), Paper 72, Berkeley Law Scholarship Repository, http://scholarship.law.berkeley.edu/clrcircuit/72, accessed 1 June 2018.

[97] For a discussion of why AI challenges notions of foreseeability which are integral to tort law, see Curtis E.A. Karnow, "The Application of Traditional Tort Theory to Embodied Machine Intelligence", in *Robot Law*, edited by Ryan Calo, Michael Froomkin, and Ian Kerr (Cheltenham and Northampton, MA: Edward Elgar, 2015), 53. See also Chapter 3 at s. 2.1.3.

attempt to define what is meant by morality (which is itself the subject of much debate),[98] it is sufficient for our purposes to say that AI takes choices which would be regarded as having a moral character or outcome if they were undertaken by a human.[99]

Life is full of moral choices, and where these are of a particularly serious or consequential nature, answers are often provided for by the law—saving each individual citizen from the terrible dilemma of making a decision. For example, in most countries voluntary euthanasia (assisted suicide) is illegal. However, in the Netherlands, Belgium, Canada and Switzerland it is allowed under strictly controlled circumstances.

It might be thought that no new laws are needed for AI, because it can simply follow the promulgated laws that apply to humans in any given legal system.[100] Thus, it might be permissible for a robot in Switzerland to administer a fatal dose of drugs to a consenting patient, but not across the border in France where this would be illegal. However, the law does not take away all moral choices.

First, in many circumstances the law leaves gaps for discretion where no right or wrong answer is mandated.

Secondly, even where the law does stipulate a moral outcome, there are some circumstances in which that law (or perhaps its enforcement) might be overridden by other concerns. For instance, although assisting suicide is illegal in the UK, the Crown Prosecution Service has published guidance which provides a prosecution is less likely to be pursued if "the suspect was wholly motivated by compassion".[101]

---

[98] Bernard Gert and Joshua Gert, "The Definition of Morality", *The Stanford Encyclopaedia of Philosophy* (Spring 2016 Edition), edited by Edward N. Zalta, https://plato.stanford.edu/archives/spr2016/entries/morality-definition/, accessed 1 June 2018.

[99] For instance, the decision to murder the children of a rival would be considered reprehensible if carried out by a human. Such behaviour is regularly seen in the animal world and is rarely considered to attract moral condemnation. See Anna-Louise Taylor, "Why Infanticide Can Benefit Animals", *BBC Nature*, 21 March 2012, http://www.bbc.co.uk/nature/18035811, accessed 1 June 2018.

[100] For proposals along these lines, see Oren Etzioni, "How to Regulate Artificial Intelligence", *The New York Times*, 1 September 2017, https://www.nytimes.com/2017/09/01/opinion/artificial-intelligence-regulations-rules.html, accessed 1 June 2018.

[101] Director of Public Prosecutions, "Suicide: Policy for Prosecutors in Respect of Cases of Encouraging or Assisting Suicide", February 2010, updated October 2014, https://www.cps.gov.uk/legal-guidance/suicide-policy-prosecutors-respect-cases-encouraging-or-assisting-suicide, accessed 1 June 2018.

Thirdly, applying human moral requirements to AI may be inappropriate given that AI does not function in the same manner as a human mind. Laws set certain minimum moral standards for humanity but also take into account human frailties. Many of the benefits of AI result from it operating differently from humans, avoiding unconscious heuristics and biases that might cloud our judgement.[102] The Director of the Digital Ethics Lab at Oxford University, Luciano Floridi, has written:

> AI is the continuation of intelligence by other means. ... It is thanks to this decoupling that AI can colonise tasks whenever this can be achieved without understanding, awareness, sensitivity, hunches, experience or even wisdom. In short, it is precisely when we stop trying to reproduce human intelligence that we can successfully replace it. Otherwise, AlphaGo would have never become so much better than anyone at playing Go.[103]

In philosopher Philippa Foot's famous "Trolley Problem"[104] thought experiment, participants are asked what they would do if they saw a train carriage (a trolley), heading down railway tracks, towards five workmen who are in the train's path and would not have a chance to move before being hit. If the participant does nothing, the train will hit the five workmen. Next to the tracks is a switch, which will move the trolley onto a different spur of tracks. Unfortunately on the second spur is another workman, who will also be hit and killed if the train carriage is directed down that set of tracks. The participant has a choice: act, and divert the trolley so that it hits the one person, or do nothing and allow the trolley to kill five.[105]

---

[102] As we explore in later chapters, the theoretical ability for AI to avoid human bias does not obviate the need to ensure that those humans originally programming AI or providing their seed data sets do not accidentally or intentionally imbue AI with human fallibilities or prejudice.

[103] Luciano Floridi, "A Fallacy that Will Hinder Advances in Artificial Intelligence", *The Financial Times*, 1 June 2017, https://www.ft.com/content/ee996846-4626-11e7-8d27-59b4dd6296b8, accessed 1 June 2018. See also Nate Silver, *The Signal and the Noise: Why So Many Predictions Fail—But Some Don't* (London: Penguin, 2012), 287–288.

[104] Philippa Foot, *The Problem of Abortion and the Doctrine of the Double Effect* in *Virtues and Vices* (Oxford: Basil Blackwell, 1978) (the article originally appeared in the *Oxford Review*, Number 5, 1967).

[105] See Judith Jarvis Thompson, "The Trolley Problem", *Yale Law Journal*, Vol. 94, No. 6 (May, 1985), 1395–1415.

The most direct analogy to the Trolley Problem for AI is the programming of self-driving cars.[106] For instance: if a child steps into the road, should an AI car hit that child, or steer into a barrier and thereby kill the passenger? What if it is a criminal who steps into the road?[107] The parameters can be tweaked endlessly, but the basic choice is the same—which of two (or more) unpleasant or imperfect outcomes should be chosen?

Aspects of the Trolley Problem are by no means unique to autonomous vehicles. For instance, whenever a passenger gets into a taxi, they delegate such decisions to the driver. Moreover, vehicles are often designed in a manner which strikes a balance between protection for pedestrians and other road users, and the safety of the passengers within that vehicle. Design features of cars such as curved bonnets

---

[106] In this book, the terms "self-driving" and "autonomous" when used in relation to vehicles refer to the delegation by humans of certain decision-making functions featuring in driving. Broadly speaking, these fall into three areas: (i) Decisions as to the destination of travel; (ii) Decisions as to the route to be taken; and (iii) Granular decisions as to how a car should travel on the road—such as its reaction to obstacles, what speed to use, when to overtake and so on. Autonomy type (i) does not tend to occur at present. Autonomy types (ii) and (iii) do, however. The Trolley Problem dilemma operates most vividly within problem (iii), although as explained below, certain "moral" trade-offs may be involved in deciding how to route a journey (within autonomy type [ii]). A further point concerning autonomous vehicles is that—on an individual level—they may be following set instructions without any degree of choice-making. However, these will be autonomous in the relevant sense so long as the software within the vehicles, which may come from a single central hub and be sent to the individual vehicles via the Internet, contains features which would qualify as AI within this book's definition. See, for example, Joel Achenbach, "Driverless Cars Are Colliding with the Creepy Trolley Problem", *Washington Post*, 29 December 2015, https://www.washingtonpost.com/news/innovations/wp/2015/12/29/will-self-driving-cars-ever-solve-the-famous-and-creepy-trolley-problem/?utm_term=.30f91abdad96, accessed 1 June 2018; Jean-François Bonnefon, Azim Shariff, and Iyad Rahwan, "The Social Dilemma of Autonomous Vehicles", *Cornell University Library Working Paper*, 4 July 2016, https://arxiv.org/abs/1510.03346, accessed 1 June 2018.

[107] The scenario involving a criminal pedestrian was posed by researchers at MIT, in their "Moral Machine" game, which is described by its designers as "A platform for gathering a human perspective on moral decisions made by machine intelligence, such as self-driving cars. We show you moral dilemmas, where a driverless car must choose the lesser of two evils, such as killing two passengers or five pedestrians. As an outside observer, you judge which outcome you think is more acceptable", "Moral Machine", *MIT Website* http://moralmachine.mit.edu/, accessed 1 June 2018.

might protect pedestrians, at the expense of those within the vehicle.[108] However, although it is true that decisions are sometime delegated to *human* service providers, and trade-offs exist in other areas of design, AI is unique in that it will engender the delegation of important trade-offs to non-human decision-makers.[109]

Recognising the novel moral issues raised by AI, Germany was the first country to create a set of ethical rules applicable to autonomous vehicles. In the introduction to a report on Automated and Connected Driving by the German Ministry of Transport's Ethics Commission, the Commission encapsulates the problem as follows:

> What technological development guidelines are required to ensure that we do not blur the contours of a human society that places individuals, their freedom of development, their physical and intellectual integrity and their entitlement to social respect at the heart of its legal regime?[110]

The Commission set 15 "Ethical rules for automated and connected vehicular traffic", including a requirement that: "The protection of individuals takes precedence over all other utilitarian considerations". In keeping with Germany's attitude to human dignity, set down in Article 1(1) of its Constitution, the ninth rule provides: "In the event of unavoidable accident situations, any distinction based on personal features

---

[108] Tso Liang Teng and V.L. Ngo, "Redesign of the Vehicle Bonnet Structure for Pedestrian Safety", *Proceedings of the Institution of Mechanical Engineers, Part D: Journal of Automobile Engineering*, Vol. 226, No. 1 (2012), 70–84.

[109] Many commentators have pointed out the applicability of the Trolley Problem to self-driving cars, but beyond articulating the issue, few have actually suggested a legal or moral answer. See, for example, Matt Simon, "To Make Us All Safer, Robocars Will Sometimes Have to Kill", *Wired*, 17 March 2017, https://www.wired.com/2017/03/make-us-safer-robocars-will-sometimes-kill/, accessed 1 June 2018; Alex Hern, "Self-Driving Cars Don't Care About Your Moral Dilemmas", *The Guardian*, 22 August 2016, https://www.theguardian.com/technology/2016/aug/22/self-driving-cars-moral-dilemmas, accessed 1 June 2018; Jean-François Bonnefon, Azim Shariff, and Iyad Rahwan, "The Social Dilemma of Autonomous Vehicles", *Science*, Vol. 352, No. 6293 (2016), 1573–1576; Noah J. Goodall, "Machine Ethics and Automated Vehicles", in *Road Vehicle Automation*, edited by Gereon Meyer and Sven Beiker (New York: Springer, 2014), 93–102.

[110] "Ethics Commission at the German Ministry of Transport and Digital Infrastructure", 5 June 2017, https://www.bmvi.de/SharedDocs/EN/Documents/G/ethic-commission-report.pdf?__blob=publicationFile, accessed 1 June 2018.

(age, gender, physical or mental constitution) is strictly prohibited. It is also prohibited to offset victims against one another".[111]

Moral issues arise not just for autonomous vehicles, but also in many other uses of AI. An AI system designed to assist with triage and prioritisation of patients in an accident and emergency ward may have to make moral choices as to which patient ought to be treated sooner, for instance when deciding whether to favour the elderly or younger patients. Indeed, any allocation of resources by AI between competing demands raises similar issues. An autonomous weapon may have to decide whether to fire a weapon at an enemy when the enemy is surrounded by civilians, taking the risk of causing collateral damage in order to eliminate the target.[112]

A common objection to the Trolley Problem or its variants being applied to AI is to say that humans are very rarely faced with extreme situations where they must choose between, for example, killing five schoolchildren or one member of their family. However, this objection confuses the individual example with the underlying philosophical dilemma. Moral dilemmas do not arise only in life and death situations. To this extent, the Trolley Problem is misleading in that it could encourage people to think that AI's moral choices are serious, but rarely arise. In fact, all decisions involving choice and discretion will involve the weighing up of one or more values against others so as to arrive at an answer.[113] Inevitably a decision to do one thing rather than another will involve the privileging of certain principles over others. For instance, an AI car might be programmed with a tendency to avoid certain areas when transporting its passengers from A to B—thereby leading to de facto social exclusion and marginalisation. This is a much more subtle aspect of the choices to which we delegate AI, but nonetheless one with profound consequences.

---

[111]The proper mechanism for designing such moral rules is discussed at length in Chapter 7.

[112]See Kenneth Anderson and Matthew Waxman, "Law and Ethics for Robot Soldiers", *Columbia Public Law Research Paper* No. 12-313, *American University WCL Research Paper* No. 2012-32 (2012), http://papers.ssrn.com/sol3/papers.cfm?abstract_id=2046375, accessed 1 June 2018.

[113]See, for example, Ugo Pagallo, *The Law of Robots: Crimes, Contracts and Torts* (New York: Springer, 2013), "such as autonomous lethal weapons or certain types of robo-traders, truly challenge basic pillars of today's legal systems", xiii.

There may be a moral element involved when AI recommends news stories, books, songs or films—on the basis that these shape how we see the world and what actions we take. An angry or disaffected person who is repeatedly recommended violent films might be encouraged to commit violent acts; a person harbouring racist tendencies might find these exacerbated if she is shown sources which tend to support this world view.[114] Recent controversy over political processes such as the 2016 US election and the UK's "Brexit" referendum has demonstrated the potential power of the information on social media to create a feedback loop reinforcing various predilections and prejudices. Such information is increasingly chosen—and even generated—by AI.

In order to make the moral choices highlighted above, AI must necessarily engage with unclear laws and competing principles and be aware of their outcomes. This is the essence of acting as a moral (and more importantly for present purposes) a legal agent. As set out below, the increasing unpredictability of AI renders it ever more difficult to tether each decision AI takes to humans through a traditional chain of causation.

3.2   Independent Development

In this book, AI capable of "independent development" means a system which has at least one of the following qualities: (a) the capability to learn from data sets in a manner unplanned by AI system's designers; and (b) the ability of AI systems to themselves develop new and improved AI systems which are not mere replications of the original "seed" program.[115]

---

[114]These issues also bring into play the questions of free speech protection for AI discussed in Chapter 3.

[115]To be clear, it is not inherent in the nature of AI as defined in this book that it should *also* have the above qualities allowing for independent development. However, assuming that AI technology in the medium term follows similar trends to those in recent years, techniques which incorporate independent development, such as deep learning, will continue to feature in AI. The idea of AI adapting independently is, however, a key part of others' definitions of AI. For instance, Pei Wang suggests that intelligence is "the ability for a system to adapt to its environment while working with insufficient knowledge and resources". Pei Wang continued: "Being adaptive means to behave according to experience, and such a system can be useful for situations where the behaviors of the system cannot be predetermined by its designer". Pei Wang, "The Risk and Safety of AI", *NARS: An AGI Project*, https://sites.google.com/site/narswang/EBook/topic-list/the-risk-and-safety-of-ai, accessed 1 June 2018. See also Pei Wang, *Rigid Flexibility: The Logic of Intelligence* (New York: Springer, 2006).

3.2.1 Machine Learning and Adaptation

A machine learns whenever it changes its structure, program or data, in such a manner that its expected future performance improves.[116] In 1959, Arthur Samuel, a pioneer in AI and computer gaming, is said to have defined machine learning as the "[f]ield of study that gives computers the ability to learn without being explicitly programmed".[117]

In the 1990s, an expert in the field of "evolvable hardware", Adrian Thompson, used a program which foreshadowed today's machine learning AI to design a circuit that could discriminate between two audio tones. He was surprised to find that the circuit used fewer components than he had anticipated. In a striking early example of adaptive technology, it transpired that the circuit had made use of barely perceptible electromagnetic interference created as a side effect between adjacent components.[118]

Today, machine learning can be categorised broadly as supervised, unsupervised, or reinforcement. In supervised learning, the algorithm is given training data which contains the "correct answer" for each example.[119] A supervised learning algorithm for credit card fraud detection could take as input a set of recorded transactions, and for each individual datum (i.e. each transaction), the training data would contain a flag that says if it is fraudulent or not.[120] In supervised learning, specific error messages are crucial, as opposed to feedback which merely tells the system that it was mistaken. As a result of this feedback, the system

---

[116] Nils J. Nilsson, *Introduction to Machine Learning: An Early Draft of a Proposed Textbook* (2015), https://ai.stanford.edu/~nilsson/MLBOOK.pdf, accessed 1 June 2018.

[117] Samuel is widely quoted as the source of this definition, but where, when and indeed whether he ever wrote or said it remains obscure. See, for example, Andres Munoz, "Machine Learning and Optimization", *Courant Institute of Mathematical Sciences* (2014), 1, https://www.cims.nyu.edu/~munoz/files/ml_optimization.pdf, accessed 1 June 2018.

[118] Report: Evolvable hardware, "Machines with Minds of Their Own", *The Economist*, 22 May 2001, http://www.economist.com/node/539808, accessed 1 June 2018.

[119] Andrew Ng, "CS229 Lecture Notes: Supervised Learning", Stanford University, http://cs229.stanford.edu/notes/cs229-notes1.pdf, accessed 1 June 2018. Semi-supervised learning is similar to supervised learning, save that not all of the training data is labelled.

[120] Jean Francois Puget, "What Is Machine Learning?" IBM DeveloperWorks, 18 May 2016, https://www.ibm.com/developerworks/community/blogs/jfp/entry/What_Is_Machine_Learning?lang=en, accessed 1 June 2018.

generates hypotheses about how to categorise future unlabelled data, which it updates based on the feedback it is given each time. Although human input is required to monitor and provide feedback, the novel aspect of a supervised learning system is that its hypotheses about the data, and their improvements over time, are not pre-programmed.

In unsupervised learning, the algorithm is presented with data but not given any labels or feedback. Unsupervised learning systems function by grouping the data into clusters of similar features. Unsupervised learning is particularly exciting from the perspective of independent development because it can, in the words of the eminent cognitive scientist Margaret Boden, "be used to *discover* knowledge": the programmers do not need to know anything about the patterns in the data—the system can find these and draw inferences all by itself.[121] Zoubin Ghahramani, Chief Scientist of Uber, explains:

> It may seem somewhat mysterious to imagine what the machine could possibly learn given that it doesn't get any feedback from its environment. However, it is possible to develop [a] formal framework for unsupervised learning based on the notion that the machine's goal is to build representations of the input that can be used for decision making, predicting future inputs, efficiently communicating the inputs to another machine, etc. In a sense, unsupervised learning can be thought of as finding patterns in the data above and beyond what would be considered pure unstructured noise.[122]

A particularly vivid example of unsupervised learning was a program that, after being exposed to the entire YouTube library, was able to recognise images of cat faces, despite the data being unlabelled.[123] This process is

---

[121] Margaret Boden, *AI: Its Nature and Future*, (Oxford: OUP, 2016), 47. Emphasis original. See also Zoubin Ghahramani, "Unsupervised Learning", *Gatsby Computational Neuroscience Unit, University College London*, 16 September 2004, http://mlg.eng.cam.ac.uk/zoubin/papers/ul.pdf, accessed 1 June 2018.

[122] Ibid, Ghahramani, 3.

[123] Quoc V. Le et al. "Building High-Level Features Using Large Scale Unsupervised Learning", in *Acoustics, Speech and Signal Processing (ICASSP), 2013 IEEE International Conference*, 2013. The underlying neurons were trained using the techniques of "model parallelism" and asynchronous SGD (defined below). The authors concluded: "Our work shows that it is possible to train neurons to be selective for high-level concepts using entirely unlabeled data. In our experiments, we obtained neurons that function as detectors for faces, human bodies, and cat faces by training on random frames of YouTube videos. These neurons naturally capture complex invariances such as out-of-plane and scale invariances".

not limited to frivolous uses such as feline identification: its applications include genomics as well as in social network analysis.[124]

Reinforcement learning, sometimes referred to as "weak supervision", is a type of machine learning which maps situations and actions so as to maximise a reward signal. The program is not told which actions to take, but instead has to discover which actions yield the most reward through an iterative process: in other words, it learns through trying different things out.[125] One use of reinforcement learning involves a program being asked to achieve a certain goal, but without being told how it should do so.

In 2014, Ian Goodfellow and colleagues including Yoshua Bengio at the University of Montreal developed a new technique for machine learning which goes even further towards taking humans out of the picture: Generative Adversarial Nets (GANs). The team's insight was to create two neural networks and pit them against each other, with one model creating new data instances and the other evaluating them for authenticity. Goodfellow et al. summarised this new technique as: "...analogous to a team of counterfeiters, trying to produce fake currency and use it without detection, while the discriminative model is analogous to the police, trying to detect the counterfeit currency. Competition in this game drives both teams to improve their methods until the counterfeits are indistiguishable from the genuine articles".[126] Yann LeCun, Director of AI Research at Facebook, has described GANs as "the most interesting

[124] Andrew Ng, "Unsupervised Learning", *Coursera Stanford University Lecture Series on Machine Learning*, https://www.coursera.org/learn/machine-learning/lecture/olRZo/unsupervised-learning, accessed 1 June 2018.

[125] Richard S. Sutton and Andrew G. Barto, *Reinforcement Learning: An Introduction*, Vol. 1, No. 1 (Cambridge, MA: MIT Press, 1998), 4.

[126] Ian J. Goodfellow, Jean Pouget-Abadie, Mehdi Mirza, Bing Xu, David Warde-Farley, Sherjil Ozair, Aaron Courville, Yoshua Bengio, "Generative Adversarial Nets", arXiv:1406.2661v1 [stat.ML] 10 Jun 2014, accessed 16 August 2018. Goodfellow later explained that his insight came about following a drunken argument with colleagues. See Cade Metz, "Google's Dueling Neural Networks Spar to Get Smarter, No Humans Required", *Wired*, 4 November 2017, https://www.wired.com/2017/04/googlesdueling-neural-networks-spar-get-smarter-no-humans-required/, accessed 16 August 2018.

idea in the last 10 years in [machine learning]", and as a technique which "opens the door to an entire world of possibilities".[127]

The above forms of machine learning—particularly those towards the fully unsupervised end of the spectrum—indicate AI systems' ability to develop independently from human input and to achieve complex goals.[128]

Programs which utilise techniques of machine learning are not directly controlled by humans in the way they operate and solve problems. Indeed, the great advantage of such AI is that it does not approach matters in the same way that humans do. This ability not just to think, but to think differently from us, is potentially one of the most beneficial features of AI.

Chapter 1 described how the ground-breaking program AlphaGo used reinforcement learning to defeat a human champion player at the notoriously complex game "Go". In October 2017, DeepMind announced a further milestone: researchers had created an AI system which was able to master Go without access to any data on games played by humans. Previous iterations of Go playing software, including the program which defeated Grandmaster Lee Sedol in 2016, had learned their skills via scanning and analysing millions of moves contained in vast data sets of games played by humans.[129] AlphaGo Zero, as the 2017 program was called, had a different method: it learned entirely without human input. Instead, it was provided only with the rules and, within a number of hours, had mastered the game to such an extent that after just three days of self-training it was able to beat the previous version of AlphaGo by 100 games to 0. DeepMind explained the new methodology as follows:

---

[127] Yann LeCun, "Answer to Question: What are Some Recent and Potentially Upcoming Breakthroughs in Deep Learning?", *Quora*, 28 July 2016, https://www.quora.com/What-are-some-recent-and-potentially-upcoming-breakthroughs-in-deep-learning, accessed 16 August 2018.

[128] Andrea Bertolini, "Robots as Products: The Case for a Realistic Analysis of Robotic Applications and Liability Rules", *Law Innovation and Technology*, Vol. 5, No. 2 (2013), 214–247, 234–235.

[129] See Chapter 1 at s. 5 and FN 111. A subsequent iteration of AlphaGo, "AlphaGo Master" beat Ke Jie, at the time the world's top-ranked human player, by three games to nil in May 2017. See "AlphaGo at The Future of Go Summit, 23–27 May 2017", *DeepMind Website*, https://deepmind.com/research/alphago/alphago-china/, accessed 16 August 2018.

It is able to do this by using a novel form of reinforcement learning, in which AlphaGo Zero becomes its own teacher. The system starts off with a neural network that knows nothing about the game of Go. It then plays games against itself, by combining this neural network with a powerful search algorithm. As it plays, the neural network is tuned and updated to predict moves, as well as the eventual winner of the games.[130]

AlphaGo Zero is an excellent example of the capability for independent development in AI. Though the other versions of AlphaGo were able to create novel strategies unlike those used by human players, the program did so on the basis of data provided by humans. Through learning entirely from first principles, AlphaGo Zero shows that humans can be taken out of the loop altogether soon after a program's inception. The causal link between the initial human input and the ultimate output is weakened yet further.

DeepMind say of AlphaGo Zero's unexpected moves and strategies: "These moments of creativity give us confidence that AI will be a multiplier for human ingenuity, helping us with our mission to solve some of the most important challenges humanity is facing".[131] This may be so, but with such creativity and unpredictability comes attendant dangers for humans, and challenges for our legal system.

3.2.2 AI Generating New AI

Some AI systems are able to edit their own code—the equivalent of a biological entity being able to change its DNA. One example of this is a program built in 2016 by a team of researchers from Microsoft and Cambridge University, which used neural networks and machine learning

---

[130] Silver et al., "AlphaGo Zero: Learning from Scratch", DeepMind Website, 18 October 2017, https://deepmind.com/blog/alphago-zero-learning-scratch/, accessed 1 June 2018. See also the paper published by the DeepMind team: David Silver, Julian Schrittwieser, Karen Simonyan, Ioannis Antonoglou, Aja Huang, Arthur Guez, Thomas Hubert, Lucas Baker, Matthew Lai, Adrian Bolton, Yutian Chen, Timothy Lillicrap, Fan Hui, Laurent Sifre, George van den Driessche, Thore Graepel, and Demis Hassabis, "Mastering the Game of Go Without Human Knowledge", *Nature*, Vol. 550 (19 October 2017), 354–359, https://doi.org/10.1038/nature24270, accessed 1 June 2018.

[131] Silver et al., "AlphaGo Zero: Learning from Scratch", DeepMind Website, 18 October 2017, https://deepmind.com/blog/alphago-zero-learning-scratch/, accessed 1 June 2018.

in order to augment its own ability to solve mathematical problems in increasingly sophisticated ways.[132]

The Microsoft/Cambridge program derived data from multiple sources, including other programs. Some commentators described this as "stealing" code.[133] This approach has been used in other instances, such as Prophet, a patch generation system that "works with a set of successful human patches obtained from open source software repositories to learn a probabilistic, application independent model of correct code".[134] Where the source used by AI to learn and develop is other AI, the causation and authorship of any new code generated can become yet more obscure.

Several papers published in 2016 showed that it is possible to train an AI network to learn to learn, a process known as "meta-learning". Specifically, AI engineers created neural networks which then learned independently to perform a complex technique: stochastic gradient descent (SGD).[135] SGD is particularly useful in machine learning because it optimises the system's ability to perform a function with only a small

---

[132] Matej Balog, Alexander L. Gaunt, Marc Brockschmidt, Sebastian Nowozin, and Daniel Tarlow, "Deepcoder: Learning to Write Programs", *Conference Paper, International Conference on Learning Representations 2017*, https://openreview.net/pdf?id=ByldLrqlx, accessed 1 June 2018. See also Alexander L. Gaunt, Marc Brockschmidt, Rishabh Singh, Nate Kushman, Pushmeet Kohli, Jonathan Taylor, and Daniel T. Terpret, "A Probabilistic Programming Language for Program Induction", *Cornell University Library Working Paper*, abs/1608.04428, 2016, http://arxiv.org/abs/1608.04428, accessed 1 June 2018.

[133] Matt Reynolds, "AI Learns to Write Its Own Code by Stealing from Other Programs", *New Scientist*, 22 February 2017, https://www.newscientist.com/article/mg23331144-500-ai-learns-to-write-its-own-code-by-stealing-from-other-programs/, accessed 1 June 2018. See also, for criticism of the description as "stealing", Dave Gershgorn, "Microsoft's AI Is Learning to Write Code by Itself, Not Steal It", *Quartz*, 1 May 2017, https://qz.com/920468/artificial-intelligence-created-by-microsoft-and-university-of-cambridge-is-learning-to-write-code-by-itself-not-steal-it/, accessed 1 June 2018.

[134] Fan Long and Martin Rinard, "Automatic Patch Generation by Learning Correct Code", in *Proceedings of the 43rd Annual ACM SIGPLAN-SIGACT Symposium on Principles of Programming Languages*, 298–312, http://people.csail.mit.edu/rinard/paper/popl16.pdf, accessed 1 June 2018.

[135] Marcin Andrychowicz, Misha Denil, Sergio Gomez, Matthew W. Hoffman, David Pfau, Tom Schaul, Brendan Shillingford, and Nando de Freitas, "Learning to Learn by Gradient Descent by Gradient Descent", arXiv:1606.04474v2 [cs.NE], https://arxiv.org/abs/1606.04474, accessed 1 June 2018. See also Sachin Ravi and Hugo Larochelle, Twitter, "Optimisation as a Model for Few-Shot Learning", *Published as a Conference Paper at ICLR 2017*, https://openreview.net/pdf?id=rJY0-Kcll, accessed 1 June 2018.

number of training samples, rather than attempting to review all the data available to it.[136]

The advent of meta-learning is significant because it shows that with minimal levels of human input at the outset, AI can acquire for itself techniques which then enable it to continue to learn, improve and adapt. Technology journalist and author Carlos E. Perez summed up these developments:

> So not only are researcher[s] who hand optimize gradient descent solutions out of business, so are folks who make a living designing neural architectures! This is actually just the beginning of Deep Learning systems just bootstrapping themselves... This is absolutely shocking and there's really no end in sight as to how quickly Deep Learning algorithms are going to improve. This meta capability allows you to apply it on itself, recursively creating better and better systems.[137]

As noted in Chapter 1, various companies and researchers announced in 2017 that they had created AI software which could itself develop further AI software.[138]

In May 2017, Google demonstrated a meta-learning technology called AutoML. Google CEO Sundar Pichai explained in a presentation that "[t]he way it works is we take a set of candidate neural nets, think of these as little baby neural nets, and we actually use a neural net to iterate through them until we arrive at the best neural net".[139]

Two final points on AI's capacity for independent development: *First*, the list of achievements above comes with a "best before" date. Even within the field of machine learning, developments will no doubt continue after this book's publication. *Second*, although the present section

---

[136] Andrew Ng, Jiquan Ngiam, Chuan Yu Foo, Yifan Mai, Caroline Suen, Adam Coates, Andrew Maas, Awni Hannun, Brody Huval, Tao Wang, and Sameep Tando, "Optimization: Stochastic Gradient Descent", Stanford UFLDL Tutorial, http://ufldl.stanford.edu/tutorial/supervised/OptimizationStochasticGradientDescent/, accessed 1 June 2018.

[137] Carlos E. Perez, "Deep Learning: The Unreasonable Effectiveness of Randomness", *Medium*, 6 November 2016, https://medium.com/intuitionmachine/deep-learning-the-unreasonable-effectiveness-of-randomness-14d5aef13f87, accessed 1 June 2018.

[138] See Chapter 1 at s. 5.

[139] See also Sundar Pichai, "Making AI Work for Everyone", *Google Blog*, 17 May 2017, https://blog.google/topics/machine-learning/making-ai-work-for-everyone/, accessed 1 June 2018.

has concentrated predominantly on forms of machine learning, this is only because they are the dominant techniques at the time of writing. As mentioned in Chapter 1, in the future other AI technologies may bring about even greater independence. One constant remains: with each advance in the automation of AI's development, it becomes yet further removed from human input.[140]

3.3   Why AI Is Not Like Chemicals or Biological Products

It might be objected that AI is not the only man-made entity capable of independent development. A bacterium created in a laboratory might be able to adapt to different environments (such as whether it is hosted in a human or an animal) and/or to change its form over time, in response to stimuli such as new antibiotics.

Current legal systems have strategies for dealing with liability arising from such chemical or bio-engineered products, even in situations where these might continue to develop once they have been released from the laboratory. In the EU, questions of liability arising from chemical or biological products that change once released into the environment may be addressed the EU Product Liability Directive which imposes a strict liability regime broadly speaking on the "producer"[141] of a defective product.[142] The EU also utilises a range of prophylactic and ongoing

---

[140]At present, many AI systems require a significant amount of human fine-tuning, especially when they are produced by companies interested in achieving striking results even at a high cost in terms of resources. The philosophical paradigm of decreasing human involvement may therefore be slowed by practical and economic constraints.

[141]Under art. 3 of the EU Product Liability Directive 85/374/EC, "producer" is defined widely to include any person who, by putting his/her name, trademark or other distinguishing feature on the product, presents himself/herself as the producer; any importer which has imported the defective product, component or raw material into the European Union market; and any supplier (e.g. the retailer, distributor or a wholesaler) if the producer cannot be identified.

[142]For further discussion of the potential use of such product liability regimes as a means of establishing responsibility for AI, see Chapter 3 at s. 2.2.

processes for monitoring the safe use and development of such products.[143] More general rules such as negligence can penalise a legal person whenever they did something they ought not to have done, or failed to do something which they ought to have done (which might include the reckless release of a dangerous substance into the environment).

However, the main difference between AI and other products which can develop and change is AI's ability to take into account and interact with laws and rules when it undergoes such changes. Bacteria and viruses are not legal agents because they cannot interact with rules beyond anything more basic than perhaps an imperative to reproduce. Though AI at present may operate using simplistic reward or error functions that could be closer to bacteria than human level reasoning, at a theoretical level it is possible for AI to engage with an unlimited number of aims and operate within various parameters and constraints. The combination of the ability to take decisions and to take those decisions based on their predicted effect within a system of rules and norms is what renders an entity an agent.

4 Conclusions on the Unique Features of AI

AI is unlike other technologies, which are essentially fixed and static once human input has ended. A bicycle will not re-design itself to become faster. A baseball bat will not independently decide to hit a ball or smash a window.

No legal system ascribes all responsibility for human actions—at least where they are carried out by adults with normal mental faculties—to their parents, their teachers or their employers. At a certain age or level of maturity, humans are treated as being independent agents who are held responsible for their own deeds. A person's tendency to undertake certain

---

[143] See, for a summary, European Medicines Agency, "The European Regulatory System for Medicines: A Consistent Approach to Medicines Regulation Across the European Union" (2014), http://www.ema.europa.eu/docs/en_GB/document_library/Leaflet/2014/08/WC500171674.pdf, accessed 1 June 2018.

actions may have been shaped by their upbringing but this does not mean that parents are forever tethered to their children. In developmental psychology, the threshold is called the "age of reason". In law, it is known as the "age of majority".[144] We are approaching this point for AI.[145]

[144] See T.E. James, "The Age of Majority", *American Journal of Legal History*, Vol. 4, No. 1 (1960), 22, 33, which notes that this concept has been applied across different cultures for millennia.

[145] To put it another way, this is the moment at which the AI entity becomes a "Träger von Rechten". See Andreas Matthias, *Automaten als Träger von Rechten* (Berlin: Logos, 2010); Andrea Bertolini, "Robots as Products: The Case for a Realistic Analysis of Robotic Applications and Liability Rules", *Law, Innovation and Technology*, Vol. 5, No. 2 (2013), 214–247, 223. See also Peter M. Asaro, "The Liability Problem for Autonomous Artificial Agents", *Ethical and Moral Considerations in Non-human Agents, 2016 AAAI Spring Symposium Series*. See also David C. Vladeck, "Machines Without Principals: Liability Rules and Artificial Intelligence", *Washington Law Review*, Vol. 89 (2014), 117–150, esp. at 124–129.

CHAPTER 3

Responsibility for AI

If AI is unique as a legal phenomenon, the next question is what we should do about it.

This chapter and the two following respond in three stages. Chapter 3 discusses responsibility for AI, in terms of both liability for harm and how to account for positive output such as the authorship of creative works. Chapter 4 addresses potential moral justifications for granting AI rights. Chapter 5 then brings together themes raised in 3 and 4: arguing that a legal person is formed from "a bundle of rights and obligations", and proposing legal personality for AI as an elegant and pragmatic solution to the issues raised.

Various legal mechanisms could be used to determine who or what is responsible for AI when it causes harm or creates something of value. There is no single "silver bullet" answer. This chapter will explore how existing laws from legal systems around the world might be applied. Those which follow discuss what changes could be needed.

1 PRIVATE AND CRIMINAL LAW DISTINGUISHED

Most legal systems split criminal and private law. AI can lead to consequences in both.

© The Author(s) 2019
J. Turner, *Robot Rules*, https://doi.org/10.1007/978-3-319-96235-1_3

Private law refers to the legal relationship between people and involves the creation, alteration, and destruction of rights.[1] Many private law relations are voluntary to begin with. For instance, people can usually choose whether or not to enter into a contract, but once they have done so it will be legally binding.

In private law, rights and obligations usually come in pairs: a liability for one party is a claim to another.[2] Deterrence of wrongful conduct is a common aim to both private and criminal laws. Another purpose of private law is to ensure rights are vindicated and that parties are compensated for harm.[3] The usual remedy in private law is a payment of money to the innocent party, although other remedies can require the defendant to undertake or desist from a particular act.[4]

Criminal law is arguably a society's most powerful weapon against those who transgress. Criminal laws are usually enforced by the state and apply whether or not individual perpetrators have expressly agreed to be bound by them. Criminal laws have various purposes, including to signify the state's disapproval of certain conduct, retribution, deterrence and protection of society as a whole.[5] If a person commits a crime, then they will usually be punished, typically by imprisonment and/or a fine. Some legal systems still practise "corporal punishment", which involves the infliction of pain, mutilation or even death on a criminal.[6]

[1] Private law is sometimes also called "civil law". However, this terminology can be confusing because the term "civil law" can also be used to describe legal systems that are founded upon one great codification (such as the Code civil in France or the Bürgerliches Gesetzbuch in Germany) and in which judicial precedent does not play the same crucial role as in Common law systems. See further Chapter 6 at Sections 3.1 and 6.3.2.

[2] For a formal account of this model, see the discussion of Hohfeld's incidents in Chapter 4 at s. 1.1.

[3] Gary Slapper and David Kelly, *The English Legal System* (6th edn. London: Cavendish Publishing), 6.

[4] Courts can also order parties to do or not do certain things, for example, a court may require a party to cease production of a mobile phone containing technology which has been unlawfully copied from another company.

[5] For discussion, see H.L.A. Hart, *Punishment and Responsibility: Essays in the Philosophy of Law* (Oxford: Oxford University Press, 2008). See also *American Legal Institute Model Penal Code*, as Adopted at the 1962 Annual Meeting of The American Law Institute at Washington, DC, 24 May 1962, para. 1.02(2) for a slightly expanded list of aims along the same lines.

[6] John H. Farrar and Anthony M. Dugdale, *Introduction to Legal Method* (London: Sweet & Maxwell, 1984), 37.

Designation as a crime is a community's most emphatic denunciation of conduct.[7] For this reason, the requirements to be found guilty tend to exceed those for private law in terms of individual culpability or blameworthiness. There may in criminal law also be a higher burden of proof needed to show guilt. Unlike civil law liability in tort or contract, being found guilty of a crime will usually have lasting effects on an individual. A conviction can lead to both social stigma and permanent legal disabilities. In some jurisdictions, criminals are barred from voting and other basic civil rights.[8] Indeed, when the USA imposed a general ban on slavery in the Thirteenth Amendment to the Constitution, it was nonetheless preserved "as a punishment for crime whereof the party shall have been duly convicted".

2   PRIVATE LAW

Private law obligations relating to AI are most likely to arise from two sources[9]: civil wrongs[10] and contract.[11] Civil wrongs occur where the legal rights of one party are infringed.[12] If Damien throws a television

---

[7] Evidence of Lord Denning, *Report of the Royal Commission on Capital Punishment, 1949–1953* (Cmd. 8932, 1953), s.53.

[8] See, for example, "Felon Voting Rights", *National Conference of State Legislatures*, http://www.ncsl.org/research/elections-and-campaigns/felon-voting-rights.aspx, accessed 1 June 2018; Hanna Kozlowska, "What would happen if felons could vote in the US?", *Quartz*, 6 October 2017, https://qz.com/784503/what-would-happen-if-felons-could-vote/, accessed 1 June 2018.

[9] For an attempt at a systematic categorisation of obligations under English law, see *English Private Law*, edited by Andrew Burrows (3rd edn. Oxford: Oxford University Press, 2017).

[10] For a classical statement, see Frederick Pollock, *The Law of Torts: A Treatise on the Principles of Obligations Arising from Civil Wrongs in the Common Law* (5th edn. London: Stevens & Sons, 1897), 3–4.

[11] The distinction between civil liability arising from tort and contract may be traced at least as far as Roman law. The Institutes of Gaius (compiled c. 170 AD) stipulate that obligations could arise under two headings: *ex delicto* and *ex contracto*. The Institutes of Justinian (compiled in the sixth century AD) added two more categories, namely *quasi ex delicto* and *quasi ex contractu*. The latter are outside the scope of the present work. For discussion, see Lord Justice Jackson, "Concurrent Liability: Where Have Things Gone Wrong?", *Lecture to the Technology & Construction Bar Association and the Society of Construction Law*, 30 October 2014, https://www.judiciary.gov.uk/wp-content/uploads/2014/10/tecbarpaper.pdf, accessed 1 June 2018.

[12] Civil wrongs are referred to in some systems as "delicts" or "torts". The source of the latter is the Latin *torquere*, to twist, which became in Medieval Latin *tortum*: a wrong or injustice. In the French Civil Code, the relevant Chapter is entitled: *Des délits et des quasi-délits*.

out of a hotel room window and injures Charles, an unlucky pedestrian walking down the street outside, Damien has done a civil wrong to Charles by interfering with his right to walk down the street in peace and/or his right to bodily integrity. Charles will be able to seek damages from Damien under private law.[13] Contract is based on agreement. If Evelyn agrees to sell a new car to Frederica, but instead delivers a second-hand model, then Frederica may sue Evelyn for breaching their agreement. The corollary is that if Evelyn delivers a new car as promised but Frederica refuses to pay for it, then Evelyn may also sue Frederica based on their exchange of promises.

Within civil wrongs, liability can arise in a number of different ways. Important categories for present purposes include negligence, strict and product liability, and vicarious liability. We will discuss these in turn.

2.1   Negligence

Negligence is conduct which fails to conform to a required standard.[14] In a famous UK case *Donoghue v. Stevenson*, a producer of bottled ginger ale was required to pay compensation to a woman who fell ill after opening a bottle which contained a dead snail.[15] The producer was held to have a duty of care to whoever might reasonably be expected to open the bottle, even though there was no direct contract between them.[16] The judgement explained: "…you must take reasonable care to avoid acts or omissions which you can reasonably foresee would be likely to injure your neighbour". Neighbours were defined as "persons who are so closely and directly affected by my act that I ought reasonably to have them in contemplation".[17]

---

[13] There may well also be criminal consequences in this type of situation.

[14] Donal Nolan and John Davies, "Torts and Equitable Wrongs", in *English Private Law*, edited by Burrows (3rd edn. Oxford: Oxford University Press, 2017), 934.

[15] [1932] A.C. 562. See also Percy Winfield, "The History of Negligence in the Law of Torts", *Law Quarterly Review*, Vol. 42 (1926), 184, an art. which pre-dated the *Donoghue* judgement by some six years.

[16] Duties in contract and tort can, however, exist concurrently. See the judgement of the UK House of Lords in *Henderson v. Merrett Syndicates* [1994] UKHL 5.

[17] Ibid., 580–581.

Similar rules apply across many different types of legal systems, including those of France,[18] Germany[19] and China.[20]

2.1.1 How Would the Law of Negligence Apply to AI?

If harm is caused, the *first* question is whether anyone was under a duty not to cause, or to prevent that harm. The owner of a robot lawnmower might be under a duty towards anyone in the vicinity of that lawnmower. This would include, for example, a duty to take reasonable care to ensure that the AI lawnmower does not stray into the next-door neighbour's garden and decapitate their prize-winning roses.

The *second* question is whether the duty was breached. If the owner of the lawnmower has taken reasonable precautions in the circumstances then he will be exonerated, even if the lawnmower caused harm. If the

---

[18] Art. 1382 of the French Civil Code provides that "any act of man, which causes damages to another, shall oblige the person by whose fault it occurred to repair it". art. 1383 states: "One shall be liable not only by reason of one's acts, but also by reason of one's imprudence or negligence". The precise standard to which a person is held when considering whether they are at fault is not defined. However, it appears that much like the common law standard of negligence, fault is an error of conduct measured against the standard of a reasonable man. British Institute of International and Comparative Law, "Introduction to French Tort Law", https://www.biicl.org/files/730_introduction_to_french_tort_law. pdf, accessed 1 June 2018. All translations of the French Civil Code herein are those of Prof. Georges Rouhette with the assistance of Dr. Anne Rouhette-Berton http://www. fd.ulisboa.pt/wp-content/uploads/2014/12/Codigo-Civil-Frances-French-Civil-Code-english-version.pdf, accessed 1 June 2018.

[19] German law has an equivalent provision in s. 823 of the German Civil Code: "A person who, intentionally or negligently, unlawfully injures the life, body, health, freedom, property or another right of another person is liable to make compensation to the other party for the damage arising from this", https://www.gesetze-im-internet.de/bgb/__823. html, accessed 1 June 2018.

[20] Tort Law of the People's Republic of China, 2009. For an English translation of the text, see World Intellectual Property Organisation website, http://www.wipo.int/edocs/ lexdocs/laws/en/cn/cn136en.pdf, accessed 1 June 2018. For discussion, see Ellen M. Bublick, "China's New Tort Law: The Promise of Reasonable Care", *Asian-Pacific Law & Policy Journal*, Vol. 13, No. 1 (2011), 36–53, 44. Bublick writes: "To an outsider, the American notion of reasonable care for the safety of others seems compatible with the Chinese concept of 'harmony,' particularly if the legal focus on reasonable care for the safety of others is seen as creating a norm that generates moral and cultural power in its own right, not just when sanctions are imposed after a breach".

neighbour decides to borrow the lawnmower without the owner's permission and the neighbour uses it on her own garden, where it causes damage, then the owner would have a strong argument that the damage was not caused by his breach of any duty.

The *third* question is whether the breach of duty caused the damage. If the lawnmower was, through the owner's negligence, rolling towards the neighbour's garden but immediately before it damaged any flowers, a car ran off the road and destroyed the neighbour's rosebed, then the lawnmower owner might have breached his duty to keep the machine under control but the damage would have not been caused by this breach because of the car driver's intervening act.

A *fourth* question in some legal systems is whether the damage was of a type or extent which was reasonably foreseeable. The cost of replacing the roses is likely to be foreseeable, but a loss of prize money from a particularly lucrative rose-growing competition that the neighbour would otherwise have entered may not be.

The owner is not the only person who might be under a duty of care in the above situation. This might also apply to the designer of the AI, or the person (if any) who taught or trained it. For instance, if the design of the AI contained a fundamental flaw (let's say it interpreted children as weeds to be destroyed), then the designer might have breached a duty to design a robot safely.

2.1.2 Advantages of Negligence
Duty Can Be Adapted Depending on Circumstance
The level of duty can expand and shrink according to context.[21] This means the law of negligence can take into account the shifting uses to which AI might be put.[22] As we move along the spectrum from narrow

---

[21] See, for example, the English case *Bolton v. Stone* [1951] AC 850, HL in which the court set out the factors to be taken into account in determining liability.

[22] The levels of precautions which a person is required to take so as to avoid causing harm to others can vary from system to system. In the UK, the approach is slightly less mechanical, in that certain other factors can serve to adjust the duties. The UK courts take into account positive externalities arising from dangerous conduct, as well as potential negative externalities. If an action is socially desirable, then this may reduce the duty to take precautions, notwithstanding the risk of damage eventuating. See *Watt v. Hertfordshire CC* [1954] 1 WLR 835. See also the US Court of Appeals in *United States v. Carroll Towing Co.* 159 F.2d 169 (2d. Cir. 1947).

AI which can only be used for one task to general AI which is multi-purpose, this feature of the law of negligence will become increasingly useful.

As a rule of thumb, the chance of harm occurring can be multiplied by the gravity of potential harm to arrive at a calculation of what precautions should be taken.[23] When transporting nuclear waste, a high level of precaution is justified because although the chances of a leak may be very low, the danger is extreme. Sometimes courts will also take into account the potential benefit to society from an activity: beneficial but risky activities are likely to be given more lenient treatment than a dangerous activity of no public benefit. For instance, the police are less likely to be held liable for negligent driving when pursuing a criminal than a joyrider would be, because there is a social advantage to the former activity but not the latter.[24]

These features are helpful in that the producers, operators and owners of AI systems which are capable of causing great harm would be required to take the most precautions. As such, negligence can (at least in theory) avoid creating restrictive rules which might dampen innovation and development unnecessarily.

Flexibility as to Whom Duty Is Owed

There is no set list of people who might claim in negligence. This is useful because the people with whom AI interacts may change over time and may not be predictable at the outset. Moreover, many of the people who are potentially affected by anything that the AI does will have no prior contractual relationship with the AI's creator, owner or controller.[25] For

---

[23] See, for example, *United States v. Carroll Towing Co.* 159 F.2d 169 (2d. Cir. 1947).

[24] See, for example, the judgement of the UK Supreme *Court in Robinson v. Chief Constable of West Yorkshire Police* [2018] UKSC 4.

[25] In some systems, contractual and tort liability can be concurrent though. See, for instance, the position in the UK: *Henderson v. Merrett* [1995] 2 AC 145. In France, contractual and tort claims are non-cumulative, except in cases of professional negligence, under art 1792 of the French Civil Code. See Simon Whittaker, "Privity of Contract and the Law of Tort: The French Experience", *Oxford Journal of Legal Studies*, Vol. 16 (1996), 327, 333–334. In Germany, liability may be concurrent in contract and Tort. See Lord Justice Jackson, "Concurrent Liability: Where Have Things Gone Wrong?" *Lecture to the Technology & Construction Bar Association and the Society of Construction Law*, 30 October 2014, https://www.judiciary.gov.uk/wp-content/uploads/2014/10/tecbarpaper.pdf, accessed 1 June 2018, 6 and the sources cited therein.

example, an AI-enabled delivery drone might come into contact with all sorts of people and things on the way to its destination, especially if it is able to design its own route and to adapt it without human input.

Duty Can Be Voluntary or Involuntary

A duty giving rise to potential liability might be undertaken deliberately or it might arise out of a person's dangerous activities. If Juan wants to practise juggling knives whilst walking down the street, then he will come under a duty of care to passers-by regardless of whether he wants to or not.

As noted above, contractual liability requires that parties *agree* to be liable. If people were only liable for AI when they decided to be, this would lead to gaps in protections for third parties who stand to be affected by activities involving AI. The non-voluntary aspect of negligence is helpful in that it encourages subjects in any given legal system to have a greater regard to all other participants than they might otherwise do if they were purely seeking a profit-maximising objective. In other words, the possibility of negligence liability can cause subjects to take into account the externalities of their actions and indeed to price these into their calculations (at least to the extent that such risk can be accurately calculated).

2.1.3 Shortcomings of Negligence
How Do We Set Standards for AI's Behaviour?

The key question in negligence is generally whether the defendant acted in the same way as the average, reasonable person in that situation. In old English cases, judges illustrated this idea by asking whether a fictional "man on the Clapham Omnibus" might have done the same thing.[26]

However, problems arise when the reasonable person test is applied to humans using AI, all the more so to AI itself.

One option would be to ask what the reasonable designer or user of the AI might have done in the circumstances.[27] For example, it may be reasonable to set a car to operate in a fully autonomous mode on a relatively clear motorway, but not in a hectic urban environment.[28] Designers might supply AI with "health warnings" stipulating what is and is not advisable. This may be a workable short-term solution, but

---

[26] *McQuire v. Western Morning News* [1903] 2 KB 100 at 109 per Lord Collins MR.

[27] Ryan Abbot, "The Reasonable Computer: Disrupting the Paradigm of Tort Liability", *The George Washington Law Review*, Vol. 86, No. 1 (January 2017), 101–143, 138–139.

[28] See, for example, s. 3(2) of the UK Automated and Electric Vehicles Act 2018.

it runs into difficulties where there is no human operator of the AI on whom liability could be easily fixed.[29] Moreover, using AI in the wrong way is only one source of potential harm. An AI entity designed for a specific purpose might still cause harm through some form of unforeseeable development even when it is used in that field. One example of AI causing harm through an attempt to carry out its stipulated goal is the intelligent toaster which burns a house down in a quest to make as much toast as possible.[30] The more unpredictable the manner of failure, the more difficult it will be to hold the user or designer responsible without resorting to a form of strict liability.

In order to get around these issues, Ryan Abbot has proposed that if a manufacturer or retailer can show that an autonomous computer, robot or machine is safer than a reasonable person, then the supplier should only be liable in negligence rather than strict liability for harm caused by the autonomous entity.[31] Abbott's negligence test would focus on the AI's "act instead of its design, and in a sense, it would treat a computer tortfeasor as a person rather than a product".[32] Abbot argues that negligence would be determined according to the standard of the "reasonable computer", on the basis that "[i]t should be more or less possible to determine what a computer would have done in a particular situation".[33] Abbot contemplates establishing this standard by "considering the industry customary, average, or safest technology".[34]

In practice, applying a "reasonable computer" standard may be very difficult. A reasonable human person is fairly easy to imagine. The

---

[29] This is a solution tentatively suggested by Hubbard in F. Patrick Hubbard, "'Sophisticated Robots': Balancing Liability, Regulation, and Innovation", *Florida Law Review*, Vol. 66 (2015), 1803, 1861–1862.

[30] See also Nick Bostrom's "paperclip machine" thought experiment, discussed in Chapter 1 at s. 6.

[31] As to strict liability, see the following section. It will be assumed for present purposes that Abbot's definition of "autonomous" covers substantially the same entities AI within this book.

[32] Ryan Abbot, "The Reasonable Computer: Disrupting the Paradigm of Tort Liability", *The George Washington Law Review*, Vol. 86, No. 1 (January 2017), 101–143, 101.

[33] Ibid.

[34] Ibid., 140. Some efforts are currently underway at the level of standard-setting bodies such as the International Standards Organisation to establish general rules on these features, so at a minimum the agreement and articulation of such standards will be a prerequisite for Abbot's scheme to work. For nascent efforts along these lines, see for instance the International Standards Organisation proposal: "ISO/IEC JTC 1/SC 42: Artificial Intelligence", *Website of the ISO*, https://www.iso.org/committee/6794475.html, accessed 1 June 2018. See also Chapter 7 at s. 3.5.

law's ability to set an objective standard of behaviour takes as its starting point the idea that all humans are similar. More precisely, the law assumes that we have a certain set of capabilities and limitations arising from our shared physiology. Some humans may be braver, cleverer or stronger than others, but when setting the negligence standard these variations do not matter. AI, on the other hand, is heterogeneous in nature: there are many different techniques for creating AI and the variety is likely only to increase in the future as new technologies are developed. Applying the same standard to all of these very different AI entities may be inappropriate.

Finally, certain applications of the reasonableness test in negligence are bound up with the way that *humans* operate in the world, in a manner which may not be applicable to artificial entities. For example, in UK law, a doctor will not be liable in negligence if she adopts a treatment accepted at the time as proper by a responsible body of medical opinion, even of other medical professionals would disagree.[35] It is an open question whether this test would be applied to a medical AI, which one might reasonably expect to be not just as safe as a doctor, but even safer, much in the same way that we expect autonomous vehicles to be safer than those driven by humans.[36]

Reliance on Foreseeability

The law of negligence relies on the concept of foreseeability. It is used in establishing both the range of potential claimants by asking: "was it foreseeable that this person would be harmed?" and the recoverable harm by asking: "what type of damage was foreseeable?" As noted in Chapter 2, the actions of AI are likely to become increasingly unforeseeable, except

---

[35] See *Bolam v. Friern Hospital Management Committee* [1957] 2 All ER 118, as modified by *Bolitho (Administratrix of the Estate of Patrick Nigel Bolitho (deceased)) v. City and Hackney Health Authority* [1997] 4 All ER 771. In addition to being accepted by a body of medical practitioners, the practice must not be in the opinion of the court, unreasonable, illogical or indefensible.

[36] For discussion of this problem in medical liability, see Shailin Thomas, "Artificial Intelligence, Medical Malpractice, and the End of Defensive Medicine", *Harvard Law Bill of Health blog*, 26 January 2017, http://blogs.harvard.edu/billofhealth/2017/01/26/artificial-intelligence-medical-malpractice-and-the-end-of-defensive-medicine/ (Part I), and http://blogs.harvard.edu/billofhealth/2017/02/10/artificial-intelligence-and-medical-liability-part-ii/ (Part II), accessed 1 June 2018.

perhaps at a very high level of abstraction and generality.[37] In consequence, holding a human responsible for any and all actions of AI would become less focussed on the human's fault (usually the hallmark of negligence) and more like a system of strict, or product liability—which are discussed further below.

2.2   Strict and Product Liability

Strict liability exists where a party is held liable regardless of their fault. It is controversial: by abandoning any mental requirements for liability, strict liability cuts against fundamental notions of human agency—namely the ability to understand consequences and plan for them.[38] Justifications for strict liability include to ensure that the victim is properly compensated, to encourage those engaged in dangerous activities to take precautions,[39] and to place the costs of such activities on those who stand to benefit most.[40]

---

[37] See Curtis E.A. Karnow, "The Application of Traditional Tort Theory to Embodied Machine Intelligence", in *Robot Law*, edited by Ryan Calo, Michael Froomkin, and Ian Kerr (Cheltenham and Northampton, MA: Edward Elgar, 2015), 53.

[38] See, for example, H.L.A. Hart, "Legal Responsibility and Excuses", in *Determinism and Freedom in the Age of Modern Science*, edited by Sidney Hook (New York: New York University Press, 1958). Hart's criticisms are of criminal law strict liability, but the same criticisms can be made to civil law.

[39] In English law, the paradigm example of such strict liability is the rule in *Rylands v. Fletcher* (1866) L.R. 1 Ex. 265; (1868) L.R. 3 H.L. 330.

[40] See, for example, Justice Frankfurter in *United States v. Dotterweich* 320 U.S. 277 (1943): "Hardship there doubtless may be under a statute which penalizes the transaction though conscious wrongdoing may be totally wanting. Balancing relative hardships, Congress has preferred to place it upon those who have at least the opportunity of informing themselves of the existence of conditions imposed for the protection of consumers before sharing in illicit commerce, rather than to throw the hazard on the innocent public who are totally helpless". Another justification for strict liability is that in order to live in a broadly fair society, a person who seeks gain must assume the potential price of risks associated with that gain. Tony Honoré has called this "outcome responsibility". See Tony Honoré, "Responsibility and Luck: The Moral Basis of Strict Liability", *Law Quarterly Review*, Vol. 104 (October 1988), 530–553, 553. Stapleton has expressed this idea as follows: "Perhaps [strict liability] can be accounted for solely by the pragmatic interest in ease of adjudication which is achieved by adopting the (impossible) target of a perfect production-line norm across all units, but it seems more likely that such a widespread consensus also has a moral dimension - a view that enterprise should pay its way for this bad luck, even if unavoidable". Jane Stapleton, *Product Liability* (London: Butterworths, 1994), 189.

"Product liability" refers to a system of rules which establish who is liable when a given product causes harm. Often, the party held liable is the "producer" of that product, though intermediate suppliers may be included as well.[41] The focus is on the defective status of a product, rather than an individual's fault.[42] These regimes became popular in the second half of the twentieth century,[43] particularly in response to increasingly complex supply chains, as well as highly publicised scandals involving mass-produced defective goods—most notably the "morning sickness" drug *Thalidomide*, which caused severe physical handicaps in children.[44]

The remainder of this section will focus on two of the most developed systems of product liability[45]: the EU's Products Liability Directive

---

[41] Under the Products Liability Directive, the producer is "...the manufacturer of a finished product, the producer of any raw material or the manufacturer of a component part and any person who, by putting his name, trademark or other distinguishing feature on the product presents himself as its producer". Products Liability Directive, art. 2(1).

[42] See Products Liability Directive, art. 1.

[43] Though there were some judicial moves towards this position in the earlier part of the twentieth century. For a notable example, see Justice Traynor's concurring opinion in the US case *Escola v. Coca Cola Bottling Co.* 24 Cal. 2d 453, 461, 150 P.2d 436, 440 (1944). In 1931, Justice Cardozo had noted "The assault upon the citadel of privity is proceeding in these days apace". *Ultramares Corp. v. Touche*, 255 N.Y. 170, 180, 174 N.E. 441, 445 (1931).

[44] See, for example, the UK's investigation into the issue: "Lord Chancellor's Department: Royal Commission on Civil Liability and Compensation for Personal Injury", better known as the "Pearson Commission" LCO 20, which was established in 1973 and reported in 1978 (Cmnd. 7054, Vol. I, Chapter 22). Its terms of references included to consider the liability for death or personal injury "...through the manufacture, supply or use of goods or services". See also The Law Commission and the Scottish Law Commission, Liability for Defective Products (June 1977) Cmnd. 6831; Strasbourg Convention on Products Liability in Regard to Personal Injury and Death, Council of Europe, 27 January 1977. See also Ontario Law Reform Commission, Report on Product Liability (Ministry of the Attorney-General, 1979).

[45] There are some differences between them but the following analysis will concentrate on the shared features which appear to be common to these and other systems around the world which have been based on them. Though for such a comparison, see, for example, Lord Griffiths, Peter de Val, and R.J. Dormer, "Developments in English Product Liability Law: A Comparison with the American System", *Tulane Law Review*, Vol. 62 (1987–1988), 354.

of 1985[46] (the "Products Liability Directive") and the US Restatement (Third) of Torts: Products Liability.[47]

In the EU, the test for defectiveness is somewhat open-ended. A product is defective when "it does not provide the safety which a person is entitled to expect, taking all circumstances into account, including (a) the presentation of the product; (b) the use to which it could reasonably be expected that the product would be put; (c) the time when the product was put into circulation".[48] The US Third Restatement adopts a slightly more structured approach.[49] Defects subject to the regime must fall into at least one of three categories[50]: (a) design; (b) instruction or warnings; and/or (c) manufacturing.

---

[46] Council Directive 85/374/EEC 25 July 1985 on the approximation of the laws, regulations and administrative provisions of the Member States concerning liability for defective products (hereafter the Products Liability Directive). As a Directive, this piece of legislation is not directly binding on individuals but rather must be transposed by individual Member States. See, for example, the Consumer Products Act 1987 in the UK; art. 1386 (1–18) in the French Civil Code.

[47] Restatement (Third) of Torts: Products Liability paras. 12–14, at 206, 221, 227 (1997). The USA does not have Federal laws of product liability. Instead, these matters are addressed state by state. The Restatement on Products Liability is an attempt by the American Law Institute to compile existing jurisprudence on the area. See Mark Shifton, "The Restatement (Third) of Torts: Products Liability-The Alps Cure for Prescription Drug Design Liability", *Fordham Urban Law Journal*, Vol. 29, No. 6 (2001), 2343–2386. For discussion see Lawrence B. Levy and Suzanne Y. Bell, "Software Product liability: Understanding and Minimizing the Risks", *Berkeley Tech. L.J.*, Vol. 5, No. 1 (1990), 2–6; Michael C. Gemignani, "Product Liability and Software", *8 Rutgers Computer & Tech. L.J.*, Vol. 173, (1981), 204, esp. at 199 *et seq.* and at FN 70.

[48] Ibid., art. 6(1).

[49] David G. Owen, *Products Liability Law* (2nd edn. St. Paul, MN: Thompson West, 2008), 332 *et seq.*

[50] Two of the UK Law Commissioners whose report preceded the legislation implementing the Defective Products Directive, the Consumer Protection Act 1987, suggested that the UK courts would be likely to adopt a similar approach to the tripartite categorisation to defects. See Lord Griffiths, Peter de Val, and R.J. Dormer, "Developments in English Product Liability Law: A Comparison with the American System", *Tulane Law Review*, Vol. 62 (1987–1988), 354. The UK courts have been somewhat more reticent about adopting the US approach wholesale though: *A and Others v. National Blood Authority and another* [2001] 3 All ER 289, in which Burton J preferred the terminology "standard" and "non-standard", rather than "manufacturing defect" and "design defect". It is questionable how much difference this change in terminology makes in practice though.

These types of rules are by no means unique to the USA and Europe; for instance, the People's Republic of China Product Quality Law 1993 (amended 2000) provides similarly that products shall be free from any unexpected dangers threatening the safety of people and property.[51] Another example is Japan's Product Liability Act (Law no. 85 of 1994).[52]

2.2.1 How Would Product Liability Apply to AI?

Suppose Alpha Ltd designs AI optical recognition technology for autonomous vehicles and supplies that technology to Bravo Plc, which uses it in its cars. Unknown to all parties, the technology cannot distinguish between certain shades of blue paint and the sky. When driving a new car he has purchased from Bravo Plc, Charlie engages the autonomous driving mode. A truck painted sky blue crosses the path of the vehicle, which does not recognise the obstacle. Charlie is killed instantly when his car hits the truck.[53] Charlie's family might be able to make a claim against Alpha Ltd, as the original producer of the AI (in addition to Bravo Plc, as the more immediate supplier). In fact, Charlie's family could pursue a supplier at *any* level of the supply chain, including those of constituent parts or raw materials, so long as they were part of the faulty product.

2.2.2 Advantages of Product Liability

Certainty

Product liability regimes specify in advance which party is to be held responsible. This is especially helpful to victims. The victim does not have to seek contribution from multiple different parties in proportion to their relative fault. Instead, once the supplier or producer of AI has been located, they are liable to the victim for 100% of the damages. The

---

[51] Ellen Wang and Yu Du, "Product Recall: China", *Getting the Deal Through*, November 2017, https://gettingthedealthrough.com/area/31/jurisdiction/27/product-recall-china/, accessed 1 June 2018.

[52] Discussed in Fumio Shimpo, "The Principal Japanese AI and Robot Strategy and Research Toward Establishing Basic Principles", *Journal of Law and Information Systems*, Vol. 3 (May 2018).

[53] For a similar real-life fact pattern, see Danny Yadron and Dan Tynan, "Tesla Driver Dies in First Fatal Crash While Using Autopilot Mode", *The Guardian*, 1 July 2016, https://www.theguardian.com/technology/2016/jun/30/tesla-autopilot-death-self-driving-car-elon-musk, accessed 1 June 2018.

onus is on the supplier or producer to seek out any other liable parties and to sue them for a contribution where appropriate.

From the perspective of the supplier or producer of AI, the certainty of their primary liability has a value in that it allows for more accurate actuarial calculations. The risk of damages can therefore be priced into the eventual cost of products, as well as provided for in the accounting forecasts of companies and in investor disclosure such as "risk factors" in a prospectus.

Encourages Caution and Safety in AI Development

Strict product liability could encourage AI developers to design products with rigorous safety and control mechanisms. Even in a situation where the AI will develop in unforeseeable ways, the designer or producer of the AI may still be identified as the best-placed person to understand and control risks.[54]

Michael Gemignani wrote the following in 1981 of computers. The same principles arguably apply with even more force to AI:

> While the computer is still in its infancy, it may prove to be as beneficial, or as potentially harmful, as atomic power. If imposition of strict liability in tort would make the manufacturers of computer hardware and software more careful and more thoughtful in their race to develop an ultimate product, that alone would justify its application.[55]

2.2.3 Shortcomings of Product Liability
Is AI a Product or a Service?

Product liability regimes are so-called because they relate to products not services. Many commentators have assumed that product liability regimes will apply to AI without examining the important preliminary question of whether it is a good or a service.[56] In the EU, products are defined

---

[54] Horst Eidenmüller, "The Rise of Robots and the Law of Humans", *Oxford Legal Studies Research Paper* No. 27/2017, 8.

[55] Michael C. Gemignani, "Product Liability and Software", *Rutgers Computer & Technology Law Journal*, Vol. 173, 204 (1981), 204.

[56] See, for instance, Andrea Bertolini, "Robots as Products: The Case for a Realistic Analysis of Robotic Applications and Liability Rules", *Law Innovation and Technology*, Vol. 5, No. 2 (2013), 214–247, 238–239; Jeffrey K. Gurney, "Sue My Car Not Me: Products Liability and Accidents Involving Autonomous Vehicles", *University of Illinois Journal of Technology Law and Policy* (2013), 247–277, 257; and Horst Eidenmüller, "The Rise of Robots and the Law of Humans", *Oxford Legal Studies Research Paper* No. 27/2017, 8.

as "all movables" in Article 2 of the Product Liability Directive, which suggests that the regime applies only to physical goods. Consequently, a robot may be covered but some cloud-based AI may not be.

There have been debates in the past as to whether information contained in media such as books or maps is to be considered a "product" for the purpose of product liability. In the 1991 US case *Winter v. G. P. Putnam's Sons*,[57] the defendant published a book called *The Encyclopedia of Mushrooms*, which wrongly said that a poisonous mushroom was edible. Predictably, someone ate that mushroom and became critically ill. The US Court of Appeals for the 9th Circuit ruled that the information in the book was not a product for the purposes of the product liability regime. The court did say as an aside that "[c]omputer software that fails to yield the result for which it was designed" is to be treated as a product and therefore subject to product liability law. However, given that the judgement was from 1991, it seems reasonable to assume that the court was referring to traditional computer programs rather than those with AI capabilities.

The problem of bringing AI within product liability regimes applies outside the EU and USA. Fumio Shimpo, a member of the Japanese Government's Cabinet Office Advisory Board on AI, writes "[for] an example of the current legal dilemma, I will refer the reader to an accident involving a robot which was caused by inaccurate information or software defect malfunction. At present, the questioning of the product liability of the information itself, which was the main cause of this accident, is outside the range of the current Japanese Product Liability Act".[58]

To the extent that AI generates bespoke advice or output based on individualised input from a user, it would seem more closely to resemble the paradigm of a service rather than a product. In the light of this uncertainty, the European Commission (one of the three law-making bodies within the EU's governing institutions) promulgated an

---

[57]938 F.2d 1033 (9th Cir. 1991). See also *Alm v. Van Nostrand Reinhold, Co.*, 480 N.E.2d 1263 (Ill. App. Ct. 1985): a book on construction that led to injuries. In *Brocklesby v. United States* 767 F.2d 1288 (9th Cir. 1985), the court held a publisher of an instrument approach procedure for aircraft strictly liable for injuries incurred due to the faulty information.

[58]Fumio Shimpo, "The Principal Japanese AI and Robot Strategy and Research Toward Establishing Basic Principles", *Journal of Law and Information Systems*, Vol. 3 (May 2018).

Evaluation Project of the Products Liability Directive, which was completed in July 2017. The Evaluation's aims included "[to]...assess if the Directive is fit-for-purpose vis-à-vis the new technological developments such as the Internet of Things and autonomous systems".[59] It investigated matters including "whether apps and non-embedded software or the Internet of things based products are considered as 'products' for the purpose of the Directive"; and "whether an unintended, autonomous behaviour of an advanced robot could be considered a 'defect' according to the Directive".

Respondents included consumers, producers, public authorities, law firms, academics and professional associations.[60] The results were published in May 2017.[61] In response to the question "According to your experience, are there products for which the application of the Directive on liability of defective products is or might become uncertain and/or problematic?", 35.42% of respondents said "yes, to a significant extent", and a further 22.92% of respondents said "yes, to a moderate extent". When asked to name the products which might give rise to such issues, 35.42% of respondents named both those "performing automated tasks based on algorithms and data analysis (e.g. cars with parking assistance)" and those "performing automated tasks based on self-learning algorithms (Artificial Intelligence)".[62] At the time of writing, the European Commission is still formulating a response to this issue[63] but as matters

[59] European Commission, "Evaluation of the Directive 85/374/EEC concerning liability for defective products", http://ec.europa.eu/smart-regulation/roadmaps/docs/2016_grow_027_evaluation_defective_products_en.pdf, accessed 1 June 2018.

[60] Results of the public consultation on the rules on producer liability for damage caused by a defective product, 29 April 2017, http://ec.europa.eu/docsroom/documents/23470, accessed 1 June 2018.

[61] "Brief factual summary on the results of the public consultation on the rules on producer liability for damage caused by a defective product", 30 May 2017, GROW/B1/HI/sv(2017) 3054035, http://ec.europa.eu/docsroom/documents/23471, accessed 1 June 2018.

[62] Ibid., 26–27.

[63] The Commission has announced that "[b]y mid-2019 the Commission will also issue guidance on the interpretation of the Product Liability Directive in the light of technological developments, to ensure legal clarity for consumers and producers in case of defective products". See European Commission, "Press Release: Artificial intelligence: Commission outlines a European approach to boost investment and set ethical guidelines", *Website of the European Commission*, 25 April 2018, http://europa.eu/rapid/press-release_IP-18-3362_en.htm, accessed 1 June 2018.

stand it seems increasingly clear that the Products Liability Directive will need to be reformed if its coverage is to extend to AI in a predictable manner, or indeed at all.

Assumes Products to be Static Once Released

Product liability regimes operate on the assumption that the product does not continue to change in an unpredictable manner once it has left the production line. AI does not follow this paradigm.

Based on the assumption of products being static, the US and EU systems are subject to a number of defences which may prove overly permissive when applied to producers of AI. In the EU, the carve-outs from liability include:

> … having regard to the circumstances, it is probable that the defect which caused the damage did not exist at the time when the product was put into circulation by him or that this defect came into being afterwards; or … that the state of scientific and technical knowledge at the time when he put the product into circulation was not such as to enable the existence of the defect to be discovered….[64]

If products liability applies to AI at all, it is probable that producers will increasingly be able to take advantage of the above safe havens, thereby lessening the protections available to consumers.[65]

2.3   *Vicarious Liability*

Legal systems have a variety of mechanisms which create responsibility for one person, the "principal", for actions undertaken by another

---

[64]Products Liability Directive, art. 7. For the relationship between these defences and those available under the US system, see Lord Griffiths, Peter de Val, and R.J. Dormer, "Developments in English Product Liability Law: A Comparison with the American System", *Tulane Law Review*, Vol. 62, (1987–1988), 354, 383–385.

[65]See also art. 6(2) of the Directive: "A product shall not be considered defective for the sole reason that a better product is subsequently put into circulation". This may have been a reasonable rule for traditional industrial products, but seems ill-suited for software where everyone rightly expects constant security updates, patches, bug fixes, etc. This is not a problem unique to AI, but it is especially pertinent for programs which *by their nature* learn and improve over time.

person, the "agent".[66] In ancient times, several civilisations had highly developed criteria for determining the situations in which a master would be held liable for the acts of his slave.[67] With the demise of slavery and the rise of industrial economies from the late eighteenth century onwards, at least some of the legal relationships originally developed for slavery came to be adapted and reapplied.[68]

Vicarious liability can arise today in employer–employee relationships (which, tellingly, are still sometimes called master–servant situations).[69] Vicarious liability is also applied where one party takes responsibility for the acts of others such as parents or teachers.[70] The broad drafting of Article 1384 of the French Civil Code is particularly well adapted to both human and non-human relationships: "A person is liable not only for the damages he causes by his own act, but also for that which is caused by the acts of persons for whom he is responsible, or by things which are in his custody".[71]

The paradigm situation of legal responsibility is that every person is responsible for their own free, willing and informed actions. Vicarious liability is an exception to this standard in that an agent can cause harm, but someone else (the principal) will be held responsible for them having done so. This does not mean that the agent will be completely exonerated. Usually, the agent will also be liable for their harmful acts, but

[66]This technique is described in various different contexts as liability through agency, employment or vicarious liability, but broadly speaking they reflect the same central idea. For consistency, they will be referred to collectively as vicarious liability.

[67]For Roman law, see William Buckland, *The Roman Law of Slavery: The Condition of the Slave in Private Law from Augustus to Justinian* (Cambridge: Cambridge University Press, 1908). For Islamic law, see the discussion in Muhammad Taqi Uusmani, *An Introduction to Islamic Finance* (London: Kluwer Law International, 2002), 108.

[68]See Evelyn Atkinson "Out of the Household: Master-Servant Relations and Employer Liability Law", *Yale Journal of Law & the Humanities*, Vol. 25, No. 2, art. 2 (2013).

[69]See *Lister v. Hesley Hall Ltd* [2001] UKHL 22, in which a boarding house for children was found vicariously liable for abuse of children carried out by one of its employees, the warden.

[70]See, for example, art. 1384 of the French Civil Code: "A person is liable not only for the damages he causes by his own act, but also for that which is caused by the acts of persons for whom he is responsible, or by things which are in his custody…".

[71]The French language original is as follows: «*On est responsable non seulement du dommage que l'on cause par son propre fait, mais encore de celui qui est causé par le fait des personnes dont on doit répondre, ou des choses que l'on a sous sa garde.*»

the victim may choose to pursue a claim against their principal on the grounds that the latter has deeper pockets. After having paid the victim, the principal can usually go on to pursue the agent for damages by way of contribution.[72]

Though the two concepts are similar, vicarious liability differs from strict liability in that not every act of the agent will render their principal liable. For vicarious liability, first there has to be a relationship between the principal and agent which falls into the recognised categories set out above (e.g. employment). Second, the wrongful act must usually take place within the scope of that relationship.[73] The UK Supreme Court recently held in *Mohamud v. WM Morrison Supermarkets plc*[74] that a petrol station owner was vicariously liable for the actions of its employee who subjected a customer to a vicious and racist assault after the customer had asked to use a printer. Crucial to this liability for the supermarket was the fact that there was a "close connection" between the assault and the employee's employment, despite the fact that the assault clearly breached the terms of the employee's contract.[75]

In addition, some legal systems (such as Germany) require also that for there to be vicarious liability, there has to be a wrongful act by the agent. So, if the agent did not act wrongfully (e.g. for want of foreseeability), there is no vicarious liability of the principal.

2.3.1 How Would Vicarious Liability Apply to AI?

A police force which uses a patrol robot might be vicariously liable in circumstances where that robot assaults an innocent member of the public during its patrol.[76] Even if they did not create the AI system which the

---

[72] In the UK, such claims are made pursuant to the Civil Liability (Contribution) Act 1978.

[73] This applies even to parents and children. The French Civil Code provides in art. 1384: "(Act of 5 April 1937) The above liability exists, unless the father and mother or the craftsmen prove that they could not prevent the act which gives rise to that liability".

[74] [2016] UKSC 11.

[75] Ibid., [45]–[47].

[76] This may not be far away. It was reported in June 2017 that the Dubai police had employed a robotic patrol robot: Agence France-Presse, "First Robotic Cop joins Dubai police", 1 June 2017, http://www.telegraph.co.uk/news/2017/06/01/first-robotic-cop-joins-dubai-police/, accessed 1 June 2018. In reality, the "robot" does not appear to use AI, but rather acts more as a mobile computer interface which allows humans to seek information and report crimes. Nonetheless, it is apparent from examples such as this that people are increasingly accepting of the prospect of roles such as police officers being undertaken by AI/ robots.

robot uses, the police force might be deemed most immediately respon-sible for the conduct of the robot and/or deriving a benefit from the ·robot. The assault may not have been desired or permitted by the police force, but it occurred within the scope of the robot's assigned role. In a sense, the robot would be in a similar situation to a slave—namely an intelligent agent whose acts might be ascribed to a principal, without that agent being treated as a full legal person in itself.

2.3.2 Advantages of Vicarious Liability
Recognition of AI Agency
Vicarious liability strikes a balance between acknowledging the independ-ent agency of AI and holding a currently recognised legal person liable for its acts. Whereas negligence and product liability tend to characterise AI as· an object rather than an agent, vicarious liability is not so limited. For this reason, unilateral or autonomous actions of AI which are not foreseeable do not necessarily operate so as to break the chain of causa-tion between the person held liable and the harm. The vicarious liability model is therefore better suited to the unique functions of AI which dif-ferentiate it from other man-made entities.

2.3.3 Shortcomings of Vicarious Liability
No Clarity on the Relationship Needed
The fact that vicarious liability is usually limited to a certain sphere of activities undertaken by the agent is both an advantage and a drawback. It means that not every act of an AI will necessarily be ascribable to the AI's owner or operator. As such, the further AI strays from its deline-ated tasks, the more likely there is to be a gap in liability. In the short to medium term, whilst (predominantly narrow) AI continues to operate within tightly limited bands, this concern is less pressing.

AI could be treated as the "student", "child", "employee", or "servant" and a human (or other legal person) as the "teacher", "par-ent", "employer" or "master". Each of these models has particular nuances as to the scope and limits of responsibility of one party for the other. However, as noted at the end of Chapter 2, at some point, the primary offender (let's say the child) is cut loose from being the responsibility of their potential principal (i.e. the parent). We would need to work out when, if ever, AI is to be cut loose from humans for legal purposes.

2.4   No-Fault Accident Compensation Scheme

A no-fault compensation scheme pays damages to victims of an accident, regardless of whether anyone else was at fault. The guaranteed nature of the damages means that as a corollary the victim will usually lose the right to sue anyone who might have caused the harm.[77]

New Zealand is the only country to operate such a scheme for all accidents.[78] It has done so since 1974, thereby removing the tort system as a means of compensation for victims and deterrence of harmful conduct. The New Zealand scheme is funded by a series of dedicated levies held in different "accounts": work, earners, non-earners, motor vehicles and (medical) treatment injuries. Money is raised from each relevant constituency by levies or taxes.[79]

New Zealand's Accident Compensation Corporation, the government body which administers the scheme, explains: "Your levies pay for treatments, visits to health providers, rehabilitation programmes and equipment that may help in your recovery… We use levies to help you in your day to day life. This may be help with childcare, at home or transport to school and work".

For people used to a system in which those cause harm may be held liable to pay damages to the victim, the idea that no one would be subject to liability for personal injuries can seem counterintuitive or even perverse. However, in at least some industries where insurance is mandatory, the New Zealand scheme is not so far in terms of economic

---

[77]The victim would be likely ton lose the right to sue the perpetrator insofar as is required to prevent the victim being compensated twice for the same harm (a phenomenon known as "double-recovery"). Accordingly, there may be an exception to this principle for exemplary damages (for extreme conduct such as deliberate and vindictive harm), where such additional damages are not provided for under the compensation scheme.

[78]An example of a more limited scheme is §§ 104, 105 Sozialgesetzbuch VII in Germany. The Sozialgesetzbuch VII introduces and regulates a mandatory public insurance for workplace accidents. It is funded through mandatory contributions by all employers. If an employee suffers a workplace accident ("Arbeitsunfall"), that employee (or their family) will be paid compensation from the mandatory insurance scheme. The employer and other co-workers who may have caused the accident negligently are, in turn, freed from liability (unless they have acted wilfully).

[79]"The levy setting process", *Website of the Accident Compensation Scheme*, https://www.acc.co.nz/about-us/how-levies-work/the-levy-setting-process/?smooth-scroll=content-after-navs, accessed 1 June 2018.

effect from jurisdictions which maintain the classical tort-based system. For example, in many countries some form of third-party insurance is required for drivers of motor vehicles. This means that if someone else is injured by a driver, then the driver's insurance company will pay any relevant damages for which the driver would otherwise be liable. It is the insurer which pays out, rather than the individual driver. The insurers are in turn funded by all drivers in the country, thereby spreading the costs of accidents through the whole of society.

2.4.1 How Would No-Fault Accident Compensation Apply to AI?

In New Zealand, if AI caused or contributed to an accident, this would be treated in exactly the same way as any other accident: no claim would need to be made against a person associated with the AI. Instead, the victim would visit a healthcare provider for treatment. The Accident Compensation Corporation would provide support and compensation to the victim. As regards revenue generation, a system adopting no-fault compensation for AI might raise a special levy from the AI industry (though defining any such industry might present its own difficulties).

2.4.2 Advantages of No-Fault Compensation

Encouragement of Safe Practices

One major objection to New Zealand's system is that it might not adequately discourage dangerous behaviour given the disconnect between the causes of the harm and the paying party. Though this criticism has intuitive appeal, there is little evidence to support the idea that the New Zealand scheme leads to more tortious acts being committed.[80] In practice, people are motivated to avoid causing harm to others by a range of social factors beyond the purely financial. In the "Haifa Kindergarten" experiment, a group of day care centres which suffered from the problem of parents failing to collect their children on time imposed a small fine each time that the parents were late. The parent absenteeism increased as soon as the fines were implemented. The reasoning for this surprising

---

[80]Donald Harris, "Evaluating the Goals of Personal Injury Law: Some Empirical Evidence", in *Essays for Patrick Atiyah*, edited by Cane and Stapleton (Oxford: Clarendon Press, 1991). Though Harris advocates replacing tort liability for personal injuries with a no-fault compensation system, he admits that the evidence supporting a link between damages liability and deterrence is inconclusive. Harris says in this regard: "the symbolic effect of tort law may greatly exceed its actual impact".

phenomenon is thought to be that a strong moral incentive to collect the children on time was replaced by a weaker financial one.[81]

The Accident Compensation Corporation seeks to shape behaviour so as to avoid harm on a prophylactic basis. Instead of using compensation and damages as deterrence, it engages in a range of preventative measures, including working with schools to teach children first aid and safety, as well as initiatives to improve health and productivity in the workplace.

Avoids Legal Questions of Liability

A no-fault compensation scheme escapes the complicated legal issues highlighted in this chapter involving causation and foreseeability of the acts of AI by avoiding them altogether. If no single person or entity needs to be held liable, then no legal theory is needed to link them to the accident. No-fault compensation combines the simplicity and certainty of a product liability or pure strict liability mechanism, but avoids their arbitrary nature by excluding any single person from paying for the harm. Instead, society as a whole (or at least the relevant industry) pays collectively.

2.4.3 Shortcomings of No-Fault Compensation

Difficulty in Scaling up

By way of indication of the scale of the scheme in New Zealand (a country of only approximately 4.7 million people),[82] in 2016 there were 1.7 million claims, which cost the scheme NZ$2.3 billion[83] (approximately US$1.16 billion).

For a small country like New Zealand, such a system is manageable. Economies of scale are possible, and the advent of "big data" processing technology may make this task yet easier. Nonetheless, it is unclear how feasible it would be to increase the scheme to a country with tens or hundreds of millions of citizens.

---

[81] See Uri Gneezy and Aldo Rustichini, "A Fine is a Price", *The Journal of Legal Studies*, Vol. 29, No. 1 (2000).

[82] "Population", *Government of New Zealand Website*, https://www.stats.govt.nz/topics/population?url=/browse_for_stats/population.aspx, accessed 1 June 2018.

[83] "Keeping You Safe", *Website of the Accident Compensation Scheme*, https://www.acc.co.nz/preventing-injury/keeping-you-safe/, accessed 1 June 2018.

Political Objections

Even if a no-fault compensation scheme was logistically and economically possible, those politicians and members of the public who are keen to see a smaller rather than a larger state on ideological grounds may well rail against the idea of having such a large and powerful government-administered program. Despite the example of New Zealand, only a handful of other countries have adopted a similar scheme in the more than 40 years since it was instigated.[84]

Whether to Limit Only to Compensation for Physical Injury

One major limitation of the New Zealand scheme is that it only covers physical (and some instances of psychological) harm to humans.

Two major areas are left out: *first*, harm to property is not covered. *Secondly*, the New Zealand scheme does not cover financial loss which is not directly related to physical harm (known as "pure economic loss").

The vast and increasing range of AI's applications means that harm which it causes will not be limited merely to physical accidents. If an AI trading program invests all of a company's money in a volatile commodity/financial instrument like Bitcoin immediately before a crash, then under the New Zealand scheme there would be no compensation available to the victim. They would have to seek recourse through the various other mechanisms identified above and below, such as negligence, product liability or contract.

---

[84] For discussions of the more general merits and disadvantages of a no-fault compensation scheme, see, for example, Geoffrey Palmer, "The Design of Compensation Systems: Tort Principles Rule, OK?" *Valparaiso University Law Review*, Vol. 29 (1995), 1115; Michael J. Saks, "Do We Really Know Anything About the Behavior of the Tort Litigation System—and Why Not?" *University of Pennsylvania Law Review*, Vol. 140 (1992), 1147; Carolyn Sappideen, "No Fault Compensation for Medical Misadventure-Australian Expression of Interest", *Journal of Contemporary Health Law and Policy*, Vol. 9 (1993), 311; Stephen D. Sugarman, "Doing Away with Tort Law", *California Law Review*, Vol. 73 (1985), 555, 558; Paul C. Weiler, "The Case for No-Fault Medical Liability", *Maryland Law Review*, Vol. 52 (1993), 908; and David M. Studdert, Eric J. Thomas, Brett I.W. Zbar, Joseph P. Newhouse, Paul C. Weiler, Jonathon Bayuk, and Troyen A. Brennan, "Can the United States Afford a "No-Fault" System of Compensation for Medical Injury?" *Law & Contemporary Problems*, Vol. 60 (1997), 1.

2.5   Contract

A contract is a legally binding agreement, or set of promises.[85] Not all promises are enforceable in law: a promise to meet a friend for dinner is unlikely to have contractual force. In order to distinguish a mere promise from a contract, legal systems impose a series of requirements. These can range from formalities such as a need for contracts to be made in writing,[86] to a requirement that something of value be exchanged.[87]

2.5.1   How Would Contracts Apply to AI?
Determining Who Is Responsible
In a paradigm situation, two or more parties would enter into a formal agreement to determine who would be legally responsible for the acts of the AI in question. Typically, in return for a payment the seller of a product or service will make a series of promises (sometimes called representations and warranties) about what it is selling.[88]

Contracts can decrease as well as increase a party's liability. Clauses in an agreement may exclude liability for all or some types of harm, or put limits on what is payable. The seller of a medical AI diagnostic program may exclude liability to a hospital buying the software for harm caused where the AI misdiagnoses a patient. At the other end of the spectrum, a seller of AI could agree to pay any relevant debts incurred by the buyer (i.e. indemnify her) for *any* harm which that AI causes. In 2015, the CEO of Volvo announced that the company would accept all liability

---

[85] There is some academic debate as to whether contract should be defined exclusively in terms of an agreement or promises but this is outside the scope of the present work. See, for discussion, *Chitty on Contracts*, edited by Hugh Beale (32nd edn. London: Sweet & Maxwell Ltd, 2015), 1-014–1-024. In the Proposal for a Regulation of the European Parliament and of the Council on a Common European Sales, Law Com (2011) 635 final, art. 2 (a) defines a contract as "an agreement intended to give rise to obligations or other legal effects".

[86] Historically, this was more common but has now been abandoned. Other requirements might include stipulations as to the language of the contract and the jurisdiction to which they are subject. See Mark Anderson and Victor Warner, *Drafting and Negotiating Commercial Contracts* (Haywards Heath: Bloomsbury Professional, 2016), 18.

[87] In some systems, the requirement for something of value to pass is known as "consideration".

[88] However, it can also be the case that a contract, and indeed contractual terms, will be deemed to have been agreed by the parties as a result of their relationship. When a person buys a crate of apples, there is usually an implied term that those apples will not be full of maggots.

for harm caused by its cars when they are operating autonomously.[89] It is hard to say whether the CEO's statement was intended to have contractual effect. However, in a seminal English case, *Carlill v. Carbolic Smoke Ball Company*,[90] a company's boast on a promotional poster that it would pay £100 to anyone who used their product and was not cured of 'flu, was held to be binding. Volvo might end up being held to its promise.

Can AI Conclude a Contract in Its Own Right?

Suppose you are buying a new sofa online. You see a sofa you like, being sold by a vendor called SOFASELLER1. You pay the purchase price and the sofa is delivered. Would it matter if SOFASELLER1 was an AI system?

Where an AI system is contracting on behalf of a further principal, in the capacity of an agent then it seems likely in many situations that the contract will be effective. Indeed, this is how much trading occurs online, where automated programs are mandated to buy, sell and bid on behalf of people and companies. Fumio Shimpo points out that not all such contracts will be binding under Japanese Law; if the AI fails to identify itself as such and entices a person to enter into a contract, then such contract might be deemed "equivalent to a mistake of an element (Article 95 of the Japanese Civil Code)", and potentially rendered ineffective.[91]

There are many automated contractual systems operating today—from consumer sales to high-frequency trading of financial instruments. At present, these all conclude contracts on behalf of recognised legal people. That may not always need to be the case. Blockchain technology is a system of automated records, known as distributed ledgers. Its uses can include chains of "self-executing" contracts, which can be executed without any need for human input. This technology has already given rise to novel and uncertain questions as to liability arising from a

---

[89] Kirsten Korosec, "Volvo CEO: We Will Accept All Liability When Our Cars Are in Autonomous Mode", *Fortune*, 7 October 2015, http://fortune.com/2015/10/07/volvo-liability-self-driving-cars/, accessed 1 June 2018.

[90] [1892] EWCA Civ 1.

[91] Fumio Shimpo, "The Principal Japanese AI and Robot Strategy and Research toward Establishing Basic Principles", *Journal of Law and Information Systems*, Vol. 3 (May 2018).

particular blockchain system in which all parts are interconnected.[92] In a situation where AI concludes a contract without direct or indirect instructions from a principal, it remains unclear how a legal system would address liability arising from such an agreement; the AI would require legal personality to be able to go to court to enforce such contract—the possibility of which is discussed further in Chapter 5.

United Nations Convention on the Use of Electronic Communications in International Contracts

There have already been some attempts to create special laws to account for the role of computers in concluding contracts. Article 12 of United Nations Convention on the Use of Electronic Communications in International Contracts 2005 provides:

> A contract formed by the interaction of an automated message system and a natural person, or by the interaction of automated message systems, shall not be denied validity or enforceability on the sole ground that no natural person reviewed or intervened in each of the individual actions carried out by the automated message systems or the resulting contract.

Legal commentators Čerkaa, Grigienėa and Sirbikytė have contended that Article 12: "states that a person (whether a natural person or a legal entity) on whose behalf a computer was programmed should ultimately be responsible for any message generated by the machine". On this basis, they argue that Convention is an appropriate tool for the determining responsibility for AI, in the absence of other direct regulation, because "[s]uch an interpretation complies with a general rule that the principal of a tool is responsible for the results obtained by the use of that tool since the tool has no independent volition of its own".[93]

---

[92]Dirk A. Zetzsche, Ross P. Buckley, and Douglas W. Arner, "The Distributed Liability of Distributed Ledgers: Legal Risks of Blockchain", *EBI Working Paper Series* (2017), No. 14; "Blockchain & Liability", *Oxford Business Law Blog*, 28 September 2017, https://www.law.ox.ac.uk/business-law-blog/blog/2017/09/blockchain-liability, accessed 1 June 2018.

[93]Paulius Čerkaa, Jurgita Grigienėa, Gintarė Sirbikytėb, "Liability for Damages Caused By Artificial Intelligence", *Computer Law & Security Review*, Vol. 31, No. 3 (June 2015), 376–389.

However, Article 12 does not stand for the proposition that the afore-mentioned academics suggest.[94] Article 12 is expressed as a negative proposition: computer-generated contracts are not to be denied validity solely because of a lack of review. The academics reverse this by suggest-ing a positive proposition, requiring that *every* computer has a person responsible—thereby transforming the meaning of Article 12. Even if Article 12 did fix responsibility for AI on the "person on whose behalf it was programmed", application of this provision is likely to become increasingly problematic the more that AI is able to learn and develop independently of its original inception, and thereby act as an agent in its own right.[95]

2.5.2 *Advantages of Contractual Liability*
Respect for Parties' Autonomy
Contracts give legal expression to human agency and choice. For this reason, in many economies and legal systems, freedom of contract is treated as a paramount value.[96]

Unlike the various other schemes described above where policy decisions as to risk allocation are taken either by judges or legislators, contract allows parties to exercise their autonomy so as to allocate risk between them. It can be assigned a price, and that price can be reflected in the transaction. In theory, this should lead to resources being allo-cated most efficiently according to market forces.

---

[94] However, the conclusion they point to was apparently reached by UNCITRAL in its deliberations, though does not formally form part of the convention. This is noted in the materials accompanying the published version of the Convention, which states at 70: "UNCITRAL also considered that, as a general principle, the person (whether a natural person or a legal entity) on whose behalf a computer was programmed should ultimately be responsible for any message generated by the machine (see A/CN.9/484, paras. 106 and 107)". See http://www.uncitral.org/pdf/english/texts/electcom/06-57452_Ebook.pdf, accessed 1 June 2018.

[95] See also the discussion of s. 9(3) of the UK Copyright, Designs and Patents Act, dis-cussed at s. 4.1, which contains similar language.

[96] See, for example, Robert Joseph Pothier, *Treatise on Obligations, or Contracts*, trans-lated by William David Evans (London: Joseph Butterworths, 1806); James Gordley, *The Philosophical Origins of Modern Contract Doctrine* (Oxford: Clarendon Press, 1993), Chapter 6.

2.5.3 *Shortcomings of Contractual Liability*
Contracts Only Apply to a Limited Set of Parties
The main disadvantage of relying solely on contracts to regulate liability
for AI is that they are very limited in terms of to whom they apply (a
feature sometimes referred to as "privity"). Contracts only create rights
and obligations between the contracting parties or occasionally a limited
class of third-party beneficiaries.[97] Contracts are therefore of no use in
determining liability where there was no prior contractual agreement. A
pedestrian who is injured by a self-driving car whilst walking down the
pavement next to a road will not have agreed a contract with the designers, owners or operators of any vehicles driving past.

Secrecy
Parties to a contract may agree that its content, and even its existence,
is to be kept private as between them. This can be very helpful for commercial entities who wish to protect certain elements of their dealings
from competitors or the public. However, where contracts are private,
then this can also have negative effects in terms of minimising the signalling effect such agreements might otherwise have to other market
participants.[98] Without accurate information about what certain parties
are doing, others will find it difficult to regulate their own behaviour.
Secrecy might prevent consistent market behaviour from developing and
thereby increase the cost to parties of negotiating each individual agreement on liability from scratch.

AI companies may have strong individual incentives to hide their
agreements on responsibility for harm. Even the existence of such an
agreement might be reported in the press as suggesting that the AI is
somehow unsafe. Many systems require certain transactions to be
recorded on a public register, such as those relating to land. One solution to the secrecy issue would be for contracts concerning liability for
AI to be made public. The obvious objection to this is that it would be
enormously bureaucratic to store such details on a public register, and
commercial parties may well refuse to do so, on the basis of well-established legal principles including confidentiality and privacy. Distributed

[97]The term "privity" is derived from the Latin: *privatus* - meaning private.
[98]For an influential analysis of the signalling effect of agreements in the labour market,
see Michael Spence, "Signaling, Screening and Information", in *Studies in Labor Markets*,
edited by Sherwin Rosen (Chicago: University of Chicago Press, 1981), 319–358.

ledger technology such as blockchain offers one option as to how contracts relating to AI might be made a matter of public record. However, it seems unlikely that many market participants would agree to this level of public scrutiny unless they were required to by law.

Quasi-Hidden Contracts

Contractual arrangements concerning AI will work best where arrangements are made between parties who are able to understand the obligations to which they are binding themselves, and are able to weigh up the benefits and disadvantages of the position they have taken. In reality, this often is not the case.

Members of the public enter into many different contracts on a daily basis without realising or consciously agreeing the terms. This can include accepting the conditions of carriage when we take a bus or a subway,[99] or the End User License Agreement which mobile app users generally flick past before clicking a box to signify that they accept. Many apparently "free" services are provided on the basis of quasi-hidden contracts. Users might receive a utility such as online mapping services, and in return, they signify their consent by contract to the provider recording and using their location and search data. There is occasional disquiet when the extent of such agreements on personal data is brought to the attention of consumers—as occurred in 2018 when a scandal broke over Facebook's data collection and use by third parties such as Cambridge Analytica.[100] Despite the somewhat manufactured outrage in the press, the extent to which people were signing away rights to their data secrecy would in most cases have been discoverable to any user who had looked closely enough at the terms and conditions to which they agreed as a *quid pro quo*.

Even if average consumers do not have the time or inclination to pore through dozens of pages of tightly worded legalese, there are often "safety-nets" which guarantee consumer rights against exploitative or

---

[99] See, for instance, *Parker v. South Eastern Railway Co* (1877) 2 CPD 41.

[100] Dylan Curran, "Are You Ready? Here Is All the Data Facebook and Google Have on You", *The Guardian*, 30 March 2018, https://www.theguardian.com/commentis-free/2018/mar/28/all-the-data-facebook-google-has-on-you-privacy, accessed 1 June 2018.

unfair contracts. These can include legislation which bans unfair contractual terms[101] or requires special attention to be drawn to particularly onerous terms.[102] If contracts concerning AI are to become as widespread for non-expert members of the public, it may be necessary for the law to impose limits or safeguards upon the rights that people can unwittingly sign away.

Limitations of Language

A further disadvantage of using contracts to manage responsibility for AI is that though such legal agreements are very useful for planning what should happen in circumstances predicted by the parties, they are less helpful for determining what should happen where the contract is vague or silent. Individually negotiated contracts can often result in a compromise between the parties, with the result that neither agrees on the meaning of a contentious clause.

At least for written agreements, creative drafting may be able to cater for some uncertainty, but it remains likely that the rigid nature of contracts will have some difficulty in accommodating the unpredictability of AI. Moreover, the interpretation of words is an inherently uncertain exercise.[103] Contractual disputes can be resolved by courts, but in advance of their decision any certainty will have been compromised.

2.6   Insurance

Insurance is a specific type of contract law, in which one party (the insurer) agrees either to pay certain amounts of money or, more rarely, to undertake steps to otherwise compensate another party (the insured), if certain events occur. Typically, in exchange the insured will pay a sum

---

[101] In EU countries, see, for example, the Unfair Terms in Consumer Contracts Directive (93/13/EC).

[102] Additional protection is provided by the various government and non-governmental bodies tasked with reviewing and periodically raising awareness of particularly egregious or harmful conduct undertaken by companies under the cover of contractual agreements. See, for example, the Federal Trade Commission in the USA, the Consumer Protection Association in the UK or the Consumer Rights Organisation in India.

[103] See Jacob Turner, "Return of the Literal Dead: An Unintended Consequence of Rainy Sky v. Kookmin on Interpretation?" *European Journal of Commercial Contract Law*, Vol. 1 (2013).

known as the "premium" at specified intervals, for example, monthly or annually.

Insurance is a form of risk management, whereby the insurer adopts the risk of certain events occurring, in exchange for a fee.[104] Insured parties will often pay a relatively small premium comparative to the overall amount which is to be paid out. The less likely an event, the lower the ratio of premium to payout. A householder might pay $500 a year for building's insurance, which might pay out $500,000 in the event that the building is destroyed by an insured risk, such as fire. The insurer benefits because—assuming they have got their calculations correct—the net amount of premiums it is paid will exceed the amounts of money it pays out to insured parties.[105]

2.6.1 How Would Insurance Apply to AI?

US Judge and author Curtis Karnow has suggested that best way of dealing with liability for artificial intelligence is to have an insurance scheme:

> Just as insurance companies examine and certify candidates for life insurance, automobile insurance and the like, so too developers seeking coverage for an agent could submit it to a certification procedure, and if successful would be quoted a rate depending on the probable risks posed by the agent. That risk would be assessed along a spectrum of automation: the higher the intelligence, the higher the risk, and thus the higher the premium and vice versa.[106]

Insurers could sell "third-party" policies to potential defendants to protect against claims for harms caused to others by AI. They could also

---

[104] See, generally, Kenneth S. Abraham, "Distributing Risk: Insurance", *Legal Theory, and Public Policy*, Vol. 48 (1986).

[105] "Primary layer" insurers will often pass on some or even all of the risk above a certain threshold to re-insurers, who may in turn do the same, thereby spreading such risk further through the market.

[106] Curtis E.A. Karnow, "Liability for Distributed Artificial Intelligences", *Berkeley Technology Law Journal*, Vol. 11, No. 1 (1996), 147–204, 176. Karnow may not be correct in his assessment that higher intelligence leads to more risks; at least some risks in the use of AI arise from it having not enough intelligence to recognise the costs of its actions or their wider impact. It might be more correct to say that the higher the level of responsibility which AI is accorded, the higher the risks. More intelligent AI is likely to be given more responsibility, thereby creating the link between intelligence and risk (albeit indirectly, and with the caveat that the intelligent AI may well be safer).

sell "first-party" policies to potential victims so as to ensure that they are compensated in the event that they are harmed by AI.

For most activities and industries, insurance policies are voluntary. As such, there can be gaps in coverage where an uninsured party causes harm and then disappears or is unable to satisfy claims for compensation made against it. There are some notable exceptions, such as mandatory automobile insurance,[107] which is imposed by law on the basis of the high number of car users, the frequency of car accidents and a desire on the part of policy-makers to ensure that victims have a quick and certain recourse, particularly in the event that the driver at fault is impecunious.[108] Similar policy considerations may well make it desirable for some form of AI insurance to be made mandatory, at least to cover risks to third parties.

2.6.2 Case Study: UK Automated and Electric Vehicles Act 2018

The UK Parliament enacted the Automated and Electric Vehicles Act in July 2018.[109] This legislation extends the compulsory insurance scheme for normal road vehicles in the UK to cover automated ones. Section 2 of the Act provides:

(1) Where— (a) an accident is caused by an automated vehicle when driving itself..., (b) the vehicle is insured at the time of the accident, and (c) an insured person or any other person suffers damage as a result of the accident, the insurer is liable for that damage.

---

[107] In the USA, a state-by-state list of mandatory car insurance requirements is provided at the consumer website, *The Balance*, "Understanding Minimum Car Insurance Requirements", 18 May 2017, https://www.thebalance.com/understanding-minimum-car-insurance-requirements-2645473, accessed 1 June 2018. For the position in the UK, see "Vehicle Insurance", *UK Government*, https://www.gov.uk/vehicle-insurance, accessed 1 June 2018.

[108] For early arguments in favour of such a rule, at a time when car driving was in its infancy, see Wayland H. Elsbree and Harold Cooper Roberts, "Compulsory Insurance Against Motor Accidents", *University of Pennsylvania Law Review*, Vol. 76 (1927–1928), 690; Robert S. Marx "Compulsory Compensation Insurance", *Columbia Law Review*, Vol. 25, No. 2 (February 1925), 164–193; and for a more modern perspective, see Harvey Rosenfield, "Auto Insurance: Crisis and Reform", *University of Memphis Law Review*, Vol. 29 (1998), 69, 72, 86–87.

[109] For more information on the drafting process see "Automated and Electric Vehicles Act", *Parliament Website*, https://services.parliament.uk/bills/2017-19/automatedandelectricvehicles.html, accessed 1 June 2018. See also Chapter 8 at s. 5.3.3.

The point of section 2(1) of the Act is to make clear that an insurer will be required to provide coverage for accidents caused by a vehicle when driving in autonomous mode, where that vehicle is already insured. The Act also extends mandatory insurance from covering only harm to third-parties to include the party insured (often the driver of the vehicle). This is helpful from the perspective of legal certainty and will likely encourage the development of the UK's autonomous vehicle industry. However, the Act does not resolve underlying legal questions of ultimate responsibility. Section 5(1) provides: "any other person liable to the injured party in respect of the accident is under the same liability to the insurer or vehicle owner". There is no indication as to whom these other liable parties may be. The result is that difficult questions of ultimate responsibility for AI are simply "kicked down the road".

2.6.3 Advantages of Insurance
Partial Solution to Unpredictability
The essence of insurance law is to cater for situations of uncertainty. Insurance policies cover parties against matters as diffuse as natural disasters, incurable or debilitating illness, as well as human-caused events such as sabotage or terrorism.[110] The unpredictability of AI which makes it particularly problematic for other areas of law may not be such an issue for insurers. By passing on the cost of harm to insurers for a fixed price, parties can plan for unknown risks with much greater certainty. The cost of insurance policies can therefore be written into financial predictions for investors and passed on to the end user of a good or service in the price they pay, thus spreading the burden throughout market participants.

Behaviour Channelling
Insurance typically has a channelling effect as regards the behaviour of the insured because the insurer has an interest in minimising the risk of harm. Insurers may require certain behaviour of the insured parties in order that their policy remains valid. For example, insurers of contents in a property may insist that there are locks on the doors and windows. As

---

[110]Terrorism is often excluded from main policies and provided in a supplementary policy with its own premium.

regards AI, insurers could require that insured parties adhere to certain minimum standards in design and its implementation.[111]

2.6.4 Shortcomings of Insurance
Parasitic on Underlying Liability

Insurance does not alter underlying legal liabilities. Rather, it redirects the liability to pay damages away from the person who caused harm (if any) to the insurer.[112]

If the victim of harm caused by AI would not have a right of recourse against the insured party, then the insurer would have no reason to pay the victim any money. Insurance only operates via the liability stipulated under the various other private law theories of responsibility and compensation set out above (or otherwise by specific legislative intervention—as in the Automated and Electric Vehicles Act 2018). A party may be insured for harm they negligently cause or for which they are strictly liable. This means that, from a victim's perspective, an insurance policy taken out by an AI owner/controller will only be helpful to the extent that the victim can assert a right against the insured party.

One option is for the various different candidates to each insure themselves separately. So for an autonomous vehicle, insurance might be taken out by the company which has produced the vehicle (which we will assume for the sake of this example also designed the AI), as well as the owner of the vehicle. That way, if either a passenger or another road user is injured or suffers loss as a result of a crash caused by the AI, then there is at least some certainty for the victim that they will receive a payout, and there is further certainty for the insured parties that they will pay only the premium. However, this still does not stop the different insurers from fighting as to liability between them if one pays out the victim in full then seeks contribution payments from the others—as might occur under section 5(1) of the Act.

---

[111] Chapters 7 and 8 of this book set out the potential content for such requirements.

[112] Curtis E.A. Karnow, "Liability for Distributed Artificial Intelligences", *Berkeley Technology Law Journal*, Vol. 11, No. 1 (1996), 147–204, 196.

Exceptions and Exclusions

A prudent insurer will set boundaries on its liability. It will exclude liability for harm caused by the deliberate or wilful act of the insured. A building's insurer would not pay out if the owner of the building deliberately sets it alight.[113]

Insurers might seek to exclude liability where the AI undertakes an activity outside a set range (e.g. if a delivery robot is used as a concierge). The more unpredictable the insured AI, the more difficult it will be for the insurer to assess and ultimately set a price for the likelihood of damage. Whether or not this renders insurance prohibitively expensive remains to be seen. As recent US experience in medical insurance exchanges demonstrates, it can be extremely difficult for a government to compel insurers to enter markets which they do not consider to be economically viable.[114]

3 CRIMINAL LAW

There may be significant overlap between the conduct which can give rise to civil and criminal consequences. Generally speaking, the more stringent measures available under criminal law require a higher degree of fault. Criminal liability usually requires not just a culpable act (sometimes referred to as actus reus), but also a certain mental state on the part of the defendant: the guilty mind or mens rea. Unlike tort law, which usually uses an objective mental standard (asking what a reasonable person would have done), in criminal law the focus is generally on the defendant's subjective state of mind: what did the perpetrator *actually* believe and intend to do.

---

[113]Indeed, some legal systems expressly prohibit insurance policies from covering wilful acts. For example, s. 533 of the California Insurance Code. For commentary, see James M. Fischer, "Accidental or Willful?: The California Insurance Conundrum", *Santa Clara Law Review*, Vol. 54 (2014), 69, http://digitalcommons.law.scu.edu/lawreview/vol54/iss1/3, accessed 1 June 2018.

[114]Olga Khazan, "Why So Many Insurers Are Leaving Obamacare: How Rejecting Medicaid and Other Government Decisions Have Hurt Insurance Markets", *The Atlantic*, 11 May 2017, https://www.theatlantic.com/health/archive/2017/05/why-so-many-insurers-are-leaving-obamacare/526137/, accessed 1 June 2018.

The mental requirements necessary for a crime to have been committed differ between legal systems and between different crimes themselves. Sometimes the mens rea required for guilt go beyond defendant having foreseen the consequences of her actions and require that she actually intended, desired or willed the consequences to take place.[115] Under English law, a person who throws a brick off a balcony is unlikely to be found guilty of murdering a person on whom the brick lands unless she *intended* either to cause death or serious harm.[116]

3.1   How Would Criminal Law Be Applied to Humans for the Actions of AI?

3.1.1   AI as an Innocent Agent

Where AI is deemed to have followed the instructions of a human and undertaken an act which, if carried out by a human, would be a crime, then the actions of the AI would normally be attributed to the human.[117] Provided that the human had the requisite mental state, then she will be guilty. The AI would be legally irrelevant.[118] It would be a mere tool in the hands of the perpetrator, like the knife used by a murderer. As the California Supreme Court found in *People v. Davis*: "Instruments other than traditional burglary tools certainly can be used to commit the offense of burglary... a robot could be used to enter the building".[119]

Innocent agents need not be limited to inanimate objects. An entity which is considered to have some intelligence may still be an innocent

---

[115] J.Ll.J. Edwards, "The Criminal Degrees of Knowledge", *Modern Law Review*, Vol. 17 (1954), 294.

[116] Extreme carelessness may not suffice for murder, though it could be enough for the lesser crime of "manslaughter". "Homicide: Murder and Manslaughter", website of the UK Crown Prosecution Service, http://www.cps.gov.uk/legal/h_to_k/homicide_murder_and_manslaughter/#intent, accessed 1 June 2018.

[117] For an exploration of innocent agency, see Peter Alldridge, "The Doctrine of Innocent Agency", *Criminal Law Forum*, Autumn 1990, 45.

[118] This analysis follows a structure proposed by Gabriel Hallevy in "The Criminal Liability of Artificial Intelligence Entities—From Science Fiction to Legal Social Control", *Akron Intellectual Property Journal*, Vol. 4, No. 2, art. 1. Hallevy later expanded on these ideas in two books: *Liability for Crimes Involving Artificial Intelligence Systems* (Springer, 2015), and *When Robots Kill: Artificial Intelligence Under Criminal Law* (Boston: Northeastern University Press, 2013).

[119] 958 P.2d 1083 (Cal. 1998).

agent. If an adult asks a child to pour a poisonous liquid into another person's drink when they are not watching, then the adult who provided the poison and directed the child is likely to be found guilty of a crime, even if the child would not be. This section concerns the criminal liability of *humans* for the acts of AI. Section 4.5 of Chapter 5 will cover the possibility of criminal liability for the *AI itself*.

3.1.2 Vicarious Criminal Liability of Humans

Vicarious liability in criminal law operates in a broadly similar manner to private law and is subject to the same limitations as set out above. One major difference between the two is that private law vicarious liability does not focus on the mens rea of the principal; rather, the question is on the relationship between the principal and agent. By contrast, in criminal law, the principal must normally have the mens rea necessary for the relevant crime.[120] If the mens rea requirement is merely that the principal was reckless as to harm (as opposed to intending harm), then this may not be a particularly difficult barrier for a prosecutor to overcome.

If an AI engineer creates an AI system for making toast and that machine then burns down a house, killing everyone in it, on the reasoning that "all the bread would be toasted", then the programmer may face criminal consequences for their reckless behaviour in creating such a program. Legal scholar Gabriel Hallevy describes this as "natural-probable-consequence" liability, explaining that it "seems legally suitable for situations in which an AI entity committed an offense, while the programmer or user had no knowledge of it, had not intended it, and had not participated in it".[121]

3.2 Advantages of Humans Being Criminally Responsible for AI

Criminal law functions best where it accords closely to society's moral precepts.[122] An effective system of criminal law cannot be imposed

---

[120] For a recent restatement of this principle with regard to joint enterprise criminal liability in the UK, see the joint decision of the UK Supreme Court and Judicial Committee of the Privy Council in *R v. Jogee, Ruddock v. The Queen* [2016] UKSC 8, [2016] UKPC 7.

[121] "The Criminal Liability of Artificial Intelligence Entities—From Science Fiction to Legal Social Control", *Akron Intellectual Property Journal*, Vol. 4, No. 2, art. 1, 13.

[122] See generally: Roger Cotterell, *Emile Durkheim: Law in a Moral Domain (Jurists: Profiles in Legal Theory)* (Edinburgh: Edinburgh University Press, 1999).

without reference to what a given polity *thinks* ought to be criminal. Psychological studies suggest that humans are innately retributivists: if someone has caused harm, our natural response is to seek out a person responsible who deserves to be made to suffer.[123]

3.3   Shortcomings of Humans Being Held Criminally Responsible for AI

3.3.1   Retribution Gap

Given that criminality is such a serious and often enduring sanction, it ought to be reserved for situations in which the perpetrator's wrong-doing is of a particularly blameworthy character. The big challenge as regards AI is that the more advanced it becomes, the more difficult it will be to hold a human responsible, let alone blameworthy for its acts without stretching accepted notions of causation out of recognition. Legal philosopher John Danaher has described the delta between humanity's expectations that someone will be held responsible, and our present inability to apply criminal law to AI as opening up a "retribution gap".[124]

Though, as shown above, it is quite possible to split the function of assigning responsibility from the function of paying compensation in the private law context, splitting responsibility from punishment in criminal law is far more problematic. Retributive punishment is linked to moral desert and not just pragmatic considerations.[125] Danaher cautions: "… I have noted how doctrines of command responsibility or gross negligence could be unfairly stretched so as to inappropriately blame the manufacturers and programmers. Anyone who cares about the strict requirements of retributive justice, or indeed justice more generally, should be concerned about the risk of moral scapegoating".[126]

---

[123] See, for example, Carlsmith and Darley, "Psychological Aspects of Retributive Justice", in *Advances in Experimental Social Psychology*, edited by Mark Zanna (San Diego, CA: Elsevier, 2008).

[124] John Danaher, "Robots, Law and the Retribution Gap", *Ethics and Information Technology*, Vol. 18, No. 4 (December 2016), 299–309.

[125] Anthony Duff, *Answering for Crime: Responsibility and Liability in Criminal Law* (Oxford: Hart Publishing, 2007).

[126] John Danaher, "Robots, Law and the Retribution Gap", *Ethics and Information Technology*, Vol. 18, No. 4 (December 2016), 299–309.

There are then two options: either to treat the actions of AI as "Acts of God" which have no legal consequences or to somehow find a "responsible" human. Unlike earthquakes or floods, the acts of AI are unlikely to be viewed as unfortunate but morally neutral natural disasters.

3.3.2 Over-Deterrence

The severity of criminal liability may lead to a chilling effect on progress and development of new and more powerful AI, if it is the case that programmers are potentially subject to criminal sanctions. The financial burden of compensation payments to victims of harm caused by AI can be passed on to an employer or insurer—or may even be treated simply as a business risk. Criminal liability by contrast is usually personal and it is difficult for an individual person to avoid by saying that he was merely following superior orders. Moreover, criminality has a social cost that cannot necessarily be displaced or expunged in monetary terms. If this threat hangs over programmers, then they might be less inclined to invent or release otherwise helpful technology.[127]

4   Responsibility for Beneficial Acts: AI and IP

The foregoing sections of this chapter, and indeed the majority of academic debate, have focussed on liability for *harm* caused by AI. The present section will address responsibility for beneficial acts or creations. When a human paints a picture, writes a book, invents a new medicine or designs a bridge, then most legal systems provide structures for determining ownership over that work and for protecting the author against unauthorised copying of their creation. Other laws protect commercial reputation. This is called the the law of "intellectual property" (IP).

AI is already creating new and innovative products and designs, whether in technical fields such as engineering and architecture,[128] or in industries such as art or music production.[129]

---

[127] See also Chapter 5 at s. 4.5 where this factor is discussed as a potential motivation for giving AI legal personality.

[128] See *Artificial Intelligence in Engineering Design*, edited by Duvvuru Siriam and Christopher Tong (New York: Elsevier, 2012).

[129] Bartu Kaleagasi, "A New AI Composer Can Write Music as well as a Human Composer", *Futurism*, 9 March 2017, https://futurism.com/a-new-ai-can-write-music-as-well-as-a-human-composer/, accessed 1 June 2018.

122 J. TURNER

AI systems can go even further than replicating a person's style. Researchers from Rutgers University, the College of Charleston and Facebook's AI Research Lab have created AI capable of making abstract art so convincing that human experts could not tell which works were made by AI and which by human artists.[130] Sceptics might argue that AI can never be truly "creative" in a philosophical sense, and that such programs merely synthesise and replicate existent work. The problem with this argument is that the same point could be made of virtually any human artistic or literary creation. Indeed, there is a good argument for saying that AI is even *more* creative than humans, in that all humans are restricted by our biological faculties, whereas AI is capable of "thinking" and operating in an entirely different manner. Regardless of one's philosophical position on the matter, there is already ample evidence of AI creating works which would qualify for protection under intellectual property law were they created directly by a human.[131]

Despite these advances in creative technologies, legal structures for protecting creations are lagging well behind.

4.1 Copyright

Copyright is a system of protection of original works which focusses on the creative activity of the creator when he or she composed the work in question. Most other intellectual property rights focus instead on the objective character of the subject matter regardless of how it was brought into existence. Thus, if Vincent paints a picture which does not copy from anyone else's picture or design, then he is likely to be accorded copyright protection in whatever he has painted, even if it is the same as a picture that someone else has painted (unbeknownst to Vincent). The focus of the protection for copyright is more on the creative process and less on the objective novelty of the output.

---

[130]Elgammal et al., "CAN: Creative Adversarial Networks Generating 'Art' by Learning About Styles and Deviating from Style Norms", Paper published on the eighth International Conference on Computational Creativity (ICCC), held in Atlanta, GA, 20–22 June 2017 arXiv:1706.07068v1 [cs.AI], 21 June 2017, https://arxiv.org/pdf/1706.07068.pdf, accessed 1 June 2018.

[131]For examples, see Ryan Abbott, "I Think, Therefore I Invent: Creative Computers and the Future of Patent Law", *Boston College Law Review*, Vol. 57 (2016), 1079, http://lawdigitalcommons.bc.edu/bclr/vol57/iss4/2, accessed 1 June 2018. See in particular FN 23–138 and accompanying text.

Under EU law, original literary and artistic works are covered by various copyright protections, which provide certain rights to the author.[132] A work or part of a work is regarded as original if it is the author's own intellectual creation,[133] reflecting his or her personality through an expression of free and creative choices, thereby stamping the work with his or her personal touch.[134]

Although individual words, figures or mathematical concepts as such do not qualify as an original work, a sentence or phrase may be protected if it constitutes an expression of the intellectual creation of the author through the choice, sequence and combination.[135] As noted above, AI is capable of creating original work for the purposes of this definition. Under EU law, the first owner of copyright is the author.[136] The relevant legislation and case law assumes implicitly that the author is a legal person. The ownership of an original work can be adjusted by employment or other contractual relationship, but the point remains that in legal terms, copyright ownership always assumes that the creator is also an entity capable of holding rights.[137]

Generally speaking, legal systems do not provide for copyright-protected works being created by non-humans. Andres Guadamuz wrote in the Magazine of the World Intellectual Property Organisation: "Creative works qualify for copyright protection if they are original, with most definitions of originality requiring a human author. The legislation of several jurisdictions, including Spain and Germany, appear to

---

[132] Jonathan Turner, *Intellectual Property and EU Competition Law* (2nd edn. Oxford: Oxford University Press, 2015), at para. 6.03 *et seq.*

[133] C-5/08 *Infopaq International v. Danske Dagblades judgment* paras. 34–39, CJ; C-403, 429/08 *FAPL v. QC Leisure* judgment paras. 155–156.

[134] *Eva-Maria Painer v. Standard VerlagsGmbH, Axel Springer AG, Süddeutsche Zeitung GmbH, Spiegel-Verlag Rudolf Augstein GmbH & Co KG, Verlag M. DuMont Schauberg Expedition der Kölnischen Zeitung GmbH & Co KG* (Case C-145/10).

[135] Ibid. See also *SAS Institute v. World Programming* judgement paras. 65–67, CJ. 37 C-393/09 *Bezpečnostní softwarová asociace v. Ministerstvo kultury* judgment paras. 48–50, CJ.

[136] Directive 2001/29, Arts. 2–4; Directive 2006/115, Arts. 3(1), 7, and 9(1).

[137] The recitals to Directive 2006/116/EC on the term of protection of copyright and certain related rights refer to cases where "one or more *physical* persons are identified as authors" (emphasis added)—presumably in distinction to references to "persons" elsewhere in the directive, which would refer to legal persons also.

suggest that only works created by a human can be protected by copyright".[138] The US Copyright Office has declared that it will "register an original work of authorship, provided that the work was created by a human being",[139] citing the 1884 case, *Burrow-Giles Lithographic Co. v. Sarony.*[140]

In the US case *Comptroller of the Treasury v. Family Entertainment Centers*,[141] a Maryland Court was asked to decide whether animatronic puppets that danced and sang at restaurants triggered a state tax on food "where there is furnished a performance". The court decided that the animatronic puppets were not performing:

> [A] pre-programmed robot can perform a menial task but, because a pre-programmed robot has no 'skill' and therefore leaves no room for spontaneous human flaw in an exhibition, it cannot 'perform' a piece of music ... Just as a wind-up toy does not perform for purposes of [the statute,] neither does a pre-programmed mechanical robot.[142]

Although this was a tax case, the discussion of creativity in relation to a performance could be relevant to copyright. The puppets in *Family Entertainment Centers* were not robots in the sense used in this book; as the court found, they were deterministic, pre-programmed automatons. There was no discretionary or unpredictable aspect to their

---

[138] Andres Guadamuz, "Artificial Intelligence and copyright", *WIPO Magazine*, October 2017, http://www.wipo.int/wipo_magazine/en/2017/05/article_0003.html, accessed 1 June 2018. For Spain, see Law No. 22/1987 of 11 November 1987, on intellectual property, and for Germany, see Urheberrechtsgesetz Teil 1 - Urheberrecht (§§ 1–69g), Abschnitt 3 - Der Urheber (§ 7). § 7 UrhG does not state expressly that the author of a copyrighted work has to be human being. It merely states: "The creator ('Schöpfer') is the author". It is generally understood, though, that the law supposes that only humans can "create" and thus be "creators".

[139] The Compendium of U.S. Copyright Office Practices: Chapter 300, https://copyright.gov/comp3/chap300/ch300-copyrightable-authorship.pdf, accessed 1 June 2018.

[140] 111 U.S. 53, 58 (1884). The position is supported by later US case law (e.g. *Feist Publications v. Rural Telephone Service Company, Inc.* 499 U.S. 340 (1991)) which specifies that copyright law only protects "the fruits of intellectual labor" that "are founded in the creative powers of the mind".

[141] 519 A.2d 1337, 1338 (Md. 1987), overturned on other grounds in *318 North Market Street, Inc.* et al. *v. Comptroller of the Treasury*, 554 A.2d 453 (Md. 1989).

[142] Ibid., at 1339.

performance. Based on the reasoning, it appears that the outcome of *Family Entertainment Centers* would have been different if the puppets in question used AI to adapt and perfect their performance over time.

Some legal systems have attempted to accommodate AI, or at least computer-generations, within their provisions on intellectual property.[143] For instance, the UK, Ireland and New Zealand acknowledge that different principles are required for AI than for direct human creators, but nonetheless seek to establish a causal link between the eventual creation and an initial human input. The UK Copyright, Designs and Patents Act 1998 (CDPA) provides at section 9(3):

> In the case of a literary, dramatic, musical or artistic work which is computer-generated, the author shall be taken to be the person by whom the arrangements necessary for the creation of the work are undertaken.[144]

Section 178 of the CDPA provides that a computer-generated work is one that "is generated by computer in circumstances such that there is no human author of the work". This provision does not allow for AI itself to be considered the author. Instead, it engenders a two-stage analysis: the first stage is to identify whether there is a human author. If a human author cannot be found, the second stage is to identify the person "by whom the arrangements necessary for the creation of the work are undertaken". Where the work is generated by an AI entity, disputes may arise at both stages.

As to the first, there may be issues including how far the inputs must be related to the outputs so as to classify the person who provided those inputs as the author. As to the second, it is unclear how one would

---

[143] For discussions of how computer-generated creations might be addressed particularly in US copyright law, as well as a proposal for a general scheme applicable to AI-generated works, see Annemarie Bridy, "Coding Creativity: Copyright and the Artificially Intelligent Author", *Stanford Technology Law Review* (2012), 1. See also Ralph D. Clifford, "Intellectual Property in the Era of the Creative Computer Program: Will the True Creator Please Stand Up?" *Tulane Law Review*, Vol. 71 (1997), 1675, 1696–1697; and Pamela Samuelson, "Allocating Ownership Rights in Computer-Generated Works", *University of Pittsburgh Law Review*, Vol. 47 (1985), 1185.

[144] New Zealand and Ireland both use the same language. See Copyright Act of 1994, 2 (New Zealand); Copyright and Related Rights Act 2000, Part I, 2 (Act. No. 28/2000) (Ireland).

identify the person who "made the arrangements". It could be the person who built the system, the person who trained it or the person who fed it these specific inputs.[145] Matters are complicated yet further if one or more of these parties is another AI entity.

4.2   Case Study: The "Monkey Selfie" Case

In 2014, a crested macaque monkey (or rather a charity which claimed to be acting on behalf of that monkey) demanded copyright in a "selfie" (self-portrait) which it had taken using a professional photographer's camera.[146] The monkey, Naruto, was named as a plaintiff in a case in the Northern District of California, against the photographer, David Slater.[147] It was reported in late 2017 that the photographer had settled with the monkey's representatives,[148] after more than two years of costly

---

[145] Toby Bond, "How Artificial Intelligence Is Set to Disrupt Our Legal Framework for Intellectual Property Rights", *IP Watchdog*, 18 June 2017, http://www.ipwatchdog.com/2017/06/18/artificial-intelligence-disrupt-legal-framework-intellectual-property-rights/id=84319/, accessed 1 June 2018. See also Burkhard Schafer et al., "A Fourth Law of Robotics? Copyright and the Law and Ethics of Machine Coproduction", *Artificial Intelligence and Law*, Vol. 23 (2015), 217–240; Burkhard Schafer, "Editorial: The Future of IP Law in an Age of Artificial Intelligence", *SCRIPTed*, Vol. 13, No. 3 (December 2016), via: https://script-ed.org/wp-content/uploads/2016/12/13-3-schafer.pdf, accessed 1 June 2018.

[146] Guadamuz, Andrés, "The Monkey Selfie: Copyright Lessons for Originality in Photographs and Internet Jurisdiction", *Internet Policy Review*, Vol. 5, No. 1 (2016), https://doi.org/10.14763/2016.1.398. http://policyreview.info/articles/analysis/monkey-selfie-copyright-lessons-originality-photographs-and-internet-jurisdiction, accessed 1 June 2018.

[147] *NARUTO, a Crested Macaque, by and through his Next Friends, People for the Ethical Treatment of Animals, Inc., Plaintiff-Appellant, v. DAVID JOHN SLATER; BLURB, INC., a Delaware corporation; WILDLIFE PERSONALITIES, LTD., a United Kingdom private limited company*, No. 16-15469 D.C. No. 3:15-cv-04324- WHO, https://assets.documentcloud.org/documents/2700588/Gov-Uscourts-Cand-291324-45-0.pdf, accessed 1 June 2018.

[148] Jason Slotkin, "'Monkey Selfie' Lawsuit Ends With Settlement Between PETA, Photographer", *NPR*, 12 September 2017, https://www.npr.org/sections/thetwo-way/2017/09/12/550417823/-animal-rights-advocates-photographer-compromise-over-ownership-of-monkey-selfie, accessed 1 June 2018.

legal battle,[149] which Slater said had left him broke.[150] Slater was reportedly required by the settlement agreement to donate 25% of the earnings from his book to charities "that protect the habitat of Naruto and other crested macaques in Indonesia", as the animal charity described it.[151]

In April 2018, despite the parties having settled out of court, the US Court of Appeals for the 9th Circuit chose to rule on the matter nonetheless and concluded that the relevant Copyright Act made no provision for animals to sue. There the story ended for Naruto's selfie rights claim. Interestingly, the Court of Appeals left open the possibility of animals "asserting" constitutional rights in other contexts, noting that animals still had constitutional standing to bring claims in a Federal Court, following a precedent set in a previous case involving dolphins and whales.[152]

The *Naruto* case demonstrates the jurisprudential difficulties which arise when a "creative" act is carried out by a non-human entity. Although the eventual conclusion of the courts was that the relevant statute did not extend to protecting the intellectual property animals or other entities without legal personality, the wider question is whether it *should* do.

---

[149] Monkey selfie case: Judge rules animal cannot own his photo copyright, *The Guardian*, 7 January 2016, https://www.theguardian.com/world/2016/jan/06/monkey-selfie-case-animal-photo-copyright, accessed 1 June 2018. David Slater announced in 2017 that he was "broke" as a result of the court case, despite having ultimately prevailed. Julia Carrie Wong, "Monkey Selfie Photographer Says He's Broke: 'I'm Thinking of Dog Walking", *The Guardian*, 13 July 2017, https://www.theguardian.com/environment/2017/jul/12/monkey-selfie-macaque-copyright-court-david-slater, accessed 1 June 2018.

[150] Ibid.

[151] Meagan Flyn, "Monkey Loses Selfie Copyright Case. Maybe Monkey Should Sue PETA, Appeals Court Suggests", *The Washington Post*, 24 April 2018, https://www.washingtonpost.com/news/morning-mix/wp/2018/04/24/monkey-loses-selfie-copyright-case-maybe-monkey-should-sue-peta-appeals-court-suggests/?utm_term=.afe1b1b181d6, accessed 1 June 2018.

[152] *NARUTO, a Crested Macaque, by and through his Next Friends, People for the Ethical Treatment of Animals, Inc., Plaintiff-Appellant, v. DAVID JOHN SLATER; BLURB, INC., a Delaware corporation; WILDLIFE PERSONALITIES, LTD., a United Kingdom private limited company*, No. 16-15469 D.C. No. 3:15-cv-04324- WHO, http://cdn.ca9.uscourts.gov/datastore/opinions/2018/04/23/16-15469.pdf, accessed 1 June 2018, citing at p. 11 *Cetacean Community*, 386 F.3d at 1171.

4.3   Patents and Other Protections

Copyright is not the only type of intellectual property law to be challenged by AI. Patents are a form of local monopoly granted over a particular invention. A classic example of an invention protectable by patent is a new pharmaceutical drug. By contrast to copyright's emphasis on the state of mind of the creator, the criteria required for protection vary between systems but generally speaking patents will be granted if an application is made for an invention which is new, non-obvious and of some potential use, regardless of the process by which they came into existence.[153] However, as with the "creativity" issue, current laws do not accommodate AI as the inventor of patents.[154]

The difference between copyright and patent protection is particularly important where AI is involved. It may be easier for AI to create subject matter protectable by patents (albeit not hold them) than for AI to create subject matter protectable by copyright.

Other tests apply to the creation and enforcement of IP rights known as trademarks (which protect branding) and designs (which protect the appearance of products). Like patents, the conditions for protection of these two categories are objective. After being exposed to a data set featuring furniture from many other companies (as well as perhaps other sources of inspiration, such as nature or art), it is quite conceivable that an AI system might create an entirely new design, let's say for a chair. The AI system might even acquire a reputation for making innovative furniture. Both of the above are in theory capable of protection under IP law, at least when created or developed by humans.

Without either a new rule for ascribing AI's works or discoveries to an existing legal person, such as a human or a company, current laws are manifestly unsuitable for accommodating and safeguarding AI's creations. This lacuna in legal protection might in turn discourage the development of creative AI in circumstances where the original developers are unsure who, if anyone, would own its creations.

---

[153] For the US rules, see, 35 U.S.C. paras. 101–02, 112 (2000). In the European system, the criteria are that the invention must be "new, involve an inventive step and are susceptible of industrial application". Art. 52 European Patent Convention.

[154] Ryan Abbot, "Everything is Obvious", 22 October 2017, https://papers.ssrn.com/sol3/papers.cfm?abstract_id=3056915, accessed 1 June 2018.

5 FREE SPEECH AND HATE SPEECH

The freedom to express ideas, within certain limits, is protected by many legal systems. In the USA, there is the First Amendment to the Constitution; in Europe, there is Article 10 of the European Convention on Human Rights. Similar protections exist under the constitutions of South Africa,[155] India[156] and other countries.

If AI can generate content which, if spoken or written by a human, would qualify for free speech protection, the question arises whether the AI's speech should be granted the same protections. In order to address this question, it is first necessary to investigate the reasons underpinning legal protections for free speech. Toni Masaro and Helen Norton describe the compendium of reasons for protecting free speech (in the USA) as follows:

> ...there is no unifying theory of the First Amendment. The most influential theories have been clustered into arguments based on democracy and self-governance, a marketplace of ideas model, and autonomy.[157]

Motivations like "autonomy" seem to be linked to conceptions of individual human dignity, which do not at present apply to AI.[158] However, as regards instrumentalist values such as the "marketplace of ideas", there does not seem to be any reason why society would derive less benefit from a new idea generated by AI than it would from a new idea generated by a human.[159]

---

[155] Constitution of South Africa, s. 16.

[156] Constitution of India, art. 19.

[157] Toni M. Massaro and Helen Norton, "Siri-ously? Free Speech Rights and Artificial Intelligence", *Northwestern University Law Review*, Vol. 110, No. 5, 1175, citations omitted.

[158] Though see Chapter 4 for discussion of when AI might justify such protection in its own right.

[159] At present, AI lacks the consciousness required for it to be deemed worthy of non-instrumentalist protections, but as shown in Chapter 4, this may not always be the case.

Not all speech is protected, and in most systems, some speech is prohibited. Where speech is deemed injurious to another person, then it can lead to private law liability in libel or slander. Where it is thought harmful to religion, it can lead to criminal blasphemy charges. Speech insulting to the royal family or head of state in some countries can lead to charges under *lese-majeste* rules.[160] Other laws may prohibit speech which incites violence. In short, there is a myriad of complex legal principles across the world which both protect and constrain what a person may say. In some countries, these protections are not limited to human persons. The US Supreme Court has confirmed that corporations are entitled to have their freedom of speech protected.[161] The question of how such rights and restrictions might apply to AI remains undetermined.

These are not just hypothetical problems. Comedian Stephen Colbert helped design a Twitter bot called "@realhumanpraise": a program which pairs epithets from a film review website with Fox News personalities, with sometimes scurrilous results.[162] Though @realhumanpraise may not use AI, it is certainly conceivable that an AI-powered program might be used to similar (if not more offensive) effect. Where the relevant laws require some form of intent, as well as the harmful speech, then it seems difficult for a human to be held liable for the "speech" of the AI system. This is especially so where the combination of words and ideas used is not foreseeable.

Mr. Colbert's program was intended as satirical, but many have raised concerns as to the possibility for automatically generated Internet content to shape human opinions and even elections. One prominent example is the alleged use of "Twitter Bots" by individuals and organisations aligned to Russia, to shape opinions in matters including the 2016 US

---

[160] "Lese-majeste Explained: How Thailand Forbids Insult of Its Royalty", *BBC Website*, http://www.bbc.co.uk/news/world-asia-29628191, accessed 1 June 2018.

[161] *Citizens United v. Federal Election Commission*, 558 U.S. 310 (2010).

[162] Ross Luipold, "Colbert Trolls Fox News By Offering @RealHumanPraise On Twitter, and It's Brilliant", *Huffington Post*, 5 November 2013, http://www.huffingtonpost.co.uk/entry/colbert-trolls-fox-news-realhumanpraise_n_4218078, accessed 1 June 2018.

election[163] and the UK's Brexit vote.[164] It is not clear whether AI has yet played any role in the generation of messages apparently designed to polarise voters, but the possibility is obvious.

In November 2015, Victor Collins was found dead in the hot tub of another man, James Bates. Mr. Bates was accused of murder. His Amazon Echo, a home speaker device incorporating an AI virtual assistant, was potentially a key "witness" to the alleged crime, and the Arkansas local police issued a warrant asking Apple to divulge data from the relevant period. In a February 2017 court filing, Amazon cited US First Amendment freedom of speech protections—not just for the human voice commands which may have been heard by the AI device, but also for the device's responses. Amazon abandoned this argument a month later, but the episode again called into question whether the AI was entitled to have its speech protected.[165]

As with protections for free speech, the intentions or even the identity of the speaker of "harmful" speech may be of far less important than the content. Should a racist message be seen as any less problematic because it is generated by AI rather than a human? In one notable public relations disaster, Microsoft's flagship AI chatbot "Tay", which was apparently modelled to speak like a "teen girl" was rapidly decommissioned after it began sending racist, neo-Nazi, conspiracy-theory-supporting and sexualised messages.[166]

[163] Samuel C. Woolley, "Automating Power: Social Bot Interference in Global Politics". *First Monday*, Vol. 21, No. 4 (2016).

[164] Alexei Nikolsky and Ria Novosti, "Russia Used Twitter Bots and Trolls 'to Disrupt' Brexit Vote", *The Times*, 15 November 2017. See also Brundage, Avin et al., *The Malicious Use of Artificial Intelligence: Forecasting, Prevention, and Mitigation*, February 2018, https://img1.wsimg.com/blobby/go/3d82daa4-97fe-4096-9c6b-376b92c619de/downloads/1c6q2kc4v_50335.pdf, accessed 1 June 2018.

[165] Rich McCormick, "Amazon Gives up Fight for Alexa's First Amendment Rights After Defendant Hands Over Data", *The Verge*, 7 March 2017, https://www.theverge.com/2017/3/7/14839684/amazon-alexa-firstamendment-case, accessed 20 August 2018. The case was *State of Arkansas v. James A. Bates* Case No. CR-2016-370-2.

[166] Helena Horton, "Microsoft Deletes 'Teen Girl' AI After It Became a Hitler-Loving Sex Robot Within 24 hours", *The Telegraph*, 24 March 2016, http://www.telegraph.co.uk/technology/2016/03/24/microsofts-teen-girl-ai-turns-into-a-hitler-loving-sex-robot-wit/, accessed 1 June 2018. It should be noted that Tay did not generate the content unprompted; various computer programmers swiftly discovered how to game its algorithms to cause it to generate offensive content. See Chapter 8 at s. 3.2.2 for discussion of how the program was corrupted.

Because current rules protecting and prohibiting speech are focussed on shaping the actions of humans, there remains a gap as to how the speech of AI is to be regulated. One option for AI-generated hate speech is to penalise the publisher on a strict liability basis (e.g. public social networks such as Facebook, Instagram or Twitter). A law enacted by Germany against social media hate speech (from any source) which is communicated by a social network has already been criticised by some as overstepping the mark.[167] Moreover, it is not always certain that AI speech will be conducted via the medium of such a provider. In any case, until a solution is chosen the law will remain unclear, and potential loopholes for harmful speech will persist.[168]

6   Conclusions on Responsibility for AI

The aim of this chapter has been to demonstrate the ways in which established legal mechanisms might address responsibility for AI. Running through each is a tension as to whether AI should be treated as an object, a subject, a thing or a person. Current laws can and will in the short term continue to determine responsibility for AI in the ways set out above. The bigger question is whether society's aims would be better served by reformulating our relationship with AI in a more radical fashion. The following chapters consider some of the changes we might make.

---

[167] Yascha Mounk, "Verboten: Germany's Risky Law for Stopping Hate Speech on Facebook and Twitter", *New Republic,* 3 April 2018, https://newrepublic.com/article/147364/verboten-germany-law-stopping-hate-speech-facebook-twitter, accessed 1 June 2018.

[168] Toni M. Massaro and Helen Norton, "Siri-ously? Free Speech Rights and Artificial Intelligence", *Northwestern University Law Review,* Vol. 110, No. 5.

CHAPTER 4

Rights for AI

Why do we protect the rights of others? Moral arguments concentrate on it being somehow "wrong" to harm the entity concerned.[1] Pragmatic arguments proceed on the basis that protecting others is helpful to those who are prevented from doing harm. Moral grounds are an end in themselves and pragmatic ones are a means to an end. The two justifications can apply independently from each other, but they are not mutually exclusive. For instance, it is morally unacceptable to wound another human being without just cause; in addition, it is sensible not to cause wanton harm to a human, lest they (or their family or friends) seek revenge against the wrongdoer.

This chapter concentrates predominantly on the moral reasons why some AI systems might one day be deemed worthy of protection. Moral rights are addressed before legal ones because, as will be suggested below, recognition of the former tends to predate (and indeed to precipitate) the latter. Chapter 5 will discuss additional pragmatic reasons for protecting AI rights, as well as endowing it with responsibilities. Together, these justifications might form the basis for granting AI legal personality, though this is far from the only way of protecting AI's moral rights.

Granting rights to robots might sound ridiculous. However, protecting AI in this way could be in accordance with widely–held moral precepts. Chapter 4 aims to answer three questions: What do we mean

---

[1] The terms moral and ethical are used interchangeably in this book.

J. Turner, *Robot Rules*, https://doi.org/10.1007/978-3-319-96235-1_4

by rights? Why do we grant rights to other entities? And could AI and robots qualify for rights by virtue of the same principles? In so doing, we will seek to challenge common preconceptions about why certain entities deserve rights, and not others.

1   WHAT ARE RIGHTS?

1.1   Hohfeld's Incidents

The word "right" is used in many different contexts: workers' rights, animal rights, human rights, a right to life, to water, to free speech, to equal treatment, to privacy, to property and so on. But without clarifying what is meant by a "right", we risk talking at cross-purposes.

This book adopts the approach of legal theorist Wesley Hohfeld, who separated rights into four categories, or "incidents": privileges, powers, claims and immunities.[2] In addition to distinguishing between the different types of rights, Hohfeld's other key insight was to pair each category of right in a reciprocal relationship with a right held by another person. The four categories listed above correspond to the following: duty, no-claim, liability, disability. Thus, if Person A has a claim to something, Person B must have a liability to provide Person A with that thing.

There are three advantages to Hohfeld's categorisation. *First* it is exhaustive of the various different "rights" mentioned in common parlance as well as legal treatises. *Secondly*, it acknowledges the differences between various types of rights.[3] *Thirdly*, Hohfeld's model explains

---

[2] In formal terms, these incidents can be expressed as follows: Privileges: A has a privilege to φ if and only if A has no duty not to φ. Claims: A has a claim that B φ if and only if B has a duty to A to φ. Powers: A has a power if and only if A has the ability to alter her own or another's Hohfeldian incidents. Immunities: B has an immunity if and only if A lacks the ability to alter B's Hohfeldian incidents. See Leif Wenar, "Rights", *The Stanford Encyclopaedia of Philosophy*, edited by Edward N. Zalta (Fall 2015 Edition), https://plato. stanford.edu/archives/fall2015/entries/rights/, accessed 1 June 2018.

[3] The philosopher Isaiah Berlin divided liberties into positive and negative categories (i.e. freedoms *to*, and freedoms *from*). See Isaiah Berlin, "Two Concepts of Liberty", in *Four Essays on Liberty* (Oxford: Clarendon Press, 1969), 121–154. This categorisation is helpful up to a point in terms of providing intellectual clarity to the nature of liberties, but—at least in terms of the positive liberties—Hohfeld's categorisation is more helpful for present purposes because of its emphasis both on the holder of rights and those interacting with that holder.

how the different categories of rights interact with each other, and with those of other people. Hohfeld's framework demonstrates that rights are social constructs. The correlatives to each right show that they do not exist in a vacuum. Rather, they are held against other people or entities. For instance, it would not make sense for a person marooned alone on a desert island to claim that she has a right to life, because there is no one else against whom she can claim that right. To hold rights, therefore, is to coexist with others capable of upholding or infringing upon those rights.

This social feature of rights underscores the importance of considering how humans (as well as other entities already afforded rights, like corporations and animals) are to live alongside AI as it becomes increasingly prevalent. As science journalist and author John Markoff has written, we will need to ask ourselves whether robots are to become "our masters, slaves, or partners".[4]

Following Markoff's lead, this chapter asks whether AI can or should be treated as a "moral patient", namely the subject of certain protections from the actions of "moral agents". As explained in Chapter 2 at Section 2.1, agency involves a party being capable of understanding and acting on certain rules and principles. In moral terms, below a certain level of mental sophistication, a human's actions are not deemed to be blameworthy. Nonetheless, a young child who lacks moral agency is still entitled to be protected as a moral patient. Moral agency and moral patiency can coincide, but it is not necessary for this to be the case.[5]

1.2   Rights as Fictions

The social nature of rights is connected to another feature: they are communal inventions which do not have any independent, objective existence beyond our collective imagination. Like companies, countries and laws themselves, rights are collective fictions, or as Harari calls them,

---

[4] John Markoff, "Our Masters, Slaves, or Partners"? in *What to Think About Machines That Think*, edited by John Brockman (New York and London: Harper Perennial, 2015), 25–28.

[5] See John Danaher, "The Rise of Robots and the Crisis of Moral Patiency", *AI & Society*, (November 2017), 1–8.

"myths".[6] Their form can be shaped to any given context. Certainly, some rights are treated as more valuable than others, and belief in them may be more widely shared, but there is no set quota of rights which prevents new ones from being created and old ones from falling into abeyance.

Jenna Reinbold has said of the drafting of Universal Declaration of Human Rights: "...the first Commission on Human Rights undertook its work in a way that smacks of the time-honored logic of mythmaking - a logic wherein language is set to the task of unequivocally presenting a vision of the world as well as a set of mandates appropriate to the maintenance of that world".[7]

None of this means rights are unimportant. To the contrary, they make life meaningful and allow societies to function effectively. Describing rights as fictions or constructs is by no means pejorative; when used in this context, it does not entail duplicity or error.[8] It simply means that they are malleable and can be shaped according to new circumstances.[9]

Moral rights are not the same as legal rights. There are some moral relationships—for instance, the duty to tell the truth and the correlative claim right not to be lied to—which are not always protected by law.[10]

---

[6] Yuval Harari, *Sapiens: A Brief History of Humankind* (London: Random House, 2015).

[7] Jenna Reinbold, "Seeing the Myth in Human Rights", *OpenDemocracy*, 29 March 2017, https://www.opendemocracy.net/openglobalrights/jenna-reinbold/seeing-myth-in-human-rights, accessed 1 June 2018. See also Jenna Reinbold, *Seeing the Myth in Human Rights* (Philadelphia: University of Pennsylvania Press, 2017).

[8] Yuval Harari, *Sapiens: A Brief History of Humankind* (London: Random House, 2015).

[9] Seeing rights as fictions does not necessarily slide into moral relativism, where any system of norms is no "better" or "worse" than any other system. The judgement of whether a norm is "good" or "bad" is a question which can only be answered by reference to some external hierarchy of moral criteria—whether utilitarian, deontological, religious and so on. On the contrary, the idea that rights are fictions is completely value-neutral. They can be good fictions or bad fictions. The view of legal rights—and indeed all laws—as having a validity which is value-neutral accords to the legal theory known as "Positivism", which holds that "[i]n any legal system, whether a given norm is legally valid, and hence whether it forms part of the law of that system, depends on its sources, not its merits". For the source of this definition, see John Gardner, "Legal Positivism: 5 1/2 Myths". *American Journal of Jurisprudence*, Vol. 46 (2001), 199. For further discussion of Positivism, see Chapter 6 at s. 1.

[10] Though misrepresentation and fraud in certain contexts can lead to civil liability and criminal charges in most if not all legal systems.

If Alfred asks Marianne whether he looks fat in his expensive new trousers, Marianne will not be held legally liable if she fails to tell the truth. Generally speaking, the law reflects and supports society's moral values but the two are not coterminous.

The present discussion is concerned primarily with the rights that actually *are* recognised by humans as well as those which have been in the past. This is a sociological exercise and is therefore capable of objective verification. The argument made here is that *if* we recognise certain moral and legal rights, then *as a matter of logical consistency* we ought to recognise others in analogous circumstances.

One of the reasons why the idea of rights for robots provokes an instinctive negative reaction for some people[11] may be an unspoken assumption that rights are a fixed quantity, like unchanging commandments written on tablets of stone. If it is accepted that rights are fictions—albeit valuable ones for the functioning of society—this objection falls away and the path is cleared for robot rights to be recognised.

2   ANIMALS: MAN'S BEST FRIENDS?

Humanity's changing attitude towards animals provides a good analogy for how we might come to see AI. The comparison with rights for animals illustrates two things: *first*, rights for animals are culturally relative, and *second*, animal rights have varied considerably over time.

2.1   Cultural Relativity of Attitudes to Animals

Rules to protect animals are not a new idea. The story of Noah saving two of each species from the great flood (which appears also in the Quran and pre-dates the Judeo-Christian Bible by several millennia)[12] might be read as a cautionary tale on the need to preserve biodiversity.[13]

---

[11] As Professor Horst Eidenmuller has observed: "Most of us probably feel uneasy when considering whether smart robots should be accorded legal personality". Horst Eidenmuller, "Robots' Legal Personality", *University of Oxford Faculty of Law Blog*, 8 March 2017, https://www.law.ox.ac.uk/business-law-blog/blog/2017/03/robots%E2%80%99-legal-personality, accessed 1 June 2018.

[12] Helge Kvanvig, *Primeval History: Babylonian, Biblical, and Enochic: An Intertextual Reading* (The Netherlands/Danvers, MA: Brill, 2011), 21–24, 243–258.

[13] Thomas L. Friedman, "In the Age of Noah", *The New York Times*, 23 December 2007, http://www.nytimes.com/2007/12/23/opinion/23friedman.html, accessed 1 June 2018.

The Book of Proverbs says "A righteous man regardeth the life of his beast".[14] However, more generally in the Judeo-Christian tradition, humanity appears to enjoy a position of dominance over all other creations. In the Old Testament, at Genesis 1:26, God says: "Let us make man in our image, after our likeness: and let them have dominion over the fish of the sea, and over the fowl of the air, and over the cattle, and over all the earth, and over every creeping thing that creepeth upon the earth".[15]

Other cultures and religions seem to accord animals a greater importance than the above. For instance, animism ascribes souls to various entities, whether sentient (including animals and insects), living (including plants, lichens and corals) and even non-living (including mountains, rivers and lakes).[16] Hindu teachings provide that an *atman* or soul can be reincarnated in many different forms, not just humans but also as various animals.[17] Indeed, several Hindu Gods have the features of animals.[18] Further, the cow is treated as a holy creature. In India, 18 states ban the slaughter of cows[19] and vigilante groups even seek to protect cows through extrajudicial violence.[20] As noted in Chapter 2, in the Japanese Shintō religion, many different creatures and objects have *kami*, which translates as "spirit", "soul" or "energy".[21]

---

[14] Proverbs 12:10, King James Bible.

[15] Genesis 1:26, King James Bible. See also Sura 93 in the Quran.

[16] See, for example, Nurit Bird-David, "Animism Revisited: Personhood, Environment, and Relational Epistemology", *Current Anthropology*, Vol. 40, No. S1, 67–91. For the first use of the term animism, see the seminal work: Edward Burnett Tyler, *Primitive Culture: Researches into the Development of Mythology, Philosophy, Religion, Language, Art, and Custom* (London: John Murray, 1920).

[17] The Hindu American Foundation, "Official Statement on Animals", *Website of the Humane Society of the United States*, http://www.humanesociety.org/assets/pdfs/faith/hinduism_and_the_ethical.pdf, accessed 1 June 2018.

[18] For instance, the Hindu God Ganesh has the head of an elephant, and Hanuman has the head or even in some depictions the entire body of a monkey.

[19] Soutik Biswas, "Is India's Ban on Cattle Slaughter 'Food Fascism'?", *BBC Website*, 2 June 2017, http://www.bbc.co.uk/news/world-asia-india-40116811, accessed 1 June 2018.

[20] Soutik Biswas, "A Night Patrol with India's Cow Protection Vigilantes", *BBC Website*, 29 October 2015, http://www.bbc.co.uk/news/world-asia-india-34634892, accessed 1 June 2018; "India Probe After 'Cow Vigilantes Kill Muslim Man'", *BBC Website*, 5 April 2017, http://www.bbc.co.uk/news/world-asia-india-39499845, accessed 1 June 2018.

[21] "Shinto at a Glance", *BBC Religions*, last updated 10 July 2011, http://www.bbc.co.uk/religion/religions/shinto/ataglance/glance.shtml, accessed 1 June 2018. See also Chapter 2 at s. 2.1.3.

It is clear from the above that animal rights are culturally relative. For those cultures which are more open to rights for animals and objects, the idea of rights for AI may be less of a philosophical leap than it is for cultures which focus solely or largely on the spiritual welfare of humans. Several writers have noted the greater willingness of the Japanese public to accept AI and humanoid robots than is the case in the West. Indeed, a 2016 Policy Paper commissioned by the European Parliament states:

> ... fear of robots is not felt in the Far East. After the Second World War, Japan saw the birth of Astro Boy, a manga series featuring a robotic creature, which instilled society with a very positive image of robots. Furthermore, according to the Japanese Shintoist vision of robots, they, like everything else, have a soul. Unlike in the West, robots are not seen as dangerous creations and naturally belong among humans.[22]

2.2   Animal Rights Through History

There is growing acceptance of the proposition that it is wrong to cause unnecessary suffering to animals.[23] This has not always been the case. Across the world, animal rights laws were minimal at best 200 years ago. Animals were regarded as property to be treated as their owners saw fit—not entities which could have rights themselves.[24] In England,

---

[22] European Parliament Directorate-General for Internal Policies, Policy Department C, Citizens' Rights and Constitutional Affairs, "European Civil Law Rules in Robotics: Study for the JURI Committee" (2016), PE 571.379, 10.

[23] See, for example, Harold D. Guither, *Animal Rights: History and Scope of a Radical Social Movement* (Carbondale and Edwardsville, IL: Southern Illinois University Press, 2009). For an influential early text on the animal rights movement, see Henry Stephens Salt, *Animal Rights Considered in Relation to Social Progress* (New York, London: Macmillan & Co, 1894). The Georgetown Law Library lists 35 countries which have anti-animal cruelty legislation. International and Foreign Animal Law Research Guide, Georgetown Law Library, http://guides.ll.georgetown.edu/c.php?g=363480&p=2455777, accessed 1 June 2018. Further materials are available at the Michigan State University Animal legal & Historical Centre website, https://www.animallaw.info/site/world-law-overview, accessed 1 June 2018.

[24] The legal basis for the determination of animals as property is traced by some to the Old Testament. For instance, William Blackstone in *Commentaries on the Laws of England* stated that "In the beginning of the world, we are informed by holy writ, the all-bountiful creator gave to man 'dominion over all the earth; and over the fifh of the fea and over the fowl of the air, and over every living thing that moveth upon the earth.' This is the only true and folid foundation of man's dominion over external things, whatever airy metaphyfical notions may have been ftarted by fanciful writers upon this fubject. The earth therfore,

in 1793, a man called John Cornish was found not guilty of any crime when he pulled out a horse's tongue. The court ruled that Cornish could be prosecuted only if there was evidence of malice towards the horse's owner.[25]

Descartes wrote that animals were merely "beast-machines", and "automata"[26] with no soul, no minds and no ability to reason.[27] It followed that we should be no more concerned with animals' squeals of pain than we are with the creaks and crashes of machinery. In moral terms, harming them was no different from tearing a piece of paper or chopping a block of wood. The modern philosopher Norman Kemp Smith described Descartes' views as a "monstrous" thesis that "animals are without feeling or awareness of any kind".[28]

However, from the seventeenth century onwards, animal rights gradually came to be protected.[29] In 1641, the General Court of Massachusetts passed the "Body of Liberties", an early charter of fundamental rights, which included a section on animals providing: "No man shall exercise any tiranny or crueltie towards any bruite creature

and all things therein, are the general property of all mankind". William Blackstone, *Commentaries on the Laws of England* (12th edn. London: T. Cadell, 1794), Book II, 2–3.

[25] Discussed in Simon Brooman Legge, *Law Relating to Animals* (London: Cavendish Publishing Ltd., 1997), 40–41.

[26] Renee Descartes, *Oeuvres de Descartes*, edited by Charles Adam and Paul Tannery (Paris: Cerf, 1897–1913), Book V, 277.

[27] A. Boyce Gibson, *The Philosophy of Descartes* (London: Methuen, 1932), 214; E.S. Haldane and G.T.R. Ross, *The Philosophical Works of Descartes* (Cambridge: Cambridge University Press, repr. 1969), 116. Though for a contrary view, which seeks to rehabilitate Descartes' writings on animals, see John Cottingham, "'A Brute to the Brutes?': Descartes' Treatment of Animals", *Philosophy*, No. 53 (1978), 551–559.

[28] Norman Kemp Smith, *New Studies in the Philosophy of Descartes* (London: Macmillan, 1952), 136, 140.

[29] The Puritans, under Oliver Cromwell's Protectorate, had banned certain activities such as bear baiting in the mid-seventeenth century. However, the motivation for this appears to have been more aimed at dampening human enjoyment rather than lessening animal cruelty: theatre and morris-dancing were also prohibited in this period. Muriel Zagha, "The Puritan Paradox", *The Guardian*, 16 February 2002, https://www.theguardian.com/education/2002/feb/16/artsandhumanities.highereducation, accessed 1 June 2018.

which are usuallie kept for man's use".[30] When in 1821 a UK politician, Colonel Richard Martin, first proposed a statute to protect horses, he was met with derision and even laughter in Parliament.[31] Things changed quickly though; the following year Parliament enacted the Ill Treatment of Horses and Cattle Act 1822 at Colonel Martin's behest. In 1824, the Society for the Prevention of Cruelty to Animals was founded in London, the first such organisation of its kind.[32] In 1840, it received a Royal Charter from Queen Victoria.

Throughout the nineteenth and twentieth centuries, increasing protection was granted to animals in various countries around the world.[33] The American Society for the Prevention of Cruelty to Animals was established in 1866; in the UK, major pieces of animal rights legislation enacted after 1822 included the Cruelty to Animals Act 1876 and the Protection of Animals Act 1911; India passed the Prevention of Cruelty to Animals Act in 1960.[34]

Advocates for animals continue to push outwards the boundaries of their protection, whether through supporting changes in legislation or developments in case law. In 2004, a US Court of Appeals held that "all of the world's whales and dolphins", claiming for damages allegedly caused by the US Navy's use of sonar, had standing under Article III of

[30] "Massachusetts Body of Liberties" (1641), published in *A Bibliographical Sketch of the Laws of the Massachusetts Colony From 1630 to 1686* (Boston: Rockwell and Churchill, 1890). Full text available at: http://www.mass.gov/anf/docs/lib/body-of-liberties-1641. pdf, accessed 1 June 2018. The Body of Liberties continued: "If any man shall have occasion to leade or drive Cattel from place to place that is far of, so that they be weary, or hungry, or fall sick, or lambe, It shall be lawful to rest or refresh them, for competant time, in any open place that is not Corne, meadow, or inclosed for some peculiar use".

[31] Other Members of Parliament referred to Martin mockingly as "Humanity Dick". Simon Brooman Legge, *Law Relating to Animals* (London: Cavendish Publishing Ltd., 1997), 42.

[32] "History", *RSPCA Website*, https://www.rspca.org.uk/whatwedo/whoweare/history, accessed 1 June 2018.

[33] For a summary of developments, see generally Simon Brooman Legge, *Law Relating to Animals* (London: Cavendish Publishing Ltd., 1997).

[34] The Prevention of Cruelty to Animals Act, 1960 Act No. 59 OF 1960. Text available at *Michigan State University Animal Legal and Historical Centre Website*, https://www.animallaw.info/statute/cruelty-prevention-cruelty-animals-act-1960, accessed 1 June 2018.

the US Constitution to bring such claims.[35] The whales and dolphins' claim was rejected because the specific statute under which they claimed did not contain any substantive protections they were able to assert. The door was left open to such protections being provided though. The Court of Appeals held:

> '[I]f Congress and the President intended to take the extraordinary step of authorizing animals as well as people and legal entities to sue, they could, and should, have said so plainly.' In the absence of any such statement in the [the relevant statutes], we conclude that the Cetaceans do not have statutory standing to sue.[36]

Extensions of animal rights law can be controversial. As Hohfeld's structure shows, granting rights to one group, in this case animals, entails a restriction on another group—usually humans. This means that there is often resistance from those who stand to lose out. When in 2004 the UK Government introduced legislation to ban foxhunting, much of the rural population objected, seeing this move as an attack by urbanites on their way of life. This in turn sparked a constitutional crisis in which the elected part of the legislature, the House of Commons, invoked a rarely used mechanism[37] to overrule the non-elected part, the House of Lords.[38] Approximately 200,000 people demonstrated in London against the proposed foxhunting ban.[39]

From this brief historical survey, it can be seen that human attitudes towards animal rights have varied greatly over time and continue to develop.

[35] *Cetacean Community v. Bush*, 386 F.3d 1169 (9th Cir. 2004) at 1171. See also the "monkey selfie" case, *NARUTO, a Crested Macaque, by and through his Next Friends, People for the Ethical Treatment of Animals, Inc., Plaintiff-Appellant, v. DAVID JOHN SLATER; BLURB, INC., a Delaware corporation; WILDLIFE PERSONALITIES, LTD., a United Kingdom private limited company*, No. 16-15469 D.C. No. 3:15-cv-04324- WHO, discussed in Chapter 3 at s. 4.2.

[36] *Cetacean Community v. Bush*, 386 F.3d 1169 (9th Cir. 2004) at 1179.

[37] This instance was only the ninth time such a step had been taken since 1911.

[38] This was done pursuant to the Parliament Acts 1911 and 1949. The legality of this course was challenged and ultimately upheld by the UK's highest court, the House of Lords, in *R (Jackson) v. Attorney General* [2005] UKHL 56.

[39] "Huge Turnout for Countryside March", *BBC Website*, 22 September 2002, http://news.bbc.co.uk/1/hi/uk/2274129.stm, accessed 1 June 2018.

3   How the Human Got Her Rights

The entitlements we now describe as fundamental human rights were not always thought of as beyond dispute. The idea of *universal* human rights and even the concept of human rights are both relatively recent inventions.

Slavery is one of the most extreme infringements on human rights and therefore provides a useful case study of changing attitudes. The analogy of slavery is also instructive because one can readily draw comparisons with our treatment of robots. Indeed, it is no coincidence that in *Rossum's Universal Robots*, the screenplay which brought the term robots into popular use, Karel Capek used "roboti", the Czech word for "slaves", to refer to the intelligent mechanical servants who eventually rose up against their human masters.[40]

As little as 150 years ago, in large parts of the world human slavery was legal. Similarly to animals, slaves were treated primarily as property. At the beginning of the nineteenth century, slavery was permitted under international law. The UK abolished the slave traffic throughout its colonies in 1807 and, in 1814, induced France to do likewise. In 1815, the "Powers" of Europe collectively condemned slavery at the Conference of Vienna.[41]

The transition towards abolishing slavery was not all one way. In the infamous *Dred Scott v. Sandford* case of 1857, the US Supreme Court ruled that slavery was legal, holding that when the US Constitution was drafted, "neither the class of persons who had been imported as slaves, nor their descendants, whether they had become free or not, were then acknowledged as a part of the people".[42]

---

[40] See also Chapter 1 at s. 1.

[41] These included Austria, Russia, Prussia, France, Spain, Portugal, Sweden, Denmark, the Netherlands, Switzerland, Genoa and several German States. See Mathieson, *Great Britain and the Slave Trade, 1839–1865* (London: Octagon Books, 1967); Soulsby, *The Right of Search and the Slave Trade in Anglo-American Relations, 1813–1862* (Baltimore: The Johns Hopkins press, 1933); and Leslie Bethell, *The Abolition of the Brazilian Slave Trade* (Cambridge: Cambridge University Press, 2009).

[42] 60 U.S. 393 (1857).

Nowadays, few would disagree that slavery is morally wrong.[43] The prohibition of slavery is one of the central tenets of modern international law, having the status of *jus cogens*: a norm which binds all nations whether or not they have expressly agreed to it, and from which no derogation is permitted.[44] The Universal Declaration on Human Rights 1948 provides that "no one shall be held in slavery or servitude; slavery and the slave trade shall be prohibited in all their forms".[45]

Even aside from slavery, for centuries it was seen as perfectly legitimate in many cultures to stratify the value of fellow humans by features including their gender, religion, race, nationality or even social class. In consequence, vast numbers of people were seen as being expendable throughout the twentieth century. This led to acts of great malice, such as the Holocaust, the Rwandan genocide and other such deliberate ethnic massacres. Belief that some humans were superior to others also facilitated much death and suffering through indifference, where certain groups were sacrificed supposedly to serve greater ends.

In George Orwell's *Animal Farm*, the horse Boxer lives by the maxim: "I will work harder", until he is eventually worn out by the hard labour, and sent to be killed and turned into glue at the knacker's yard.[46]

---

[43] See, for instance, the Slavery Convention of 1926, by which the signatories undertake to suppress and prevent the slave trade and to bring about, progressively and as soon as possible, the entire suppression of slavery in all its forms.

[44] M. Cherif Bassiouni, "International Crimes: Jus Cogens and Obligatio Erga Omnes". *Law and Contemporary Problems*, Vol. 59 (1996), 63. The International Law Commission Draft Code of Crimes against the Peace and Security of Mankind, adopted in 1996, listed enslavement as a crime against humanity (see art. 18(d): *Yearbook of the ILC* (1996) vol II, pt 2. This formed the basis for the Rome Statute of the International Criminal Court to list enslavement as a crime against humanity in art. 7(1)(c). The International Court of Justice has regarded protection from slavery as included in the basic rights of the human person which give rise to obligations which states owe *erga omnes*. *Barcelona Traction Case*, ICJ Rep (1970), 32.

[45] Art. 4 of the Universal Declaration on Human Rights 1948. See also art. 8 of the International Covenant on Civil and Political Rights adopted by the United nations General Assembly in 1966, art. 4 of the European Convention for the Protection of Human Rights and Fundamental Freedoms 1950, art. 6 of the American Convention on Human Rights 1969, and art. 5 of the Charter on Human and Peoples' Rights 1981.

[46] George Orwell, *Animal Farm* (London: Secker & Warburg/Penguin, 2000), 82.

Today, we treat machines the same way. When they are broken, out of date or obsolete, we discard them and sell their pieces for scrap.

Proponents of slavery evinced pseudo-scientific arguments that certain ethnicities were biologically different—and therefore inferior—to others.[47] Such theories of racial superiority and inferiority have now been debunked, but modern evolutionary biology suggests that there are in fact small but significant genetic differences between ethnicities.[48] These discoveries have not caused the world to doubt that we should give people of all ethnicities the same human rights. They show that we do not necessarily protect human rights because we have no differences, but rather *in spite of* our differences.

Although at a genetic level humans have not changed significantly in the last several thousand years, our attitudes towards rights for humans have shifted significantly during that period (a trend similar in some ways to the treatment of animals described above). By contrast, AI has only come into existence fairly recently—with significant advances in the last 10 years. In consequence, there is likely to be far greater scope for changes in societal attitudes, as AI acquires new capabilities and traits. Despite what our gut instincts might suggest at present, developments in animal and human rights indicate that societal opinions *could* shift in favour of granting AI rights. The next question is whether and if so, when we *should* grant AI rights. The following sections investigate and identify the features humanity deems worthy of protection.

4   WHY ROBOT RIGHTS?

There are three general reasons for protecting the rights of others, which could be applied to at least some types of AI and robots: *first*, the ability to suffer; *secondly*, compassion; *thirdly*, their value to humans. There is also a *fourth* specific reason for protecting AI rights: situations where humans and AI are combined.

[47] See, for instance, Samuel Cartwright's notorious art. "Diseases and Peculiarities of the Negro Race", *De Bow's Review, Southern and Western States*, Volume XI (New Orleans, 1851).

[48] Yuval Harari, *Sapiens: A Brief History of Humankind* (London: Random House, 2015), 13–19, describing the 'interbreeding theory' of human evolution. For an example of another contemporary theory on the differences between races, see, for instance, Nicholas Wade, *A Troublesome Inheritance: Genes, Race and Human History* (London: Penguin, 2015).

4.1   The Argument from Pain: "Suffer Little Robots"

4.1.1 Consciousness and Qualia

One reason for protecting rights is to increase happiness and decrease suffering. This was what the utilitarian philosopher John Stuart Mill referred to as the "felicific calculus". This is not necessarily limited to humans but rather encompasses an unlimited range of entities. As Jeremy Bentham wrote in 1789:

> The day may come, when the rest of the animal creation may acquire those rights which never could have been withholden from them but by the hand of tyranny... The French have already discovered that the blackness of the skin is no reason why a human being should be abandoned without redress to the caprice of a tormentor... The question is not, Can they reason? nor, Can they talk? but, Can they suffer?[49]

The argument that an entity should have rights because it can suffer seems to assume that the entity in question is aware, or conscious of itself suffering. Otherwise, there is no *they* or *it* which can be said to suffer.

Consciousness therefore becomes a prerequisite for protection based on the ability to suffer. For the avoidance of doubt, this section is not seeking to suggest that all or even some AI is conscious and as such deserving of protection. The point is simply that *if* an AI system was to acquire this quality, *then* it should qualify for some moral rights.

An agreed definition of consciousness is a matter which continues to elude philosophers, neurologists and computer scientists. One popular definition, which this book adopts, is to say that consciousness describes "the way things seem to us", an experience referred to more formally as *qualia*.[50] It is suggested that consciousness as *qualia* can be broken down into three stages as follows:

---

[49] Jeremy Bentham, *An Introduction to the Principles of Morals and Legislation* (Oxford: Clarendon Press, 1907), Chapter XVII, Of the Limits of the Penal Branch of Jurisprudence, FN 122.

[50] Daniel Dennett said that qualia have four properties: (1) ineffable—meaning they cannot be communicated; (2) intrinsic—meaning they do not change depending on the relationship of the qualia to other things; (3) private—meaning they cannot be compared between experiencing entities; (4) directly or immediately apprehensible in consciousness—meaning, as Louis Armstrong once said of defining jazz "If you got to ask, you ain't never going to know". Daniel Dennett, "Quining Qualia", in *Consciousness in Contemporary Science*, edited by A.J. Marcel and E. Bisiach (Oxford: Oxford University Press, 1988).

For an entity to be conscious, it must be capable of (i) sensing stimuli, (ii) perceiving sensations and (iii) having a sense of self, namely a conception of its own existence in space and time.

Sensations are the raw data of the external world that are observed or felt by an entity. The first stage is achieved by even rudimentary technology which does not constitute AI. Any sensor—whether of light, heat, humidity, electromagnetic signals or any other stimuli—meets this low threshold. Clearly, today's AI systems and robots are able to take in such raw data.

The second level is perception, meaning applying some form of analysis or rule to the data which allows it to be sensibly interpreted. This jump from sense to perception occurs when we see certain visual patterns and infer that there is a certain three-dimensional object. A person may *sense* a series of lines, joining each other in various ways, but she *perceives* that she is looking at a table.[51] Perceptions do not necessarily equate to reality. A person may think she is looking at a table when in fact she is observing an optical illusion. Likewise, we may perceive the sun "rising", but in fact what we have observed is our field of vision alerting as the earth rotates on its axis whilst orbiting the sun. Technology journalist Hal Hodson sums up the difference between levels one and two of consciousness:

> Even though cameras can capture more data about a scene than the human eye, roboticists are at a loss as to how to stitch all that information together to build a cohesive picture of the world.[52]

The perception stage of consciousness appears to be present in various instances of AI. As Dr. Bruce MacLennan of the Department of Electrical Engineering and Computer Science at the university of Tennessee explains,

---

[51] Sydney Shoemaker, "Self-knowledge and Inner Sense, Lecture I: The Object Perception Model". *Philosophy and Phenomenological Research*, Vol. 54, No. 2 (1994), 249–269.

[52] Hal Hodson, "Robot Homes in on Consciousness by Passing Self-Awareness Test", *New Scientist*, 15 July 2015, https://www.newscientist.com/article/mg22730302-700-robot-homes-in-on-consciousness-by-passing-self-awareness-test/?gwaloggedin=true, accessed 1 June 2018.

In robots, as in animals, a primary function of emotion is to make rapid assessments of external or internal situations and to ready the robot to respond to them with action or information processing... These processes will be monitored by interoceptors (internal sensors) that measure these and other physical properties (positions, angles, forces, stresses, flow rates, energy levels, power drains, temperatures, physical damage, etc.) and send signals to higher cognitive processes for supervision and control.[53]

When an AI system uses rules or principles to draw conclusions from data, it can be said to "perceive" that data, by virtue of whatever heuristic has been applied. This process of being able to radically simplify huge amounts of information by sorting it into groups known as "clusters" functions in a similar way to that in which human and (probably) animal minds seem to work, when seeking to make sense of the world.

Another example of perception stage of consciousness in AI is the use of artificial "neural nets" where initial data is taken in at one level, stimulating an "input" layer of thinking cells or neurons. Those neurons in turn stimulate a further layer, which is capable of more abstract thought processes. This keeps on going through the system, allowing the AI system to develop complex conclusions before it reaches its output. It is no coincidence that the most basic neural nets developed in the 1950s were referred to by developers as "perceptrons", which were able to develop internal "representations" of concepts.[54]

The third and final stage is for there to be an entity which knows that it is experiencing the sensations. This is the "I" in "I am feeling...". In German, this is referred to as "*ich-gefühl*", translating literally to "I-feeling".[55] For AI, as well as perhaps some living organisms, the "I" could be a "we". Consciousness can be formed by collective experience, as opposed to individual. A bee may not have a particularly strong conception of a singular self which suffers, but it certainly appears to know of a greater self—the colony or the nest—of which it is part, and which can collectively suffer or thrive. One example of collective consciousness

---

[53] Bruce MacLennan, "Cruelty to Robots? The Hard Problem of Robot Suffering", *ICAP Proceedings* (2013), 5–6, http://www.iacap.org/proceedings_IACAP13/paper_9.pdf, accessed 1 June 2018.

[54] Marvin Minsky and Sydney Papert, "*Perceptrons: An Introduction to Computational Geometry*," (Cambridge, MA and London, England: The MIT Press, 1988), Prologue.

[55] Leo A. Spiegel, "The Self, the Sense of Self, and Perception", *The Psychoanalytic Study of the Child*, Vol. 14, No. 1 (1959), 81–109, 81.

portrayed in popular culture is the Borg, an alien race in *Star Trek*: a vast collection of drones linked to a hive mind known as "the Collective".[56] Some experimental AI systems which are based on swarm intelligence displayed collectively by a number of individual "bots" may eventually develop group consciousness.[57] Without this I or we then there is no *thing* which can be said to suffer. Psychologists Daniel Kahneman and Jason Riis suggest that the human mind comprises an "experiencing self" and an "evaluating self". The experiencing self is that which lives life as a series of moments. The evaluative self then tries to make sense of those moments, by a variety of different shortcuts or heuristics.[58] Their account of a distinct "self" existing through time and remembering past experiences is an example of the third element of consciousness.

Generally speaking, the third stage of consciousness remains elusive in AI. That said, some experiments and theories relating to an off-switch or kill button for AI may provide some evidence as to how AI could acquire a sense of "self".[59] Whereas most AI executes a particular task, these experiments consider how AI might be incentivised to allow a human to turn it off (a process known as safe interruptibility).[60] The reason why this is significant to consciousness is that the AI might have some conception of its own existence in order to resist or willingly allow that existence to be terminated.

[56] "Borg", *Startrek.com*, http://www.startrek.com/database_article/borg, accessed 1 June 2018. For discussion of the Borg and human consciousness, see Jacob Lopata, "Pre-Conscious Humans May Have Been Like the Borg", *Nautilus*, 4 May 2017, http://nautil. us/issue/47/consciousness/pre_conscious-humans-may-have-been-like-the-borg, accessed 1 June 2018.

[57] See, for example, Eric Bonabeau, Marco Dorigo, and Guy Theraulaz, *Swarm Intelligence: From Natural to Artificial Systems*, No. 1 (Oxford: Oxford University Press, 1999); Christian Blum and Xiaodong Li, "Swarm Intelligence in Optimization", in *Swarm Intelligence* (Heidelberg: Springer, 2008), 43–85; and James Kennedy, "Swarm intelligence", in *Handbook of Nature-inspired and Innovative Computing* (Springer US, 2006), 187–219.

[58] Daniel Kahneman and Jason Riis, "Living, and Thinking About It: Two Perspectives on Life", *The Science of Well-Being*, Vol. 1 (2005). See also Daniel Kahneman, *Thinking, Fast and Slow* (London: Penguin, 2011).

[59] See more general discussion in Chapter 8 at s. 5.4.2.

[60] See Laurent Orseau and Stuart Armstrong, "Safely Interruptible Agents", 28 October 2016, http://intelligence.org/files/Interruptibility.pdf, accessed 1 June 2018; El Mahdi El Mhamdi, Rachid Guerraoui, Hadrien Hendrikx, and Alexandre Maure, "Dynamic Safe Interruptibility for Decentralized Multi-Agent Reinforcement Learning", *EPFL Working Paper* (2017) No. EPFL-WORKING-229332.

In a 2016 paper, researchers from the University of California at Berkeley led by Stuart Russell reported on an experiment which they called "The Off-Switch Game".[61] The starting point of this game is that AI can possess instrumentalist goals beyond those with which it has originally been programmed, which can include self-preservation.[62]

Self-preservation can be problematic if AI decides to take steps to prevent a human disabling it. Russell et al. offer a novel solution which has important effects not just for the control of AI but also its nature as a potentially conscious entity:

> Our key insight is that for R to want to preserve its off switch, it needs to be uncertain about the utility associated with the outcome, and to treat H's actions as important observations about that utility. (R also has no incentive to switch itself off in this setting.) We conclude that giving machines an appropriate level of uncertainty about their objectives leads to safer designs, and we argue that this setting is a useful generalization of the classical AI paradigm of rational agents.[63]

Russell et al. demonstrate with formal mathematical proofs that so long as an AI entity is unsure whether or not it is doing what a human wants, it will always allow itself to be turned off. The AI can function, but at every decision point it must ask whether or not it is doing the right thing, and if not, whether it should be "killed" as a sanction for its failure. In other words, the AI modelled must ask itself: "To be or not to be?".[64] Though not the focus of Russell et al., the experiment arguably

---

[61] Dylan Hadfield-Menell, Anca Dragan, Pieter Abbeel, and Stuart Russell, "The Off-Switch Game", *arXiv preprint* arXiv:1611.08219 (2016), 1.

[62] See, for example, Stephen Omohundro, "The Basic AI Drives", in *Proceedings of the First Conference on Artificial General Intelligence* (2008).

[63] Ibid.

[64] Arguably, an excess of confidence in the "rightness" of an ultimate goal—particularly where that goal is not of a nature that is observable in the natural world—can lead to undesirable consequences in human actions, as well as those of AI. For instance, it might be said that belief-based fundamentalists, whether on the basis of religion, animal rights, nationalism, etc., suffer from an excess of confidence. This is the same flaw as a robot given a supposedly simple goal, which then causes enormous harm in fulfilling that goal at the expense of everything else in the world (as in Nick Bostrom's paperclip example, see Chapter 1 at s. 6). By the same token a person who is uncertain whether he will go to heaven, even if he kills ten unbelievers, is less likely to become a suicide bomber. A little uncertainty can do a lot of good.

indicates one route for AI to display the third element of consciousness identified above.[65]

There is more than one route to the third element of consciousness. Another avenue might be along the lines of that suggested by Hod Lipson and colleagues in a paper published in the journal *Science* in 2006. Lipson and colleagues showed how a four-legged robot which had no prior knowledge of its own appearance or capabilities was able to learn to move through continuous self-modelling.[66]

Many will object that neither of the two experiments described above truly indicate consciousness in a metaphysical sense, or even under this book's own definition. However, so long as it is acknowledged that (a) consciousness is an objective property which is capable of being defined and observed; and (b) consciousness is not restricted to humans, then the possibility remains open for AI to be developed which can be conscious.[67]

If and when an AI becomes conscious, the final question is whether the conscious AI can suffer. AI technology today appears capable of achieving this result—even if it is not yet conscious. Reinforcement learning involves programs analysing data, making decisions and then being informed by a feedback mechanism whether these decisions were more or less correct. The feedback mechanism qualifies results by assigning them a score according to how desirable the decision was. Each time this process takes place, the computer learns more about its task and its

---

[65] Additional evidence of AI approaching the third level of consciousness is provided by a 2015 experiment in robotics carried out by Selmer Bringsjord and colleagues, in which a robot was able to pass a "three wise men" test. This involved a robot correctly identifying that its speech function had not been disabled, without being given any other information other than the question "which pill did you receive?". See Selmer Bringsjord, John Licato, Naveen Sundar Govindarajulu, Rikhiya Ghosh, and Atriya Sen, "Real Robots that Pass Human Tests of Self-Consciousness" in *Robot and Human Interactive Communication (RO-MAN), 2015 24th IEEE International Symposium on*, pp. 498–504. IEEE, 2015.

[66] Josh Bongard, Victor Zykov, and Hod Lipson, "Resilient Machines Through Continuous Self-Modeling", *Science*, Vol. 314, No. 5802 (2006), 1118–1121.

[67] See generally the work of Stan Franklin, who argues that IDA, a piece of US Navy software, displays consciousness according to a set of (different) objective criteria—based on neuroscientist Bernard Baars' "global workspace" theory: Stan Franklin, "IDA: A Conscious Artifact?", *Journal of Consciousness Studies*, Vol. 10, No. 4–5 (2003), 47–66. See also Bernard J. Baars, *A Cognitive Theory of Consciousness* (Cambridge: Cambridge University Press, 1988); Bernard J Baars, *In the Theater of Consciousness* (Oxford: Oxford University Press, 1997).

environment, gradually honing and perfecting its abilities. Most human children discover at some point that if they touch a sharp object, it can hurt. If pain is just a signal which encourages an entity to avoid something undesirable, then it is not difficult to acknowledge that robots can experience it. In 2016, German researchers published a paper indicating that they had created a robot which could "feel" physical pain when its skin was pricked with a pin.[68]

4.1.2 Degrees of Consciousness

Consciousness is not a binary quality; it can exist in degrees.[69] It can vary on at least three levels. *First*, within a living organism, there is a spectrum of consciousness from a minimally conscious state such as deep sleep to being fully awake. For instance, in October 2013, Oxford University researchers led by Irene Tracey were able to pinpoint different degrees of consciousness during anaesthesia of a human subject.[70] *Secondly*, and

[68] Johannes Kuehn and Sami Haddadin presentation entitled, "An Artificial Robot Nervous System to Teach Robots How to Feel Pain and Reflexively React to Potentially Damaging Contacts", given at ICRA 2016 in Stockholm, Sweden, http://spectrum.ieee.org/automaton/robotics/robotics-software/researchers-teaching-robots-to-feel-and-react-to-pain, accessed 1 June 2018.

[69] See, for instance, Christof Koch and Giulio Tononi, "Can Machines Be Conscious? Yes—And a New Turing test Might Prove It", in *IEEE Spectrum Special Report: The Singularity*, 1 June 2008, http://spectrum.ieee.org/biomedical/imaging/can-machines-be-conscious, accessed 1 June 2018. Koch and Tononi write: "To be conscious, then, you need to be a single integrated entity with a large repertoire of states. Let's take this one step further: your level of consciousness has to do with how much integrated information you can generate. That's why you have a higher level of consciousness than a tree frog or a supercomputer".

[70] Róisín Ní Mhuircheartaigh, Catherine Warnaby, Richard Rogers, Saad Jbabdi, and Irene Tracey, "Slow-wave Activity Saturation and Thalamocortical Isolation During Propofol Anesthesia in Humans", *Science Translational Medicine*, Vol. 5, No. 208 (2013), 208ra148–208ra148. "Researchers pinpoint degrees of consciousness during anaesthesia", Nuffield Department of Clinical Neurosciences, 24 October 2013, https://www.ndcn.ox.ac.uk/news/researchers-pinpoint-degrees-of-consciousness-during-anaesthesia, accessed 1 June 2018. See also David Chalmers, "Absent qualia, Fading qualia, Dancing qualia", in *Conscious Experience*, edited by Thomas Metzinger (Paderborn: Exetes Schoningh in association with Imprint Academic, 1995), 256. For a similar argument, see John R. Searle, *The Rediscovery of the Mind* (Cambridge, MA: MIT Press, 1992), 66. Nicholas Bostrom critiques the "fading qualia" argument in Nicholas Bostrom, "Quantity of Experience: Brain-duplication and Degrees of Consciousness", *Mind Machines*, Vol. 16 (2006), 185–200.

again within any given species (particularly mammals which continue to develop significantly after birth), a newborn infant appears to have less consciousness than a fully developed adult.[71] *Thirdly*, consciousness may differ between species.[72]

If there are degrees of consciousness, there is no logical reason why humans in a normal waking state should occupy the pinnacle of such conscious experience. Indeed, we know that certain animals possess the ability to sense phenomena outside the bounds of human senses or even comprehension. It is generally accepted that humans are limited to five senses through which the world can be experienced: taste, sight, touch, smell and sound.[73] Bats experience the world through sonar. Other animals can sense and act on the basis of electromagnetic waves. Humans can observe such forces via other media, such as visual representations on a computer screen, but we cannot know or even accurately imagine what it is to experience them directly.[74] Owing to these additional senses, some animals are arguably *more* conscious than us, or at the very least conscious in different ways.[75]

[71] Douglas Heaven, "Emerging Consciousness Glimpsed in Babies", *New Scientist*, 18 April 2013, https://www.newscientist.com/article/dn23401-emerging-consciousness-glimpsed-in-babies/, accessed 1 June 2018.

[72] See, for example, Colin Allen and Michael Trestman, "Animal Consciousness", in *The Blackwell Companion to Consciousness*, edited by Susan Schneider and Max Velmans (Oxford: Wiley, 2017), 63–76. Colin Allen and Michael Trestman, "Animal Consciousness", *The Stanford Encyclopedia of Philosophy*, edited by Edward N. Zalta (Winter 2016 Edition), https://plato.stanford.edu/archives/win2016/entries/consciousness-animal/, accessed 1 June 2018. Nicholas Bostrom, "Quantity of Experience: Brain-duplication and Degrees of Consciousness", *Mind Machines*, Vol. 16 (2006), 185–200, 198.

[73] There are some scientists who disagree. See, for instance, Rupert Sheldrake, "The 'Sense of Being Stared at' Confirmed by Simple Experiments", *Rivista Di Biologia Biology Forum*, Vol. 92, 53–76. Anicia Srl, 1999.

[74] See Thomas Nagel, "What Is It to Be a Bat?", *The Philosophical Review*, Vol. 83, No. 4 (October 1974), 435–450.

[75] This is assuming for the sake of argument that the increase in ability to perceive one sensory stimulus is not accompanied by a reduction in another sense—for instance bats' vision is weak compared to many other animals.

Unlike humans, AI is not constrained by the finite physical space that a biological brain can occupy and the number of neurons it can contain. Just as a relatively simple computer can now undertake many more calculations in a given period of time than can even the greatest human mathematician, we cannot exclude the possibility that AI may one day acquire a greater sense of consciousness than any human and perhaps become capable of experiencing suffering of a far greater magnitude.

Susan Schneider of the University of Connecticut considers that AI might *bypass* what we consider to be consciousness and develop an entirely different form of operating. First, she notes that "[i]n humans, consciousness is correlated with novel learning tasks that require concentration, and when a thought is under the spotlight of our attention, it is processed in a slow, sequential manner... A superintelligence would surpass expert-level knowledge in every domain, with rapid-fire computations ranging over vast databases that could encompass the entire internet. It may not need the very mental faculties that are associated with conscious experience in humans". Schneider's second argument is based on physical properties. She hypothesises:

> ... consciousness may be limited to carbon substrates only. Carbon molecules form stronger, more stable chemical bonds than silicon, which allows carbon to form an extraordinary number of compounds, and unlike silicon, carbon has the capacity to more easily form double bonds...If the chemical differences between carbon and silicon impact life itself, we should not rule out the possibility that these chemical differences also impact whether silicon gives rise to consciousness, even if they do not hinder silicon's ability to process information in a superior manner.[76]

4.1.3 The Role of Scepticism

We have only a limited understanding of the human mind, let alone animal or even artificial ones. We can ask other people how they feel, and we can observe brain scans, but none of these things equate to knowing

---

[76]Susan Schneider, "The Problem of AI Consciousness", *Kurzweil Accellerating Intelligence Blog*, 18 March 2016, http://www.kurzweilai.net/the-problem-of-ai-consciousness, accessed 1 June 2018.

exactly what the other person experiences.[77] David Chalmers has termed these difficulties as "the hard problem of consciousness".[78]

The same issue applies even more so with animals. In an influential paper on consciousness, the philosopher Thomas Nagel asked "What is it to be a bat?", concluding that "reductionist" objective accounts of a subjective experience are not possible.[79] A dog may appear sad, and a chimpanzee can recoil as if in pain, but we cannot actually ask them to describe it and even if we could, we would have no way to actually know what it is like being a dog, chimpanzee or indeed a bat. Nonetheless, we continue to act as if both humans and animals are conscious, and can suffer.

This sceptical perspective on consciousness is important because it shows that even when we assume we are granting rights to others based on their capability for suffering, we cannot truly be sure of what they are feeling. It seems, therefore, that we protect the rights of others based not on what they are actually feeling, but on what we *believe* they are feeling. The following section expands on why we act in this way, and whether similar motivations might apply to robots and AI.

4.2   The Argument from Compassion

4.2.1 Evolutionary Programming and Intuition

We protect certain things because we have an emotional reaction to them being harmed. If we see a child picking a kitten up by its tail, the instinctive human reaction is to feel empathy for the kitten. Human rights have intuitive appeal because it *feels* bad to watch another person in pain. Animal rights receive similar, if not the same, support from our involuntary reactions. Why is it upsetting to see others suffer (or appear to suffer)? Understanding how others feel is one of humanity's most

---

[77] In this sense, we are still unable to solve a problem posed by John Locke, namely that although we can measure the wavelength of light reflected off a certain object, we have no way of knowing whether one person's perception of "blue" is *experienced* by another person as the colour that the first person would call "yellow". This is known as the spectral inversion thesis. John Locke, *Essay Concerning Human Understanding* (London: T. Tegg and Son, 1836), 279.

[78] For a modern exposition, see David Chalmers, "Facing Up to the Problem of Consciousness", *Journal of Consciousness Studies*, Vol. 2, No. 3 (1995), 200–219.

[79] Thomas Nagel, "What Is It to Be a Bat?", *The Philosophical Review*, Vol. 83, No. 4 (October 1974), 435–450.

powerful tools for building value and belief systems which can bind society together. Empathy is therefore another reason for creating rights. This was the view taken by Immanuel Kant:

> If a man has his dog shot, because it can no longer earn a living for him, he is by no means in breach of any duty to the dog, since the latter is incapable of judgment, but he thereby damages the kindly and inhumane qualities in himself, which he ought to exercise in virtue of his duties to mankind... a person who already displays such cruelty to animals is also no les hardened towards men.[80]

Kant even extended this theory to "duties to inanimate objects", explaining that "[t]hese allude, indirectly, to our duties to men. The human impulse to destroy things that can still be used is very immoral... Thus all duties relating to animals, other beings and things have an indirect reference to our duties towards mankind".[81]

Failing to protect the rights of other humans undermines the moral fibre of a community. It denies the basic emotional reaction we have to the perceived suffering of another. The same emotional reactions also govern our feelings towards animals, although to a lesser extent. If we treat animals with contempt, then we might start to do so with humans also. There is a link between the two because we perceive animals as having needs and sensations—even if they do not have the same sort of complex thought processes as we do. Essentially, animals exhibit features which resemble humans, and we are biologically programmed to feel empathy toward anything with those features. This phenomenon is also likely to explain why we feel greater empathy to mammals, which are anatomically closer to humans than reptiles, amphibians, insects or fish. For instance, baby mammals often have large heads and large eyes, just like human babies.[82]

---

[80] Immanuel Kant, *Lectures on Ethics*, translated by Peter Heath, edited by Peter Heath and Jerome B. Schneewind (Cambridge: Cambridge University Press, 1997), 212, (27: 459).

[81] Ibid., 27:460.

[82] M. Borgi, I. Cogliati-Dezza, V. Brelsford, K. Meints, and F. Cirulli, "Baby Schema in Human and Animal Faces Induces Cuteness Perception and Gaze Allocation in Children", *Frontiers in Psychology*, Vol. 5 (2014), 411. http://doi.org/10.3389/fpsyg.2014.00411, accessed 1 June 2018.

The emotional response of empathy is a successful evolutionary technique which allows us to collaborate with others in our species, because we can imagine what it is like to be them. Such collaboration, which (unlike in other species) can extend beyond family, tribe or colony, is one of the factors which led to the success of the human race.[83] Although there is no obvious evolutionary advantage to feeling empathy for an injured animal as opposed to an injured human, it appears that the same neural pathways are activated when we see animals are in pain.[84]

Sometimes our empathy and compassion for animals even exceed that for other humans. In 2013, scientists at Regent's University Augusta and Cape Fear Community College conducted a study in which 40% of participants said that they would save their pet rather than a human foreigner from being hit by a bus.[85] Around the world, there was outcry when Harambe, a male gorilla, was shot by zookeepers after he snatched a three-year-old boy who had wandered into his enclosure.[86]

4.2.2 Sex, Robots and Rights

The moral debates concerning robots designed for sexual activities may cast some light on the social significance of AI. If we see certain acts with robots as being unacceptable, we must ask ourselves why this is so.

Capek's film *Rossum's Universal Robots* questioned whether advanced robot slaves should have some form of civil rights, or whether they are

---

[83] Yuval Harari, *Sapiens: A Brief History of Humankind* (London: Random House, 2015), 102–110.

[84] See, for example, Claus Lamm, Andrew N. Meltzoff, and Jean Decety, "How Do We Empathize with Someone Who Is Not Like Us? A Functional Magnetic Resonance Imaging Study", *Journal of Cognitive Neuroscience*, Vol. 22, No. 2 (February 2010), 362–376, http://www.mitpressjournals.org/doi/abs/10.1162/jocn.2009.21186?url_ver=Z39.88-2003&rfr_id=ori%3Arid%3Acrossref.org&rfr_dat=cr_pub%3Dpubmed&#.WPKpwIQrLRZ, accessed 1 June 2018.

[85] Richard J. Topolski, Nicole Weaver, Zachary Martin, and Jason McCoy, "Choosing Between the Emotional Dog and the Rational Pal: A Moral Dilemma with a Tail", *Anthrozoös*, Vol. 26, No. 2 (2013), 253–263.

[86] See, for instance, Jennifer Chang, "Outrage Grows Over the Death of a Gorilla, Shot After a Child Climbed into Its Enclosure", *Quartz*, 30 May 2016, https://qz.com/695343/outrage-grows-over-the-death-of-a-gorilla-shot-to-protect-a-child-who-climbed-into-its-enclosure/, accessed 1 June 2018.

merely machines that can be harmed or destroyed at will with no moral consequences.[87]

In *Love and Sex with Robots: The Evolution of Human-robot Relationships*, David Levy surmises that "[t]he robots in the middle of this century will not be exactly like us, but close", and that "when robots reach a level of sophistication, at which they are able to engender and sustain feelings of romantic love in their humans... the social and psychological benefits will be enormous".[88] He justifies this statement on the basis that "[a]lmost everyone wants someone to love, but many people have no one. If this natural human desire can be satisfied for everyone who is capable of loving, surely the world will be a much happier place".

Others might object that something intangible is lost when such feelings are lavished on artificial entities. Joanna Bryson acknowledges the psychological tendency in humans to develop feelings for things which *look* as if they are conscious, but her proposed solution is to argue that we should therefore avoid creating robots which display conscious tendencies: "If robots ever need rights we'll have designed them unjustly".[89]

Even more difficult are questions as to whether humans should be permitted to carry out activities on robots which are prohibited on humans: Is it wrong to allow a human to play out a rape fantasy using a robot as the victim? Would anything change if the robot was aware that it was being used for activities which, if carried out on a human, would be considered morally depraved?[90]

The debates over sex robots suggest there are two reasons why we might find this technology distasteful. One is that committing degrading

---

[87] Similar issues are raised by the sympathetic depictions of sensitive and helpful androids such as Data in *Star Trek* and R2-D2 or C-3PO *Star Wars*, to the challenging examples in *Westworld* (both the 1973 film and modern TV series) as well as Alex Garland's 2015 film *Ex Machina*.

[88] David Levy, *Love and Sex with Robots: The Evolution of Human-robot Relationships* (New York and London: Harper Perennial, 2009), 303–304.

[89] Joanna Bryson, "If Robots Ever Need Rights We'll Have Designed Them Unjustly", *Adventures in NI Blog*, 31 January 2017, https://joanna-bryson.blogspot.co.uk/2017/01/if-robots-ever-need-rights-well-have.html, accessed 1 June 2018.

[90] For discussion, see Rebecca Hawes, "Westworld-style Sex with Robots: When Will It Happen—And Would it Really Be a Good Idea?", *The Telegraph*, 5 October 2016, http://www.telegraph.co.uk/tv/2016/10/05/sex-with-robots-when-will-it-happen---and-would-it-really-be-a-g/, accessed 1 June 2018.

acts with the sex robots is harmful to the robots themselves. This would rely on the arguments based on suffering outlined above. The other (and more popular) argument is that simulating immoral or illegal acts with robots harms human society in some way, by condoning or promoting an unpleasant behaviour trait: an instrumental justification. This is a similar justification to the reason why cartoons depicting child pornography are often banned—even though no child was directly harmed in the process. Evan Dashevsky, a robot ethicist from MIT's Media Lab, has summed up the second argument by giving a contemporary twist to Kant's point above:

> Would you rather live in, say, a Westworld universe filled with humans who feel free to rape and maim the park's mechanical inhabitants, or on the deck of Star Trek: The Next Generation, where advanced robots are treated as equals? The humans of one world seem a lot more welcoming than the other, don't they?[91]

4.2.3 Speciesism

Some writers and academics have suggested that to discriminate between different species when granting rights may be morally wrong. Richard D. Ryder, a psychologist and animal rights activist, termed this behaviour "speciesism", drawing deliberate parallels with racism:

> ...scientists have agreed that there is no 'magical' essential difference between humans and other animals, biologically-speaking. Why then do we make an almost total distinction morally? If all organisms are on one physical continuum, then we should also be on the same moral continuum.[92]

It is not necessary to go as far as Ryder in urging that animals, or AI should have *the same* rights as humans, but his extreme view contains an important insight: the human species is not as unique as we might think in certain respects. It is of course true that AI and robots are physically different from humans, but it should be recalled that racists and

---

[91] Evan Dashenevsky, "Do Robots and AI Deserve Rights?" *PC Magazine*, 16 February 2017, http://uk.pcmag.com/robotics-automation-products/87871/feature/do-robots-and-ai-deserve-rights, accessed 1 June 2018.

[92] Richard Ryder, "Speciesism Again: The Original Leaflet", *Critical Society*, No. 2 (Spring 2010), 81.

proponents of eugenics sought to support their arguments with scientific "evidence" on the differences between races. The probity of this science may certainly be doubted, but the more important question seems to be not whether there are physical differences between entities, but whether those differences are treated by society as important.

4.2.4 Robots and the Role of Physicality

In this chapter, readers may have noted that the term "robots" (the physical embodiment of AI) is used more frequently than elsewhere in the book. Whereas the other issues tackled in this book apply equally to embodied or unembodied AI, the granting of rights to an entity is dependent not just on that entity's own consciousness or otherwise, but also on humanity's attitudes to the entity. For the reasons given above, these attitudes can be shaped by the entity's physical form and appearance.

A significant amount of the public discussion on AI ethics to date has focussed on robots because unlike a disembodied computer program, they are easy to picture.[93] Whereas the emphasis on robots, as opposed to AI more generally, is undue in most legal contexts,[94] when it comes to rights the position is slightly different. This psychological tendency has been recognised in various studies on human reactions to robots.[95] It is this quality which Ryan Calo touched upon when he wrote that robots (as opposed to unembodied AI) are deserving of different legal

---

[93] See, for example, European Parliament Directorate-General for Internal Policies, Policy Department C, Citizens' Rights and Constitutional Affairs, "European Civil Law Rules in Robotics: Study for the JURI Committee", PE 571.379, which again focuses on "robots" as opposed to AI. The fact that robots provoke a more visceral reaction than AI is one of the reasons for this book's title.

[94] Jack Balkin criticised Calo's sole focus on robots in Jack B. Balkin, "The Path of Robotics Law" (2015). *The Circuit*. Paper 72, Berkeley Law Scholarship Repository, http://scholarship.law.berkeley.edu/clrcircuit/72, accessed 1 June 2018: "We may be misled if we insist on too sharp a distinction between robotics and AI systems, because we do not yet know all the ways that technology will be developed and deployed".

[95] See, for example, Astrid M. Rosenthal-von der Pütten, Nicole C. Krämer, Laura Hoffmann, Sabrina Sobieraj, and Sabrina C. Eimler, "An Experimental Study on Emotional Reactions Towards a Robot", *International Journal of Social Robotics*, Vol. 5 (2013), 17–34.

treatment because of their "social valence". Calo said that robots "… feel different to us, more like living agents".[96]

Real-life examples illustrate this psychological tendency. A remote-controlled machine used for defusing improvised explosive devices in Afghanistan was nicknamed "Sergeant Talon" by the soldiers in its company. They even unofficially "awarded" him three Purple Hearts, the decoration given to soldiers in the US military who are injured whilst serving.[97] Sergeant Talon was not a "robot" in the sense used in this book given that it did not utilise any AI and it was entirely under the control of a human operator.[98] However, the physical and psychological value of Sergeant Talon was clearly recognised by those working with it.

Similarly, an automated minesweeper developed at the US Los Alamos National Laboratory resembled a giant millipede and was designed to destroy mines by stepping on them, blowing off a leg or two in the process. As the machine lost one leg after another crawling across the battlefield, a colonel overseeing its work demanded that the exercise be stopped because the test was "inhumane".[99]

It is important to note that AI is not a necessary criterion in order for physical machines to generate human emotions. The two examples in the preceding paragraphs demonstrate that such reactions can arise in relation to remote-controlled mechanical entities which have no independent intelligence at all. However, it is suggested that AI-enabled entities will be all the more suited to evincing such reactions, owing to their ability to learn and improve their behaviour with a view to increasing human empathy.

The more that robots come to resemble living beings, the more it seems we will react to them as if they had feelings. In one experiment conducted by lawyer and AI ethicist Kate Darling, researchers asked people to play with toy mechanical dinosaurs called "Pleos". After an hour of play, the instructor then asked participants to hurt their Pleo

---

[96] Ryan Calo, "Robotics and the Lessons of Cyberlaw", *California Law Review*, Vol. 103, 513–563, 532.

[97] P.W. Singer, *Wired for War: The Robotics Revolution and Conflict in the 21st Century* (London and New York: Penguin, 2009), Section entitled "For the Love of a Robot".

[98] See "TALON datasheet", *QinetiQ Website*, https://www.qinetiq-na.com/wp-content/uploads/datasheet_TalonV_web-2.pdf, accessed 1 June 2018.

[99] Joel Garreau, "Bots on the Gróund", *Washington Post*, 6 May 2007, http://www.washingtonpost.com/wp-dyn/content/article/2007/05/05/AR2007050501009_2.html, accessed 1 June 2018.

with weapons they had been given. All participants refused. Even when the instructors told them they could save their own robots by "killing" someone else's, they still refused. Finally, the researchers told the participants that unless one person "killed" their Pleo, all the robots would be destroyed. Even so, only one of the participants was willing to do so.[100] Darling has used the results of this experiment to support granting rights for AI on Kantian grounds, both out of sentiment and "to promote socially desirable behaviour".[101]

4.2.5 Escaping the Uncanny Valley

There is a phenomenon in robotics known as the "uncanny valley", first identified by roboticist Masahiro Mori.[102] The uncanny valley describes the slow rise, sharp drop and then relatively fast rise in feelings of familiarity, as robots become more like humans. The uncanny valley illustrates the tendency for human observers to feel uneasy when they encounter a robot which looks and acts like a human, but is not quite accurate. This could be a product of various subtle imperfections: its jerky movements, unnerving facial expressions, a flat and monotonous voice incapable of fully capturing the human range of emotions, and so on. The point is that when we see something which looks a lot like us but is definitely not human, we feel something strange is going on. We know we are being tricked.

It is possible to create robots designed specifically to avoid this phenomenon. Partly because of a fear of falling into the uncanny valley, most robots are not designed to have exact human features (though sexbots are an exception). In the late 1990s, researchers led by Cynthia

---

[100] Kate Darling, "Extending Legal Protection to Social Robots: The Effects of Anthropomorphism, Empathy, and Violent Behavior Towards Robotic Objects", in *Robot Law*, edited by Ryan Calo, A. Michael Froomkin, and Ian Kerr (Cheltenham, UK, Northampton, MA: Edward Elgar, 2016); see also Richard Fisher, describing an experiment carried out by MIT researcher Kate Darling, in 'Is it OK to torture or murder a robot? *BBC Website*, 27 November 2013, http://www.bbc.com/future/story/20131127-would-you-murder-a-robot, accessed 1 June 2018.

[101] Kate Darling, "Extending Legal Protection to Social Robots: The Effects of Anthropomorphism, Empathy, and Violent Behavior Towards Robotic Objects", in *Robot Law*, edited by Ryan Calo, A. Michael Froomkin, and Ian Kerr (Cheltenham, UK; Northampton, MA: Edward Elgar, 2016), 230.

[102] Masahiro Mori, "The Uncanny Valley", *Energy*, translated by Karl F. MacDorman and Takashi Minato, 7(4), 33–35.

Breazeal at the MIT Artificial Intelligence Laboratory built a robot called Kismet, which was designed to recognise and simulate human emotions, by manipulating its mechanical eyes, mouth and ears.[103] The appearance of Kismet was very far from human. Instead, the builders of Kismet chose certain features of humans which our brains recognise as signalling emotions, but put them in an exaggerated and deliberately mechanical form.[104] Kismet's large eyes in relation to its other features resemble elements we naturally associate with babies and young animals. This too encourages us to feel empathy towards the robot.[105]

It might be objected that we could actively design robots so as to not evince compassion. As noted above, this is Joanna Bryson's solution to avoiding robot rights. However, it seems surprisingly easy for a robot to provoke caring feelings in humans. In the *Star Wars* films, one of the most enduringly popular characters is R2-D2. The robot looks like little more than a painted metal trashcan on wheels, with a domed colander instead of a lid. But somehow, through a combination of its beeps, trills and deft movements, R2-D2 is imbued in the mind of the audience with a distinct personality, and certainly one which can elicit sympathy.[106]

---

[103] See Website of the MIT Humanoid Robotics Group, http://www.ai.mit.edu/projects/humanoid-robotics-group/kismet/kismet.html, accessed 31 July 2017.

[104] Michael R.W. Dawson, *Mind, Body, World: Foundations of Cognitive Science* (Edmonton: AU Press, 2013), 237.

[105] Another example of this anthropomorphosizing tendency can be seen in the 2001 *Castaway* by Director Robert Zemeckis; Tom Hanks' marooned character Chuck finds companionship in a volleyball which he finds in the wreckage of his crash-landing on a desert island—"Wilson". The volleyball takes on greater significance when Chuck cuts his hand and draws a face on the ball in blood—once it has even symbolic human features, the protagonist finds it far easier to anthropomorphize the inanimate object.

[106] Animals also appear to demonstrate these tendencies. In a recent BBC nature documentary, monkeys were filmed mourning the apparent death of a mechanical baby monkey that they had accepted as being of their own species, but which was in fact just a complicated hidden camera device. See Helena Horton, "Monkeys Mourn Robot Baby in Groundbreaking New BBC Show", *The Telegraph*, 10 January 2017, http://www.telegraph.co.uk/tv/2017/01/10/monkeys-mourn-robot-baby-groundbreaking-new-bbc-show/, accessed 1 June 2018.

4.3   The Argument from Value to Humanity

4.3.1 Reciprocity of Respect

Disrespect to robots could one day endanger humans. If we take the position that the world's dominant species or entity has the power to alter the rights of all others as it sees fit and for its own interests, surely it follows that we can have no moral complaints if one day AI does the same to us. Novelist and academic C.S. Lewis made this argument in his essay "Vivisection":

> We may find it difficult to formulate a human right of tormenting beasts in terms which would not equally imply an angelic right of tormenting men.[107]

Lewis' theory bears some resemblance to the dystopic visions of superintelligent AI subjugating humanity suggested by some modern commentators. As noted in Chapter 1, such predictions often verge on hyperbole and are by no means an immediate concern.[108] However, they do add a further argument in favour of protecting AI rights. It is fallacious to assume that there is a logical connection between humanity treating AI "well" and AI doing the same for humanity, should it ultimately obtain the whip hand. However, assuming that AI is rational and will seek to preserve itself and its own interests, then adopting an attitude of mutual coexistence towards it will likely engender a similar attitude from AI towards humans—at least for so long as humanity's actions are capable of having a bearing on AI. Indeed, the assumption that humans will not wish to wantonly destroy robots forms part of the modelling employed by Russell et al. which enables them to be kept under control and yet still obedient to human commands.[109]

---

[107] C.S. Lewis, "Vivisection", *God in the Dock: Essays on Theology and Ethics* (Grand Rapids, MI: William B. Eerdmans Publishing Co, 1996).

[108] Though one academic has even gone as far as writing a "message to future AI", suggesting various instrumental reasons why a superintelligent entity (which might one day come to read the paper) ought not to destroy humanity: Alexey Turchin, "Message to Any Future AI: 'There are Several Instrumental Reasons Why Exterminating Humanity Is Not in Your Interest'", http://effective-altruism.com/ea/1hj/message_to_any_future_ai_there_are_several/, accessed 1 June 2018.

[109] Dylan Hadfield-Menell, Anca Dragan, Pieter Abbeel, and Stuart Russell, "The Off-Switch Game", *arXiv preprint* arXiv:1611.08219 (2016), 1.

4.3.2 *Inherent Value*

The law protects a range of entities and objects not because they have a particular definable use, but rather for a panoply of cultural, aesthetic and historical reasons. We refer here to these justifications collectively as "inherent" value.

We might extend such protection to Methuselah, a bristlecone pine somewhere in California's White Mountains which is said to be over 5000 years old.[110] The same type of moral reasoning might apply to the protection of a Van Gogh painting or an ancient Babylonian temple. Whether such an "inherently valuable" entity is man-made or natural does not appear to make a difference to the value which it is ascribed. The world's first cloned mammal, Dolly the sheep, was not treated with any less respect than any other sheep. In fact, owing to her unique status as the world's first *man-made* sheep, she was treated far better than natural ones.[111]

The reason why we protect these objects goes beyond the fact that they might be someone's property. Indeed, for many of the most valuable things in the world the reason why we feel they should be protected is that they are *everyone's* property; they are meaningful to all humanity.

In 2002, Germany amended its constitutional basic law to include the following provision: "Mindful also of its responsibility toward future generations, the state shall protect the natural foundations of life and animals".[112] Notably, the German Constitution stipulates that this right is protected for the benefit of "future generations", which presumably refers to humans. Thus, the recorded motivation for the protection of animal and natural life is not necessarily that life in and of itself but rather its impact on humanity.

There is a tendency to see computer programs as expendable—when one is updated then previous versions can be deleted or overwritten. However, there are sound pragmatic reasons for maintaining previous

---

[110]The exact location is a secret guarded by the US Forest Service.

[111] Roslin Institute, "The Life of Dolly", *University of Edinburgh Centre for Regenerative Medecine*, http://dolly.roslin.ed.ac.uk/facts/the-life-of-dolly/index.html, accessed 1 June 2018.

[112]Art. 20a, Basic Law of the Federal Republic of Germany. For discussion, see Erin Evans, "Constitutional Inclusion of Animal Rights in Germany and Switzerland: How Did Animal Protection Become an Issue of National Importance?", *Society and Animals*, Vol. 18 (2010), 231–250.

copies. For instance, it may be necessary for legal forensic purposes to preserve a version of an AI system at the time a relevant event took place so as to be able to make enquiries as to its functioning and thought process. Similarly, if an update or patch leads to unforeseen problems, it may well be necessary to "roll back" a program to its previous version in order to rectify the issue. Both of these motivations underline the importance to humanity of preserving types of AI in some way.

We may already be recognising the inherent value of some robots. Kismet is no longer an operational model used in experiments, but it is preserved in the MIT Museum. In London, the Science Museum held an exhibition in early 2017 dedicated to robots, which featured various different iconic designs. Although these examples both feature the physical robots, we may also wish to preserve the source code of seminal AI systems such as AlphaGo Zero for future generations to study and learn from.

4.4   The Argument from Post-humanism: Hybrids, Cyborgs and Electronic Brains

Machines and human minds are not always separate. The idea of a human augmented by AI often appears in popular culture—examples include the Cybermen in *Doctor Who*, or the Borg in *Star Trek*. In 2017, Elon Musk suggested that humans must merge with AI or become irrelevant in the AI age.[113] Shortly afterwards, he launched a new company, Neuralink which aims to achieve this goal by "developing ultra high bandwidth brain-machine interfaces to connect humans and computers".[114]

Various research projects and commercial enterprises are exploring how to use the human brain in the development of AI. Other scientists have demonstrated that tiny syringe-injectable electronics can be inserted

---

[113] Aatif Sulleyman, "Elon Musk: Humans Must Become Cyborgs to Avoid AI Domination", *Independent*, 15 February 2017, http://www.independent.co.uk/life-style/gadgets-and-tech/news/elon-musk-humans-cyborgs-ai-domination-robots-artificial-intelligence-ex-machina-a7581036.html, accessed 1 June 2018.

[114] Website of Neuralink, https://www.neuralink.com/, accessed 1 June 2018. It is based on a concept first invented by science fiction writer Iain M. Banks: "neural lace", wireless mesh which interlinks brain tissue and computer processors. See, for example, Iain M. Banks, *Surface Detail* (London: Orbit Books, 2010), Chapter 10.

into biological matter and then activated.[115] These electronics could have all sorts of applications for humans, from improving memory to processing power. In a compelling 2018 article entitled "How to become a centaur", Nicky Case argued that human and AI could combine to become greater than the sum of their parts: "Symbiosis shows us you can have fruitful collaborations even if you have different skills, or different goals, or are even different *species*. Symbiosis shows us that the world often isn't zero-sum — it doesn't have to be humans *versus* AI, or humans *versus* centaurs, or humans *versus* other humans. Symbiosis is two individuals succeeding together not despite, but *because* of, their differences".[116]

If humans can be augmented by AI, boundary issues will arise: when, if ever, might a human lose their protected status? This raises similar problems to the Roman historian Plutarch's "Ship of Theseus Paradox":

> The ship wherein Theseus and the youth of Athens returned from Crete had thirty oars, and was preserved by the Athenians down even to the time of Demetrius Phalereus, for they took away the old planks as they decayed, putting in new and stronger timber in their places, in so much that this ship became a standing example among the philosophers, for the logical question of things that grow; one side holding that the ship remained the same, and the other contending that it was not the same.[117]

This paradox, which questions the nature of continuous identity through shifting physical components, can be applied to combinations of humanity and AI. We would not deny someone their human rights if they were 1% augmented by AI. What about if 20%, 50% or 80% of their mental functioning was the result of computer processing powers? On one view, the answer would be the same—a human should not lose rights just because they have added to their mental functioning. However, consistent with his view that no artificial process can produce "strong" AI

---

[115] "Syringe-injectable Electronics", *Nature Nanotechnology*, Vol. 10 (2015), 629–636, http://www.nature.com/nnano/journal/v10/n7/full/nnano.2015.115.html#author-information, accessed 1 June 2018.

[116] Nicky Case, "How to Become a Centaur", *Journal of Design and Science*, https://jods.mitpress.mit.edu/pub/issue3-case, accessed 1 June 2018.

[117] Plutarch, *Theseus*, translated by John Dryden (The Classics, MIT), http://classics.mit.edu/Plutarch/theseus.html, accessed 1 June 2018.

which resembles human intelligence, the philosopher John Searle argues that replacement would gradually remove conscious experience.[118]

Replacement or augmentation of human physical functions with artificial ones does not render someone less deserving of rights.[119] Someone who loses an arm and has it replaced with a mechanical version is not considered less human. The same argument might be made in the future, for instance if someone suffers a brain injury causing persistent amnesia and undergoes surgery to fit a processor replacing this mental function.

The case of Phineas Gage provides a historical example of continued moral identity despite neurological change. Gage was a railway worker who suffered catastrophic brain injuries when an explosion sent an iron rod through his skull. He somehow survived but his personality was reported to be permanently altered as a result.[120] There was no suggestion though that Gage was any less of a rights-bearing citizen or human as a result. If Gage's rights were maintained following this accidental brain trauma and subsequent neurological alteration, it seems illogical for them to be reduced should such alteration take place voluntarily or even in response to an injury.

Indeed, the growing ubiquity of wearable technology blurs yet further the line between what is human and what is not. At the time of writing, humans can of course remove AI goggles, smartwatches and other personal AI devices from their bodies. There is a certain taboo about

[118] John Searle, "Minds, Brains, and Programs", *Behavioral and Brain Sciences*, Vol. 3 (1980), 417–425.

[119] By way of example of the capabilities of technology, one recent study demonstrated that brain implants can allow paralysed monkeys to walk again, by sending information wirelessly directly from the implant to electrodes near the leg muscles, thereby bypassing the injured spinal cord. David Cyranoski, "Brain Implants Allow Paralysed Monkeys to Walk", *Nature*, 9 November 2016, http://www.nature.com/news/brain-implants-allow-paralysed-monkeys-to-walk-1.20967, accessed 1 June 2018.

[120] Claudia Hammond and Dave Lee, "Phineas Gage: The Man with a Hole in His Head", *BBC News*, 6 March 2011, http://www.bbc.co.uk/news/health-12649555, accessed 1 June 2018. See also the similar case of James Brady, Presidential Press Secretary to President Reagan, shot on April 27, 1981. Discussed in Marshall S.Willick, "Artificial Intelligence: Some Legal Approaches and Implications", *AI Magazine*, Vol. 4, No. 2 (1983), 5–16, 13: "It cannot, be maintained that there is any sanctity to retaining original brain tissue, either. When James Brady was shot through his head, he lost a large amount of brain tissue. Because he remained 'alive', however, he retained all the incidents of legal personality. It has never been suggested that an attempt to regain lost brain functions by using artificial parts would entail a greater cost to his legal recognition".

voluntary surgery to integrate technology for healthy participants.[121] But this may not always be the case. It is accepted in most cultures, and even mandated in some, that humans may undergo voluntary surgery for aesthetic or religious reasons. Tattoos, piercings, circumcision and even more extreme forms of surgery are morally acceptable or even required in some cultures. It may be the case in the coming decades that the same becomes true of integrated technology. The precise boundaries between what is "human" and what is "artificial" for the purposes of ascribing rights are a matter beyond the scope of this chapter. The main point is that the distinction between human and technology may become increasingly fluid.[122]

A further biology-based route to AI through whole brain emulation does not aim to augment or update human brains, but rather to create an entirely new brain capable of intelligent thoughts, feelings and consciousness, using a combination of technology and bioengineering.[123] As noted above, owing to her status as the first cloned mammal, throughout her life Dolly the sheep was monitored and cared for by teams of scientists and veterinarians, receiving state of the art care.[124] In the same way that we treated this quasi-artificial sheep with equal or greater respect to a natural sheep, would we not do the same for an artificial human brain? This begs the question of whether it is ethically acceptable to clone a human brain in the first place. In many countries, cloning humans is heavily regulated or banned. Some have even questioned whether it was morally appropriate to clone Dolly.[125]

---

[121] "The World's Most Famous Real-Life Cyborgs", *The Medical Futurist*, http://medicalfuturist.com/the-worlds-most-famous-real-life-cyborgs/, accessed 1 June 2018.

[122] For further argument as to why technological advances should not lead to a reduction in rights, see also Nick Bostrom, "In Defence of Posthuman Dignity", *Journal of Value Inquiry*, Vol. 37, No. 4 (2005), 493–506: "From the Transhumanist standpoint, there is no need to behave as if there were a deep moral difference between technological and other means of enhancing human lives. By defending posthuman dignity we promote a more inclusive and humane ethics, one that will embrace future technologically modified people as well as humans of the contemporary kind".

[123] Anders Sandberg and Nicholas Bostrom, "Whole Brain Emulation: A Roadmap", *Technical Report #2008–3*, Future of Humanity Institute, Oxford University, www.fhi.ox.ac.uk/reports/2008-3.pdf, accessed 1 June 2018.

[124] "Dolly the Sheep", *Website of National Museums Scotland*, https://www.nms.ac.uk/explore-our-collections/stories/natural-world/dolly-the-sheep/, accessed 1 June 2018.

[125] See, for instance, John Harris, "'Goodbye Dolly?' The Ethics of Human Cloning", *Journal of Medical Ethics*, (2007), 23(6), 353–360.

A related possibility—which at present is still in the realms of science fiction—is that a human's personality or consciousness might be somehow uploaded and stored by a computer or on a network. Some scientists are already working on this idea.[126] Others, such as neuroscientist and author Roger Penrose, contend that human thinking can never be emulated by a machine.[127] If it is the case that a human mind can one day be uploaded to a computer, we will be faced with a dilemma as to whether and if so what rights it should hold. Even if the artificial mind is imperfect or rudimentary, as will likely be the case in the first iterations of any such technology, this is not necessarily a reason for denying it basic rights.

5   Conclusions on Rights for AI

Suggesting that robots deserve rights might be met with disgust or disdain. But we should remember that proponents of animal and indeed universal human rights faced exactly the same reaction at first.

Moral rights are not the same as legal rights, though protection in law often follows shortly after society has recognised a moral case for protecting something. The next chapter addresses the case for giving robots legal personality, providing additional suggestions based on pragmatism which might apply alternatively to or in addition to the ethical considerations outlined here.

If a society does decide robots should have rights, this raises further difficult questions as to what rights ought to be protected. If the justification is to reduce suffering, or the appearance of suffering, then it would seem to follow that one of the rights to be protected might involve minimising robot "suffering" except where necessary and proportionate to achieving more important aims.

In addition to minimising suffering, other rights which we might one day protect for AI may not necessarily resemble those that we protect for humans, or even animals. For example, human-centric rights bound up

---

[126] "The Immortalist: Uploading the Mind to a Computer", *BBC Magazine*, 14 March 2016, http://www.bbc.co.uk/news/magazine-35786771, accessed 1 June 2018.

[127] Roger Penrose, *The Emperor's New Mind* (Oxford: Oxford University Press, 1998). See also FN 118 above.

in social relationships such as dignity or privacy might not be appropriate for AI. In a similar vein, animals feel no shame if a human watches them mating. Instead, an AI entity's rights might include those more unique to its nature, such as a better energy supply, or to more processing power. Of course, there may well be good reasons why such potential rights for AI will be overridden—just as some human and animal rights are subordinated to more important principles—but that does not mean that AI's rights cannot exist in the first place.

These are questions which can and should be addressed via the types of consultative process outlined in Chapters 6 and 7 on building institutions for regulation.

As AI and robots become more advanced and more integrated into our societies, we will be forced to re-evaluate our notions of moral rights. If robots display the same capabilities as other protected creatures, the inquiry may switch from asking "why should we give robots rights?" to asking "why should we continue to *deny* them?"

CHAPTER 5

Legal Personality for AI

1 The Missing Link?

In October 2017, Saudi Arabia granted "citizenship" to a humanoid robot named Sophia.[1] This move was derided by commentators as a cynical media stunt and a particularly hypocritical act for a country which grants only limited rights to human women.[2] Even so, the episode is significant because it was the first time that a country purported to grant a robot or AI entity any form of legal personality in its own right. Just days after the Saudi announcement, Tokyo's Shibuya district announced that an AI system had been granted "residency".[3]

---

[1] "Sophia", *Website of Hanson Robotics*, http://www.hansonrobotics.com/robot/sophia/, accessed 1 June 2018.

[2] See, for instance, James Vincent, "Pretending to Give a Robot Citizenship Helps No One", *The Verge*, 30 October 2017, https://www.theverge.com/2017/10/30/16552006/robot-rights-citizenship-saudi-arabia-sophia, accessed 5 November 2017; Cleve R. Wootson Jr., "Saudi Arabia, Which Denies Women Equal Rights, Makes a Robot a Citizen", *Washington Post*, 29 October 2017, https://www.washingtonpost.com/news/innovations/wp/2017/10/29/saudi-arabia-which-denies-women-equal-rights-makes-a-robot-a-citizen/?utm_term=.da4c35055597, accessed 1 June 2018.

[3] Patrick Caughill, "An Artificial Intelligence Has Officially Been Granted Residency", *Futurism*, 6 November 2017, https://futurism.com/artificial-intelligence-officially-granted-residency/, accessed 1 June 2018.

© The Author(s) 2019   173
J. Turner, *Robot Rules*, https://doi.org/10.1007/978-3-319-96235-1_5

In a seminal 1992 article, Lawrence B. Solum proposed a form of legal personhood[4] for AI.[5] When that paper was written, the world was still in the midst of the second "AI Winter": a period when setbacks in AI development coupled with a lack of funding contributed to a period of relatively slow growth.[6] For the following two decades, Solum's ideas remained a mere thought experiment. Given the recent developments in the capabilities of AI and its growing use, it is now an appropriate time to reconsider this proposal.[7]

Legal personality for AI is no longer just a matter for academic debate. In February 2017, the European Parliament passed a resolution containing recommendations on Civil Law Rules on Robotics.[8] The European Parliament suggested, as one of a menu of potential solutions to the issue of liability for the acts of robots:

> ...creating a specific legal status for robots in the long run, so that at least the most sophisticated autonomous robots could be established as having the status of electronic persons responsible for making good any damage they may cause, and possibly applying electronic personality to cases where robots make autonomous decisions or otherwise interact with third parties independently.[9]

Chapter 3 demonstrated that present laws will struggle to assign responsibility for AI. Chapter 4 showed that there may be moral arguments for AI to be given some rights. The present chapter considers whether an elegant solution to one or both these issues might be to grant AI legal personality. It asks *first* whether legal personality for AI would

---

[4] In the UK, the term "legal personality" is more common, whereas in the US "personhood" seems to be preferred. In the book, the terms "personality" and "personhood" are used interchangeably.

[5] Lawrence B. Solum, "Legal Personhood for Artificial Intelligences", *North Carolina Law Review*, Vol. 70 (1992), 1231. Solum was not the only theorist to make such a proposal, see for instance, R. George Wright, "The Pale Cast of Thought: On the Legal Status of Sophisticated Androids", *Legal Studies Forum*, Vol. 25 (2001), 297.

[6] Toby Walsh, *Android Dreams: The Past, Present and Future of Artificial Intelligence* (London: Hurst, 2017), 28.

[7] See also Koops, Hildebrandt, and Jaquet-Chiffell, "Bridging the Accountability Gap: Rights for New Entities in the Information Society?", *Minnesota Journal of Law, Science & Technology*, Vol. 11, No. 2 (2010), 497–561, who make a similar observation.

[8] European Parliament Resolution with recommendations to the Commission on Civil Law Rules on Robotics (2015/2103(INL)).

[9] Ibid., para. 59(f).

be possible, and *secondly*, whether it would be desirable. The chapter closes by considering what further questions need to be resolved if AI were to be granted this status.

2   Is Legal Personality for AI Possible?

2.1   A Bundle of Rights and Obligations

Legal personality is a fiction; it is something that humans create through legal systems.[10] As such, we can decide to what it should apply and what its content should be. In an important nineteenth-century US case on separate legal personality for corporations, *Trustees of Dartmouth College v. Woodward*, Chief Justice Marshall expressed the concept as follows:

> A corporation is an artificial being, invisible, intangible, and existing only in contemplation of law. Being the mere creature of law, it possesses only those properties which the charter of its creation confers upon it, either expressly or as incidental to its very existence. These are such as are supposed best calculated to effect the object for which it was created. Among the most important are immortality, and, if the expression may be allowed, individuality; properties by which a perpetual succession of many persons are considered as the same, and may act as a single individual. They enable a corporation to manage its own affairs and to hold property without the perplexing intricacies, the hazardous and endless necessity of perpetual conveyances for the purpose of transmitting it from hand to hand. It is chiefly for the purpose of clothing bodies of men, in succession, with these qualities and capacities that corporations were invented and are in use.[11]

Instead of being a single notion, legal personality is a technical label for a bundle of rights and responsibilities.[12] Joanna Bryson,[13] Mihailis

---

[10] See also Chapter 2 at s. 2.1.1.

[11] 17 US (4 Wheat.) 518 (1819).

[12] Joanna J. Bryson, Mihailis E. Diamantis, and Thomas D. Grant, "Of, for, and by the People: The Legal Lacuna of Synthetic Persons", *Artificial Intelligence and Law*, Vol. 25, No. 3 (September 2017), 273–291, https://link.springer.com/article/10.1007%2Fs10506-017-9214-9, accessed 1 June 2018 (hereafter Bryson et al., "Of, for, and by the People"). See, more generally, Hans Kelsen, *General Theory of Law and State*, translated by Anders Wedberg, (Cambridge, MA: Harvard University Press, 1945).

[13] Joanna Bryson, an expert in AI and ethics, is a vocal critics. See, for example, Dr. Bryson's blog: *Adventures in NI*, https://joanna-bryson.blogspot.co.uk/, accessed 1 June 2018. For her posts on the topic see: https://joanna-bryson.blogspot.co.uk/2017/11/why-robots-and-

Diamantis and Thomas Grant write that legal persons are "fictive, divisible, and not necessarily accountable".[14] They observe "legal personality is an artifice", with the effect that "[l]egal people need not possess all the same rights and obligations, even within the same system".[15]

As shown in Chapter 4, legal protections for humans have changed over time and continue to shift. By way of brief examples: 2000 years ago, in Roman law the *paterfamilias* or head of a family was the subject of legal rights and obligations on behalf of the whole household, including his wife and children[16]; 200 years ago, slaves were considered to be non-persons and only subsequently granted partial rights; even today, women continue to be denied full civil rights in various legal systems across the world.[17]

The rights and obligations of non-human legal persons can also undergo development. The US Supreme Court recently (and controversially) extended constitutional freedom of speech protections to companies, enabling them to play a greater role in election campaigns.[18] There remain limits to the protections we give to legal persons compared to natural ones: in an earlier case, the US Supreme Court denied that corporations had the same right to avoid self-incrimination enjoyed by human citizens.[19]

animals-never-need-rights.html, https://joanna-bryson.blogspot.co.uk/2017/10/human-rights-are-thing-sort-of-addendum.html, and https://joanna-bryson.blogspot.co.uk/2017/10/rights-are-devastatingly-bad-way-to.html, accessed 1 June 2018.

[14] Bryson et al., "Of, for, and by the People".

[15] Ibid.

[16] Samir Chopra and Laurence White, "Artificial Agents—Personhood in Law and Philosophy", *Proceedings of the 16th European Conference on Artificial Intelligence* (Amsterdam: IOS Press, 2004), 635–639.

[17] "Six Things Saudi Arabian Women Still Cannot Do", *The Week*, 22 May 2018, http://www.theweek.co.uk/60339/things-women-cant-do-in-saudi-arabia, accessed 1 June 2018.

[18] In *Citizens United v. Federal Election Commission* 558 US 310, the US Supreme Court declared that the protections for the freedom of speech contained in the First Amendment to the US Constitution had the effect of prohibiting the government from restricting independent expenditures for communications by non-profit corporations, for-profit corporations, labour unions and other associations. See also *Burwell v. Hobby Lobby*, 573 US 134 S.Ct. 2751 (2014), holding that a closely held company could possess First Amendment freedom of religion rights.

[19] *Hale v. Henkel*, 291 US 43 (1906).

2.2 Legal "Housing" of AI Within Existing Corporate Structures

Hermit crabs are known for their ability to find empty mollusc shells and adopt these as their home. Some legal scholars have suggested in effect that an AI entity might do the same with existing legal structures.

The US legal scholar and computer programmer Shawn Bayern has argued that the US law of limited liability companies (LLCs) could be utilised to bestow legal personality on any type of autonomous system.[20] Bayern's proposal seeks to take advantage of an apparent loophole under both New York's LLC Law and the Revised Uniform LLC Act. Bayern considers it would be possible to create an LLC whose operating agreement placed it under the control of an AI system and then have every other member of the LLC withdraw, leaving the system unsupervised by humans. Beguiling though Bayern's hypothesis might be, Matthew Scherer has launched a convincing counter-argument to the effect that the relevant statutes would not be construed by courts as leaving power to control the LLC in the hands of AI because this would be contrary to legislative intention.[21]

Given that at present most AI is "narrow" in nature and fairly limited in its range of abilities, any autonomous system granted legal personality through Bayern's method is likely to lack the basic acumen necessary to take many business decision. Even if an AI entity could acquire personality, it might be subject to the same default laws as apply in the case of an LLC whose single human member suddenly became mentally incapacitated, such that they were no longer fit to manage the entity. The relevant LLC laws have not yet been tested on this point, so it remains unclear whether Bayern's proposal would be endorsed by the courts.

Nonetheless, Bayern's paper sparked discussion in several other countries as to whether their laws would operate in a similar manner to the US LLC provisions he describes. Along with Bayern, a group of legal experts from the UK, Switzerland and Germany wrote a further paper which considered how the legal systems in those countries might achieve

---

[20] Shawn Bayern, "The Implications of Modern Business-Entity Law for the Regulation of Autonomous Systems", *European Journal of Risk Regulation*, Vol. 2 (2016), 297–309.

[21] Matthew Scherer, "Is AI Personhood Already Possible Under U.S. LLC Laws? (Part One: New York)", 14 May 2017, http://www.lawandai.com/2017/05/14/is-ai-personhood-already-possible-under-current-u-s-laws-dont-count-on-it-part-one/, accessed 1 June 2018.

the housing of AI within existing corporate structures.[22] Their conclusions were that although under UK law it might be possible to house an unsupervised AI within a legal entity, German and Swiss law did not easily accommodate AI legal personality without any other controlling party. Nonetheless, Thomas Burri, one of the authors of that study, wrote subsequently "...given the present capacities of existing forms of legal entities, companies of various kinds can serve as a mechanism through which autonomous systems might engage with the legal system".[23]

It is important to distinguish between the assets of a company and its corporate form. Another way of putting this is to say that the corporate form is the container, and its assets are the contents. It is certainly already possible for a person (whether human or corporate) to own the rights to an AI system, whether through proprietorship of its software or otherwise. Indeed, such an AI system might be the only asset of that company. However, this does not mean that the AI itself has legal personality; in the same way, if a company had as its sole asset a racehorse, this would not mean that the horse was a legal person in and of itself. The problems raised in Chapter 3 in terms of assigning responsibility for AI would not be solved by creating an entity whose sole asset was the AI, because there would still be difficulty in ascribing the AI's actions to its owner.

Bayern aims to jump across the gap between container and contents by replacing the (existing) person in control of an LLC with an AI entity. However, it is questionable whether the AI entity in *control* of the LLC would be treated as having all of the LLC's liabilities. Decision-making *on behalf* of an entity is not the same as having the same legal personality as

---

[22]Shawn Bayern, Thomas Burri, Thomas D. Grant, Daniel M. Häusermann, Florian Möslein, and Richard Williams, "Company Law and Autonomous Systems: A Blueprint for Lawyers, Entrepreneurs, and Regulators", *Hastings Science and Technology Law Journal*, Vol. 9, No. 2 (2017), 135–161. One member of the group, Shawn Bayern, has written separately that it is possible in certain US states to give de facto legal personhood to an autonomous system through a permanently memberless limited liability company which in turn owns the autonomous system. Shawn Bayern, "The Implications of Modern Business-Entity Law for the Regulation of Autonomous Systems", *Stanford Technology Law Review*, Vol. 19 (2015), 93. See also Paulius Čerka, Jurgita Grigienė, and Gintarė Sirbikytė. "Is It Possible to Grant Legal Personality to Artificial Intelligence Software Systems?", *Computer Law & Security Review* Vol. 33, No. 5 (October 2017), 685–699.

[23]Thomas Burri, "How to Bestow Legal Personality on Your Artificial Intelligence", *Oxford University Law Faculty Blog*, 8 November 2016, https://www.law.ox.ac.uk/business-law-blog/blog/2016/11/how-bestow-legal-personality-your-artificial-intelligence, accessed 1 June 2018.

that entity. A human controller of an LLC does not thereby become personally liable for the LLC's debts, and, presumably, neither would the AI.

2.3 New Legal Persons

Notwithstanding the above, debates on whether *existing* corporate law can be stretched to accommodate AI are ultimately something of a sideshow. Regardless of whether legal systems could at present allow legal personality for AI, the other option is for countries to create new bespoke corporate structures.

In the recitals to its February 2017 Resolution concerning Civil Law Rules on Robotics, the European Parliament appeared to leave open the question of whether AI could be housed within recognised categories of personality or whether new ones would be needed:

> [U]ltimately, the autonomy of robots raises the question of their nature in the light of the existing legal categories or whether a new category should be created, with its own specific features and implications.[24]

When a state grants legal personality to an entity, the new legal person is recognised as a "national" of that state.[25] Under EU law, the power to create legal personality is vested in Member States, which are entitled to "lay down the conditions for the acquisition and loss of nationality".[26] Accordingly, it is a matter of national sovereignty as to what entities it decides to grant personality. As a general matter, the freedom to award nationality is one of the basic powers of all sovereign states.[27] Even if the laws of nationality are not invoked, a state is free to do anything which

---

[24] European Parliament Resolution with recommendations to the Commission on Civil Law Rules on Robotics (2015/2103(INL)), Recital AC.

[25] Thomas Burri, "Free Movement of Algorithms: Artificially Intelligent Persons Conquer the European Union's Internal Market", in *Research Handbook on the Law of Artificial Intelligence*, edited by Woodrow Barfield and Ugo Pagallo (Cheltenham: Edward Elgar, 2018). States can grant nationality to legal persons already recognised in other states (for instance by a grant of dual nationality). However, this usually applies only to natural persons, i.e. humans.

[26] *Mario Vicente Micheletti and others v. Delegación del Gobierno en Cantabria*, C-369/90, ECR 1992 I-4239, para. 10.

[27] This principle was laid down by the International Justice in the *Nottebohm case (Liechtenstein v. Guatemala)* [1955] ICJ 1.

is not prohibited under public international law[28]; there are no international rules prohibiting legal personality for AI.

The extent of countries' freedom to recognise legal persons is illustrated by the breadth of entities accorded this status around the world—from temples in India[29] to *eingetragener Vereins* (registered voluntary associations) in Germany. Burri illustrates the point as follows: "[j]ust as national legislature could determine that great apes or certain rivers are persons within the domestic legal order, it could also state so, for instance, for webpages".[30]

2.4   Mutual Recognition of Foreign Legal Persons

As soon as one country grants AI legal personality, this may have a domino effect on other nations.[31] Many countries operate a doctrine of "mutual recognition" in their conflict of law provisions whereby they will accord the same status for legal persons recognised in other countries even where that entity would not be considered a person under the local law. This occurred in the UK case, *Bumper Development Corp. v. Commissioner of Police for the Metropolis*,[32] where the Court of Appeal held that an Indian temple that was "little more than a pile of stones" could be a treated as a legal person in England because it held that status in India, even though English law had no equivalent standing for religious buildings.

EU freedom of establishment provisions require that all Member States recognise all legal persons formed in accordance with the law of at

---

[28] This proposition was established in *The Case of the S.S. "Lotus"* (France v. Turkey) (1927) P.C.I.J., Ser. A, No. 10.

[29] There are limits though: in July 2017, the Indian Supreme Court held that two sacred rivers in India did not have the same rights as humans, thereby overturning a decision by the High Court rendered in March that year, see "India's Ganges and Yamuna Rivers Are 'Not Living Entities'", *BBC News*, http://www.bbc.co.uk/news/world-asia-india-40537701, accessed 1 June 2018. The possibility remains though for such rivers to have other, non-human legal rights.

[30] Thomas Burri, "Free Movement of Algorithms: Artificially Intelligent Persons Conquer the European Union's Internal Market", in *Research Handbook on the Law of Artificial Intelligence*, edited by Woodrow Barfield and Ugo Pagallo (Cheltenham: Edward Elgar, 2018).

[31] See, for a similar argument, Shawn Bayern, "Of Bitcoins, Independently Wealthy Software, and the Zero-Member LLC", *Northwestern University Law Review Online*, Vol. 108 (2014), 257.

[32] [1991] 1 WLR 1362.

least one other Member.[33] Article 54 of the Treaty on the Functioning of the EU (the TFEU) enshrines this principle:

> Companies or firms formed in accordance with the law of a Member State and having their registered office, central administration or principal place of business within the Union shall, for the purposes of this Chapter, be treated in the same way as natural persons who are nationals of Member States.

> 'Companies or firms' means companies or firms constituted under civil or commercial law, including cooperative societies, and other legal persons governed by public or private law, save for those which are non-profit-making.

If AI is housed within an EU company in the manner contemplated by Bayern et al. above, then it may only take one country to recognise the validity of that legal structure within the EU to cause the entire bloc to do so.[34] Indeed, the breadth of the definition of "companies or firms" in Article 54 of the TFEU suggests that even new forms of AI-specific legal persons would be covered, provided that they are "profit-making".[35] If the EU were to grant such recognition, it seems plausible that other major world economies would follow suit, so as to make themselves appear as attractive as possible to footloose AI designers and entrepreneurs who may wish to take advantage of the new legal status.

2.5   Robots in the Boardroom

As noted in relation to Bayern's LLC proposal above, granting legal personality to something does not necessarily mean the entity in question

---

[33] *Überseering BV v. Nordic Construction Company Baumanagement GmbH* (2002) C-208/00, ECR I-9919.

[34] Thomas Burri, "Free Movement of Algorithms: Artificially Intelligent Persons Conquer the European Union's Internal Market", *Oxford Law Faculty Blog*, 4 January 2018, https://www.law.ox.ac.uk/business-law-blog/blog/2018/01/free-movement-algorithms-artificially-intelligent-persons-conquer, accessed 1 June 2018. See also, Thomas Burri, "Free Movement of Algorithms: Artificially Intelligent Persons Conquer the European Union's Internal Market", *Research Handbook on the Law of Artificial Intelligence*, edited by Woodrow Barfield and Ugo Pagallo (Cheltenham: Edward Elgar, 2018).

[35] There is a limited right to derogation from market freedoms under the TFEU on the basis of public policy under art. 51(1) of that treaty. However, generally speaking public policy has been construed narrowly: see, for example, *Van Duyn v. Home Office*, 41–74, ECR 1974, 1337, para. 18.

can take decisions for itself. To the contrary, non-human legal persons can usually only act through the direction of human decision-makers. A charity takes decisions through its human trustees or directors. A company takes decisions through its management board or sometimes on direct instruction from its shareholders. Company directors and shareholders can be corporations, but at the top of the chain there is invariably at least one human decision-maker. As the English judge, Sir Edward Coke put it in a seventeenth-century case:

> ... the Corporation itself is onely in abstracto, and resteth onely in intendment and consideration of the Law; for a Corporation aggregate of many is invisible, immortal, & resteth only in intendment and consideration of the Law; and therefore cannot have predecessor nor successor. They may not commit treason, nor be outlawed, nor excommunicate, for they have no souls, neither can they appear in person, but by Attorney. A Corporation aggregate of many cannot do fealty, for an invisible body cannot be in person, nor can swear, it is not subject to imbecilities, or death of the natural, body, and divers other cases.[36]

Because holding rights and taking decisions about those rights are separate functions, it would be possible for an AI to have its own legal personality but remain under the control of humans, just like any other special purpose corporate vehicle.

Once AI is capable of taking sufficiently complex decisions, it is conceivable that the need for human decision-making on company boards could be reduced or even eliminated altogether. Florian Möslein, an expert on corporate governance, predicts that "[d]ue to its rapid technological development, artificial intelligence will enter corporate boardrooms in the very near future", and that "technology will probably soon offer the possibility of artificial intelligence not only supporting directors, but even replacing them".[37] After surveying current corporate law, Möslein concludes that changes will be needed to allow AI to take major corporate decisions absent of human oversight. Directors usually have a

---

[36] *Case of Sutton's Hospital* (1612) 77 Eng Rep 960.

[37] Florian Möslein, "Robots in the Boardroom: Artificial Intelligence and Corporate Law", in *Research Handbook on the Law of Artificial Intelligence*, edited by Woodrow Barfield and Ugo Pagallo (Cheltenham: Edward Elgar, 2018).

wide power to delegate some of their duties but nonetheless remain ultimately responsible for the management of the company.[38]

In 2014, an AI system was reportedly "appointed" to the board of directors of a Hong Kong venture capital firm to assist in decision-making.[39] Particularly in industries where quantitative analysis and data science are paramount, AI may have significant advantages over humans. At the moment, AI acts as a decision-making aid to humans, but in the future the roles might be reversed.[40] Indeed in various industries, human-gathered intelligence and data are already fed into an AI system, which then generates a recommendation for humans to execute.[41]

As to whether human directors could be replaced by AI, Möslein explains that "[o]n a more general level, corporate laws usually presuppose that only 'persons' can become directors".[42] As such, the question of whether AI is entitled to take decisions for a legal entity depends in turn on whether AI has its own legal personality.

3   SHOULD WE GRANT AI LEGAL PERSONALITY?

In 2007, legal scholar and sociologist Gunther Teubner made the provocative argument that "[t]here is no compelling reason to restrict the attribution of action exclusively to humans and to social systems. Personifying other non-humans is a social reality today and a political

---

[38] Ibid. Möslein cites *Dairy Containers Ltd v. NZI Bank Ltd* [1995] 2 NZLR 30, 79. See also *In Re Bally's Grand Derivative Litigation*, 23 Del.J.Corp.L., 677, 686.

[39] "Algorithm Appointed Board Director", *BBC Website*, 16 May 2014, http://www.bbc.co.uk/news/technology-27426942, accessed 1 June 2018. The company in question is Deep Knowledge Ventures and the system is called Validating Investment Tool for Advancing Life Sciences, "VITAL".

[40] The trend of AI board appointments has continued: see, for instance, "Tieto the First Nordic Company to Appoint Artificial Intelligence to the Leadership Team of the New Data-Driven Businesses Unit", *Tieto Website*, 17 October 2016, https://www.tieto.com/news/tieto-the-first-nordic-company-to-appoint-artificial-intelligence-to-the-leadership-team-of-the-new, accessed 1 June 2018.

[41] For instance, Cade Metz, writing in Wired, describes in vivid terms how the human proxy "operator" Aja Huang followed the instructions provided to him by DeepMind's AlphaGo: Cade Metz, "What the AI Behind Alphago Can Teach Us About Being Human", *Wired*, 19 May 2016, https://www.wired.com/2016/05/google-alpha-go-ai/, accessed 1 June 2018.

[42] Ibid.

necessity for the future".[43] In order to assess this claim, we need to begin by stipulating criteria by which the merits of legal personality for AI can be assessed.

3.1   Pragmatic Justifications: Setting the Threshold

We noted at the outset of Chapter 4 that there are two potential justifications for granting AI rights: moral and pragmatic. Legal personality is neither necessary nor sufficient to protect an entity's moral rights. A company does not have any "moral" rights—just a legitimate expectation (held in reality not by the "company" but by its officers, employees and shareholders) that its legal entitlements will be respected. Conversely, animals can be said to have various moral claims, but generally speaking they lack the legal personality to advocate for these in their own name.[44]

Chapter 4 addressed several moral reasons why we might wish to grant AI rights. One vehicle for the protection of those rights could be legal personality. At least in the short to medium term, it does not appear that either technology or society is at a stage where moral rights for AI are widely recognised. Therefore, the remainder of the present chapter concentrates solely on pragmatic justifications.

In this context, we define a pragmatic solution as one where agreed aims can reliably be achieved by a particular mechanism. Bryson et al. adopt a similar approach, albeit that they argue *against* giving AI personality. Their starting point is to "specify what are the purposes of the legal system in relation to which robot legal personhood should be assessed". Bryson et al. define the basic purposes of human legal systems as follows:

> 1. to further the material interests of the legal persons it recognizes, and
> 2. to enforce as legal rights and obligations any sufficiently weighty moral rights and obligations, with the caveat that

---

[43] Gunther Teubner, "Rights of Non-humans? Electronic Agents and Animals as New Actors in Politics and Law", Lecture delivered 17 January 2007, *Max Weber Lecture Series* MWP 2007/04.

[44] Jane Goodall and Steven M. Wise, "Are Chimpanzees Entitled to Fundamental Legal Rights?", *Animal Law*, Vol. 3 (1997), 61. For extensive discussion of a similar point in relation to the environment, see Christopher D. Stone, "Should Trees Have Standing? Towards Legal Rights for Natural Objects", *Southern California Law Review*, Vol. 45 (1972), 450–501.

3. should equally weighty moral rights of two types of entity conflict, legal systems should give preference to the moral rights held by human beings.[45]

It is not clear whether the above authors consider "moral rights" to include economic interests. If not, then the statement is inaccurate because economic rights often trump moral ones: a hungry person is not at liberty to rob a supermarket.

A slightly improved version of the formula against which granting legal personality to AI is to be measured would be as follows: (a) maintaining the integrity of the legal system as a whole and (b) advancing the interests of humans. For the avoidance of doubt, the term "interests" refers to economic as well as moral claims. The act of advancing interests of humans is less narrow than "giving preference to the moral rights held by humans", because in many circumstances the interests of humans generally will be served by giving preference to a legal entity over a human, and thereby upholding the institution of separate legal personality fundamental to most advanced economies.

3.2   Filling the Accountability Gap

Solum argued that the need for legal personality for AI depends empirically on the measure of independence possessed by the AI.[46] This makes sense because, as Chapter 3 showed, traditional theories under both criminal and private law run into increasing difficulty in assigning culpability to recognised legal persons as AI becomes more independent. David Vladeck, a former Director of the Bureau of Consumer Protection at the US Federal Trade Commission, put the point as follows:

> So long as we can conceive of these machines as 'agents' of some legal person (individual or virtual), our current system of products liability will be able to address the legal issues surrounding their introduction without significant modification. But the law is not necessarily equipped to address the legal issues that will start to arise when the inevitable occurs and

---

[45] Bryson et al., "Of, for, and by the People".

[46] Lawrence B. Solum, "Legal Personhood for Artificial Intelligences", *North Carolina Law Review*, Vol. 70 (1992), 1231.

these machines cause injury, but when there is no 'principal' directing the actions of the machine. How the law chooses to treat machines without principals will be the central legal question that accompanies the introduction of truly autonomous machines, and at some point, the law will need to have an answer to that question.[47]

Without legal personality for AI, our two pragmatic aims might pull in different directions: on the one hand, in order to advance the interests of humans we might want to find a legal person responsible for harm. However, seeking a human or corporate party responsible for AI might be at the expense of the integrity of the legal system as a whole.

In Greek mythology, the bandit Procrustes was famous for placing victims on a wooden bed then cutting off the end of their limbs if they were too tall or pulling limbs out of their joints them if they were too short. In a similar way, seeking always to find an *existing* legal person responsible for all AI actions we risk damaging the coherence of the legal system. Koops, Hildebrandt and Jaquet-Chiffell comment: "For tomorrow's agents, however, applying and extending existing doctrines in these ways may stretch legal interpretation to the point of breaking".[48]

Where the chain of causation between a recognised legal person and an outcome has been broken, interposing a new AI legal person provides an entity which can be held liable or responsible. AI personality allows liability to be achieved with minimal damage to fundamental concepts of causation and agency, thereby maintaining the coherence of the system as a whole.[49]

---

[47]David C. Vladeck, "Machines Without Principals: Liability Rules and Artificial Intelligence", *Washington Law Review*, Vol. 89 (2014), 117, at 150. See also Benjamin D. Allgrove, "Legal Personality for Artificial Intellects: Pragmatic Solution or Science Fiction?" (Oxford University Doctoral Thesis, 2004).

[48]Koops, Hildebrandt, and Jaquet-Chiffell, "Bridging the Accountability Gap: Rights for New Entities in the Information Society?" *Minnesota Journal of Law, Science & Technology*, Vol. 11, No. 2 (2010), 497–561. Chapter 3 demonstrated the problems which can arise under current legal structures when attempting to assign responsibility for actions carried out by AI which are unforeseeable and sufficiently independent of human input or guidance.

[49]Curtis E.A. Karnow, "Liability for Distributed Artificial Intelligences", *Berkeley Technology Law Journal*, Vol. 11 (1996), 147. See also Andreas Matthias, "Automaten als Träger von Rechten", Plädoyer für eine Gesetzänderung (Dissertation, Humboldt Universität, 2007).

3.3   Encouraging Innovation and Economic Growth

The rights and liabilities of a company are usually separate from those of its owners or controllers.[50] Creditors of a company only have recourse to that company's own assets, a feature known as "limited liability". The limited liability of companies is a powerful tool in protecting humans from risk and thereby encouraging innovation.[51] A single human can create a company, which she owns and of which she is the sole director. She can take every decision for the company, and she alone can reap the benefits in terms of the increased value of her shareholding if the company is successful. In practice, there may be almost nothing to distinguish the company from its human owner. And yet, despite all this, corporate law allows for the company to be treated as separate from her. Even if the company goes into bankruptcy, absent any fraud or personal guarantees of its liabilities, the owner can walk away entirely unscathed.

Granting AI legal personality could be a valuable firewall between existing legal persons and the harm which AI could cause. Individual AI engineers and designers might be indemnified by their employers, but eventually creators of AI systems—even at the level of major corporates—may become increasingly hesitant in releasing innovative products to the market if the programmers are unsure as to what their liability will be for unforeseeable harm. We return to this point below when considering some of the objections raised against separate personality for AI.

Arguably, the justifications for providing such legal personality to AI are even stronger than for protecting human owners from the liability of companies. AI systems can do something that existing companies cannot: take decisions without human input. Whereas a company is merely a collective fiction for human volitions, AI by its nature has its own independent "will".[52] For this reason, legal academics Tom Allen and Robin Widdison suggest that when an agent is capable of developing its own

---

[50] The leading UK case on this principle is the House of Lords case, *Salomon v. Salomon & Co Ltd* [1897] AC 22.

[51] Paul G. Mahoney, "Contract or Concession—An Essay on the History of Corporate Law", *Georgia Law Review*, Vol. 34 (1999), 873, 878.

[52] Gunther Teubner, "Rights of Non-humans? Electronic Agents and Animals as New Actors in Politics and Law", Lecture delivered 17 January 2007, *Max Weber Lecture Series* MWP 2007/04.

strategy, it then makes sense for that agent to be held responsible for its independent actions.[53]

3.4   Distributing the Fruits of AI Creativity

Alongside issues arising from harm caused by AI, Chapter 3 also demonstrated that current legal systems are poorly suited to addressing how the fruits of AI creativity should be distributed. Existing institutions such as intellectual property protection as well as free speech and hate speech laws have not been adapted to cover situations where the creator of a meaningful output is not a recognised legal person.

Allowing AI to hold property would resolve the issues raised concerning the ownership of new intellectual property of which AI is the creator. To the extent that the relevant acts of creativity were split between humans and AI, the intellectual property rights could be shared accordingly, much as it is between multiple human creators at present.

Granting AI other civil rights, including perhaps to freedom of speech, may be justified in circumstances where that AI's speech is a valuable contribution to the marketplace of ideas, which deserves to be protected for its benefit to human society. Without such protections, the powerful could simply restrict AI's ability to generate important output and there may be no legal person in a position to complain on the AI's behalf. A corollary effect is that AI could become subject to hate speech laws, which might then be used to prevent it from engaging in harmful discourse.

3.5   Skin in the Game

In his 2017 book *Skin in the Game*, Nassim Nicholas Taleb argues that in any given social system all participants should have some kind of a vested interest to encourage them to think properly and learn from their mistakes.[54]

More than just creating a pool for compensation, personality for AI could provide an AI system with the motivation to adhere to certain rules which, otherwise, it might abandon or eschew on the grounds that they conflict with its motives. Assuming that AI is trained to value its own assets, providing AI with personality could therefore give it skin in the game.

---

[53] See Tom Allen and Robin Widdison, "Can Computers Make Contracts?", *Harvard Journal of Law & Technology*, Vol. 9 (1996), 26.

[54] Nassim Nicholas Taleb, *Skin in the Game* (London: Allen Lane, 2017).

Deterrence is a major feature of both civil and criminal law: shaping a rational actor's behaviour by signalling that undesirable consequences will follow if a particular norm is transgressed. Humans agree to live in societies and submit to various laws because on the whole it serves our interests to do so. Provided AI could be imbued with a sufficiently nuanced model of the world, the same motivating structures might in theory be applied to it.

Though an AI entity may not be swayed by the psychological and emotional aspects of wanting to be seen by its peers to act lawfully, it seems easier to conceive of an AI system acting *rationally* to avoid diminution of its assets. F. Patrick Hubbard has termed this justification a "prudential grant of personhood".[55]

3.6   Arguments Against Personality for AI

3.6.1 The "Android Fallacy"
The simplest—and least tenable—objection to AI personality rests on a mistaken conflation of the idea or personality with humanity.

Legal academic Neil Richards and roboticist William Smart argue "… one particularly seductive metaphor for robots should be rejected at all costs: the idea that robots are 'just like people'… We call this idea 'the Android Fallacy.'"[56] Richards and Smart are correct to caution against the Android Fallacy, but this does not mean that one needs to abandon the concept of AI personality. Proposals for AI legal personality rarely go as far as giving robots all of the same rights as humans, and there is no logical or legal reason why they would need to.

Arguments against AI personality often shade into the Android Fallacy. Jonathan Margolis, writing in the *Financial Times*, declared that "Rights for robots is no more than an intellectual game", and that though "AI could exceed our capabilities… its personhood is illusory".[57] However, this is more of an assertion than an argument.

---

[55] F. Patrick Hubbard, "'Do Androids Dream?': Personhood and Intelligent Artifacts", *Temple Law Review*, Vol. 83 (2010–2011), 405–474, at 432.

[56] Neil Richards and William Smart, "How Should the Law Think About Robots?", in *Robot Law*, edited by Ryan Calo, A. Michael Froomkin, and Ian Kerr (Cheltenham and Northampton, MA: Edward Elgar, 2015), 3, at 4.

[57] Jonathan Margolis, "Rights for Robots Is No More Than an Intellectual Game", *Financial Times*, 10 May 2017, https://www.ft.com/content/2f41d1d2-33d3-11e7-99bd-13beb0903fa3, accessed 1 June 2018.

Similarly, in April 2018, a group of AI experts from 14 European countries (at the time of writing numbering over 250), including computer scientists, legal scholars and CEOs of AI technology companies, warned that granting robots legal personality as contemplated in the February 2017 European Parliament Resolution would be "inappropriate" from a "legal and ethical perspective".[58] In so doing, they fell into a similar error to Margolis. The experts said further of the European Parliament's proposal:

From a technical perspective, [the proposal for electronic personhood] offers many [sic] bias based on an overvaluation of the actual capabilities of even the most advanced robots, a superficial understanding of unpredictability and self-learning capacities and, a robot perception distorted by Science-Fiction and a few recent sensational press announcements....

A legal status for a robot can't derive from the Natural Person model, since the robot would then hold human rights, such as the right to dignity, the right to its integrity, the right to remuneration or the right to citizenship, thus directly confronting the Human rights. This would be in contradiction with the Charter of Fundamental Rights of the European Union and the Convention for the Protection of Human Rights and Fundamental Freedoms.[59]

When considering whether to grant legal personality to AI, the point is not whether the potential legal person *understands* the meaning of its actions. A temple or river cannot be said to be conscious of its legal personality. Indeed, we recognise the legal personality of humans who are unaware that they have it, including young children and people in permanent comas.[60] Even though children and those of diminished mental

[58] Janosch Delcker, "Europe Divided Over Robot 'Personhood'", *Politico*, 11 April 2018, https://www.politico.eu/article/europe-divided-over-robot-ai-artificial-intelligence-personhood/, accessed 1 June 2018. For the text of the letter, see: "Open Letter to the European Commission Artificial Intelligence and Robotics", https://g8fip1kplyr33r3krz5b97d1-wpengine.netdna-ssl.com/wp-content/uploads/2018/04/RoboticsOpenLetter.pdf, accessed 1 June 2018.

[59] "Open Letter to the European Commission Artificial Intelligence and Robotics", https://g8fip1kplyr33r3krz5b97d1-wpengine.netdna-ssl.com/wp-content/uploads/2018/04/RoboticsOpenLetter.pdf, accessed 1 June 2018.

[60] David J. Calverley, "Imagining a Non-biological Machine as a Legal Person", *AI & Society*, Vol. 22 (2008), 523–537, 526.

faculties can usually only act via other representatives, nonetheless they are still legal persons. Seen in this light, there is no magic to granting AI legal personality. We are not declaring it is alive.

3.6.2 "Robots as Liability Shields"

Bryson et al. argue that "[w]e currently have a legal system that is, first and foremost, of, for, and by the (human) people. Maintaining the law's coherence and capacity to defend natural persons entails ensuring that purely synthetic intelligent entities never become persons, either in law or fact". Bryson et al. assume that "if decision makers in the system say that they are ready to consider the possibility of 'electronic personality'", then human actors will "seek to exploit that possibility for selfish ends". They continue:

> There is nothing objectionable in itself about actors pursuing selfish ends through law. A well-balanced legal system, however, considers the impact of changes to the rules on the system as a whole, particularly so far as the legal rights of legal persons are concerned. We take the main case of the abuse of legal personality to be this: natural persons using an artificial person to shield themselves from the consequences of their conduct.[61]

The liability shield criticism assumes that any instance of separate legal personality will be abused by humans on a habitual basis. To the contrary, and as noted above, it has been recognised for centuries that separate legal personality plays a valuable economic role in enabling humans to take risks without sacrificing all of their own assets.[62] Indeed, exactly the same liability shield exception might equally be raised against limited liability for companies. Surely even the most trenchant critics of AI personality would not advocate abolishing all companies, yet this is the logical conclusion of some of their arguments.

Bryson et al. cite a well-known international law case, *JH Rayner (Mincing Lane) Ltd.* v. *Department of Trade and Industry*[63] as "foreshadow[ing] the risk that electronic personality would shield some human actors from accountability for violating the rights of other legal

---

[61] Bryson et al., "Of, for, and by the People", 4.2.1.

[62] See Chapter 2 at s. 2.1.1.

[63] "International Tin Council Case, JH Rayner (Mincing Lane) Ltd. v. Department of Trade and Industry", *International Law Reports*, Vol. 81 (1990), 670.

persons, particularly human or corporate".[64] In *JH Rayner*, a number of parties had made contracts with an international organisation, the International Tin Council (ITC), whose members included various individual states. In 1972, the UK had recognised the ITC as having its own legal personality, which enabled it to, among other things, enter into contracts. Various private parties contracted directly with the ITC, which eventually defaulted on some of these agreements. It transpired that the ITC itself did not have any assets. Various disappointed parties therefore attempted to sue one of its members: the UK (via its Department of Trade and Industry), arguing that it was not a separate legal person from the ITC. The UK's House of Lords rejected the claimants' case, holding that the ITC was separate from its members.

The real difficulty for the claimants in *JH Rayner* was that they had contracted with an entity that collapsed without any assets. This is hardly a problem unique to AI; in fact, it can exist in any situation where a party incurs a liability but then lacks the ability to be able to satisfy its debts. Where a company goes into liquidation, unsecured creditors may be left out of pocket. Where an impecunious person causes harm to others, the victims may not be able to seek financial recourse against the responsible party. In short, the problems complained of by Bryson et al. are nothing new. There are various ways of addressing them including insurance, taking adequate security and economically prudent behaviour—such as not entering into a contractual relationship with a party that may be unable to satisfy its obligations for want of assets. Simply put, if the claimants in *JH Rayner* had wanted to avoid financial risk, then they should not have contracted with the ITC.

Finally, in answer to the concern that AI might be used cynically by humans to harm others with impunity, well-established rules exist to prevent this from happening in corporate law. The same principles could be applied to AI.[65] Where a company is used as a cloak for wrongdoing and its owners are seeking to exploit its corporate personality to shield themselves, the separate legal personality of the company can be ignored and liability can be fixed directly on its owners.[66] This is known as "piercing

---

[64] Bryson et al., "Of, for, and by the People", 4.2.1.

[65] See Chapter 2 at s. 2.1.1.

[66] In the UK, see, *Petrodel Resources Ltd v. Prest* [2013] UKSC 34. For the US position see *MWH Int'l, Inc. v. Inversora Murten, S.A.*, No. 1:11-CV-2444-GHW, 2015 WL 728097, at 11 (S.D.N.Y. 11 February 2015) (citing *William Wrigley Jr. Co. v. Waters*, 890 F.2d 594, 600 (2d Cir. 1989).

the corporate veil".[67] In these situations, the law acknowledges that the fiction of a company is useful only up to a point.

Indeed, the idea of a human being able deliberately to exploit an AI system's separate legal personality presupposes that the human is in sufficient control of the AI to know what it is going to do. AI personality, by contrast, is designed predominantly to cater for situations in which such control or foreseeability from a human perspective is no longer present. If a person intentionally employs AI as a vehicle to do harm to others or recklessly uses AI to achieve some other end, realising that harm to others is a likely result of such use, then the person in question may be held liable under existing regimes—both criminal and private law.

3.6.3 *"Robots as Themselves Unaccountable Rights Violators"*

A further argument raised by some critics against AI personality is that the robots *themselves* are unaccountable: "[a]dvanced robots would not necessarily have further legal persons to instruct or control them".[68] Bryson et al. argue that "[g]iving robots legal rights without counter-balancing legal obligations would only make matters worse". This may be true, but why then not give robots legal obligations?

In order to make legal personality for AI useful from the perspective of settling liability, the AI will need to be given some way of holding funds, or at least having access to a pool of assets which can be used to satisfy creditors (such as a compulsory insurance scheme). One of the shortcomings of the European Parliament proposal was that it did not make clear enough how AI personality could be used to fill any

---

[67] Lord Sumption explained in the 2013 UK Supreme Court case *Petrodel Resources Ltd v. Prest* [2013] UKSC 34 at [8]: "The separate personality and property of a company is sometimes described as a fiction, and in a sense it is. But the fiction is the whole foundation of English company and insolvency law. As Robert Goff LJ once observed, in this domain 'we are concerned not with economics but with law. The distinction between the two is, in law, fundamental': *Bank of Tokyo Ltd v. Karoon (Note)* [1987] AC 45, 64. He could justly have added that it is not just legally but economically fundamental, since limited companies have been the principal unit of commercial life for more than a century. Their separate personality and property are the basis on which third parties are entitled to deal with them and commonly do deal with them". On fiction theory, see David Runciman, *Pluralism and the Personality of the State* (Cambridge: Cambridge University Press, 2009); Martin Wolff, "On the Nature of Legal Persons", *Law Quarterly Review*, Vol. 54 (1938), 494–521; and John Dewey, "The Historic Background of Corporate Legal Personality", *Yale Law Journal*, Vol. 35 (April 1926), 655–673.

[68] Bryson et al., "Of, for, and by the People", 4.2.2.

responsibility gap. An ability for AI to hold property could have been explicitly linked to both its personality and to the ability of that AI to settle debts or pay compensation. Rights for AI in this sense (and legal protections for those rights) are a means to an end rather than end in themselves.

3.6.4 Social Dislocation and Disenfranchisement

In recent years, various commentators have observed that in addition to the traditional "right/left" economic and political divide (pursuant to which people and groups are seen as being, broadly, against or in favour of government intervention), a new gulf has emerged particularly in developed economies between groups who are "anywhere/somewhere", "open/closed"[69] or "drawbridge down/drawbridge up".[70] These categories refer to the difference in attitudes between people who favour globalisation and multiculturalism versus people who value their own local culture and economy and may be more resistant to what they perceive to be a loss of identity.

Various "shock" results in elections or polls, in particular the UK's decision to leave the EU and the election of Donald Trump in the USA are often cited as examples of this trend, whereby new coalitions across the old political spectrum formed in order to reject the established social, economic and political order—rejecting the advice of "elites" in both cases.[71] A major critique of liberal social and economic policies in the past 30–40 years is that whilst they have been seen to benefit some members of society, large parts have come to feel increasingly disenfranchised as both economic inequality and social rifts grow. David Goodhart writes of the two new groupings:

> Anywheres dominate our culture and society. They tend to do well at school… then usually move from home to a residential university in their late teens and on to a career in the professions… Such people have portable 'achieved' identities, based on educational and career success which makes them generally comfortable and confident with new places and people.

[69] David Goodhart, *The Road to Somewhere: The Populist Revolt and the Future of Politics* (Oxford: Oxford University Press, 2017).

[70] "Drawbridges Up", *The Economist*, 30 July 2016, https://www.economist.com/briefing/2016/07/30/drawbridges-up, accessed 1 June 2018.

[71] Andrew Marr, "Anywheres vs Somewheres: The Split That Made Brexit Inevitable", *New Statesman*, 7 March 2017.

Somewheres are more rooted and usually have 'ascribed' identities - Scottish farmer, working class Geordie, Cornish housewife - based on group belonging and particular places, which is why they often find rapid change more unsettling. One core group of Somewheres have been called the 'left behind' - mainly older white working class men with little education. They lost economically with the decline of well-paid jobs for people without qualifications and culturally, too, with the disappearance of a distinct working-class culture and the marginalisation of their views in the public conversation[72]

Why are these trends relevant to the question of whether to grant legal personality to AI? Though this book is not about the economic impact of AI and technological unemployment, this is undeniably a major concern for world economies and populations. White collar jobs may be increasingly threatened by AI, but nonetheless it remains likely that jobs requiring less skill and training will be replaced first, not least because those taking the relevant decisions are often skilled individuals who will not be keen to cannibalize their own jobs or those of their immediate friends and family.

Putting the two issues together, the somewhere/closed/drawbridge up group of the population may well consider it to be adding insult to injury to be told not only that an AI entity has taken their job, but *also* that the AI entity is going to be granted some form of legal rights. A new social fissure might be added to the growing list of descriptors: Technophiles versus neo-Luddites. Referring to the nineteenth-century bands who smashed machinery fearing its impact on their jobs, the latter term is not intended pejoratively. Technology writer Blake Snow describes (and advocates) "reformed Luddism", saying: "to be a reform Luddite, all you have to do is recognize the many benefits of personal technology, but do so with an untrusting eye".[73]

The Technophiles will lap up the latest AI enabled smartphone, home speaker system or smart watch. By contrast, neo-Luddites may come to view with suspicion highly expensive consumer goods, just as they do AI systems which might replace their jobs. It must be recognised that there is a tension, therefore, between the ideas advocated in this and the

---

[72] David Goodhart, *The Road to Somewhere: The Populist Revolt and the Future of Politics* (Oxford: Oxford University Press, 2017), 3.

[73] Blake Snow, "The Anti-technologist: Become a Luddite and Ditch Your Smartphone", *KSL*, 23 December 2012, https://www.ksl.com/?sid=23241639, accessed 1 June 2018.

previous chapter, which suggest moral and pragmatic reasons for protecting AI, with the need for rules on AI to march in step with society's views and expectations. Technology journalist and think-tank director Jamie Bartlett notes that signs of a more violent and destructive brand of neo-Luddism may be growing, citing riots by taxi drivers against Uber in Paris, and the burning of technology laboratories in Grenoble, Nantes and Mexico. Bartlett goes on to say of the link between technology and these wider social trends:

> I am told repeatedly in the tech startup bubble that unemployed truckers in their 50s should retrain as web developers and machine-learning specialists, which is a convenient self-delusion. Far more likely is that, as the tech-savvy do better than ever, many truckers or taxi drivers without the necessary skills will drift off to more precarious, piecemeal, low-paid work.

> Does anyone seriously think that drivers will passively let this happen, consoled that their great-grandchildren may be richer and less likely to die in a car crash? And what about when Donald Trump's promised jobs don't rematerialise, because of automation rather than offshoring and immigration? Given the endless articles outlining how "robots are coming for your jobs", it would be extremely odd if people didn't blame the robots, and take it out on them, too.[74]

Striking this balance is an ongoing challenge. Although the economic benefits to be gained from AI might first be enjoyed by those who are already highly fortunate, in turn it is to be hoped that AI will bring benefits for the whole of society. These questions of equity and distribution are outside the scope of the present work. Nonetheless, it is suggested here that the trade-off between granting AI some rights and also ensuring that the technology remains socially acceptable can be overcome, or at least managed effectively. The techniques for consultative rule-making set out in Chapters 6 and 7 aim to go some way towards bridging this gap.

---

[74] Jamie Bartlett, "Will 2018 Be the Year of the Neo-Luddite?" *The Guardian*, 4 March 2018. See also Jamie Bartlett, *Radicals: Outsiders Changing the World* (London: William Heinemann, 2018).

4 Remaining Challenges

Notwithstanding the theoretical possibility of legal personality for AI, there are still some significant challenges and issues which remain to be resolved if AI is to be granted this status.

4.1 When Does AI Qualify for Legal Personality?

We may wish to set some minimum criteria for AI personality. F. Patrick Hubbard has suggested granting legal personality to an entity if it has the following capacities: (a) an ability to interact with its environment and to engage in complex thought and communication, (b) a sense of being a self with a concern for achieving its plan for its life, and (c) the ability to live in a community with other persons based on, at least, mutual self-interest.[75] Hubbard's second criterion resembles what this book would term "consciousness".[76] If legal rights for AI are viewed from a purely pragmatic perspective, then consciousness would not be necessary. Either way, at least Hubbard's first and third criteria seem to be a good starting point as a threshold for AI personality.

The precise boundaries at which we determine that AI can or should be granted personality are a matter of legitimate debate, which can be addressed through the law-making mechanisms described in the following chapters. Further questions arise as to whether, when an entity meets the relevant tests, legal personality should be optional or compulsory. Ultimately, these are moral and political issues, rather than ones which can be resolved by legal reasoning alone.

4.2 Identification of the AI

Personality for AI or a robot presupposes that it is possible to identify that entity with reasonable certainty. This is an empirical question.

Because AI can change and adapt, it might reasonably be asked whether it is the same program from one instance to the next.[77]

---

[75] F. Patrick Hubbard, "'Do Androids Dream?': Personhood and Intelligent Artifacts", *Temple Law Review*, Vol. 83 (2010–2011), 405–474, 419.

[76] See Chapter 4 at s. 4.1.1.

[77] Curtis E.A. Karnow, "Liability for Distributed Artificial Intelligences", *Berkeley Technology Law Journal*, Vol. 11, No. 1 (1996), 147–204, 200.

However, we should not forget that the same question could be asked of humans, namely whether our identity persists as we change and develop throughout our lifetimes.[78]

The philosopher A.J. Ayer argued that human identity through time consists in the identity of our bodies.[79] However, unlike a human mind, which is (at least for now) inextricably linked to a body, a robot mind could be held in any number of different repositories—and indeed more than one simultaneously.[80]

Although robots have a physical form, at present this is merely a vehicle for the operative AI, which could in most cases be transferred to other storage or operating systems. This problem may not apply if we were to develop embodied technologies via whole brain emulation, human-computer interfaces or similar—where the intelligent software and hardware are inextricably linked. Notwithstanding future developments, currently it would not make sense to conflate an AI system with the physical hardware from which it functions. Accordingly, we must look to something inherent and identifiable in the nature of the non-physical AI as its method of identification.

The issue is particularly acute where the AI entity in question is a unit of a greater whole, as part of a "swarm" or network of AI systems. Some consumer-facing AI already has these elements. For instance, since 2009 the search algorithm used by Google has learned from both individual users as well as wider data from the entire community.[81] When a user signs into her unique Google account, her search results will reflect a combination of personalised data such as her past searches and location as well as general updates provided to all users of the platform.[82]

---

[78] See, for instance, Eric T. Olson, "Personal Identity", in *The Stanford Encyclopedia of Philosophy* (Summer 2017 Edition), edited by Edward N. Zalta, https://plato.stanford. edu/archives/sum2017/entries/identity-personal/, accessed 1 June 2018, and the sources cited therein.

[79] A.J. Ayer, *Language, Truth, and Logic* (London: Gollancz, 1936), 194.

[80] A realisation that the "female" AI protagonist, played by Scarlett Johansson, was in fact multiple entities, provided the plot twist for the film *Her* (Warner Bros, 2013).

[81] "Personalised Search for Everyone", *Official Google Blog*, 4 December 2009 https:// googleblog.blogspot.co.uk/2009/12/personalized-search-for-everyone.html, accessed 1 June 2018.

[82] Masha Maksimava, "Google's Personalized Search Explained: How Personalization Works, What It Means for SEO, and How to Make Sure It Doesn't Skew Your Ranking Reports", *Link-Assistant.com*, https://www.link-assistant.com/news/personalized-search. html, accessed 1 June 2018.

Difficult questions of identification arise when seeking to determine whether it is the same Google AI algorithm operating on person A's smartphone as on person B's computer.

In order for entities to take advantage of the benefits allowed by legal personality, it might be made a requirement that AI be registered[83] and marked with an indelible and immutable electronic identifying "stamp" corresponding to that registry, such that identification is always possible.[84] A distributed ledger or block chain system might be used to verify any register of AI and to prevent tampering with entries.

One option is for an AI system to be registered in more than one place: if an AI system takes updates from a centralised source, but is also personalised to its user, then it might be sensible for the AI to be regulated individually and collectively. Similar principles of overlapping duties apply to humans: if a lorry driver crashes and causes damage, she may be held liable in her capacity as an individual person, but she may also be held liable in her capacity as an employee of a larger entity. In the circumstances, a person harmed is likely to pursue whoever has deeper pockets (generally the employer).

If AI is to hold substantive economic rights, such as the ability to own property or hold funds, then there would need to be some way in which to link the given AI system to the method of ownership. This is one of the reasons why many countries require companies to be registered so that their identity can be verified and thereby linked to certain rights. A bank account containing money would need to be in the name of someone or something. To this extent, some form of registry for AI may be an inescapable requirement for AI to hold rights.

A human may be able to just about survive "off-grid", without a social security number and outside of the knowledge of authorities, but in developed economies this is increasingly difficult. In order to access many basic goods and services, some form of local, federal or national

---

[83] See, for example, Koops, Hildebrandt, and Jaquet-Chiffell, "Bridging the Accountability Gap: Rights for New Entities in the Information Society?" *Minnesota Journal of Law, Science & Technology*, Vol. 11, No. 2 (2010), 497–561, 516.

[84] This is the solution favoured by Karnow, who proposes a "*Registry*" for AI, which "*will certify the agent by inserting a unique encrypted warranty in the agent*". Curtis E.A. Karnow, "Liability for Distributed Artificial Intelligences", *Berkeley Technology Law Journal*, Vol. 11, No. 1, (1996), 147–204, 193 *et seq.* Allen and Widdison also propose a similar Registry, which would state the competence of registered AI agents and the limits on their liability. See also Tom Allen and Robin Widdison, "Can Computers Make Contracts?", *Harvard Journal of Law and Technology*, Vol. 9 (1996), 26.

identity verification is required. The same bottlenecking principles could be applied to AI. Thus, though not every AI entity would need to be registered, this could be made an essential prerequisite in order for that AI system to avail itself of certain pieces of legal and economic infrastructure, such as insurance, the banking system or even perhaps the Internet. As such, in order for AI systems to participate in activities in which legal liability is likely to arise, registration and licensing could be made mandatory.

4.3   What Legal Rights and Responsibilities Might AI Hold?

4.3.1 Potential Rights and Obligations

Based on the various justifications for AI personality set out in the preceding section, the rights and obligations which we may wish to grant AI include: separate legal personality and the corporate veil, the ability to own and dispose of assets, the rights to sue and be sued and the freedom to have certain expression or speech protected/prohibited.

A further power which we may want to grant AI is the ability to conclude contracts in its own name—as opposed to that of a principal. This way, AI could both oblige itself to do certain things and require counterparties to adhere to their contractual undertakings. Allowing AI to act in this manner could reinforce certainty for participants in any given market on the basis that even if a human or corporation disclaims liability for the AI's actions, the AI itself can be held responsible. As Francisco Andrade and colleagues explain:

> First, by the recognition of an autonomous consent— which is not a fiction at all—it would solve the question of consent and of validity of declarations and contracts enacted or concluded by electronic agents without affecting too much the legal theories about consent and declaration, contractual freedom, and conclusion of contracts. Secondly, and also quite important, it would "reassure the owners-users of agents", because, by considering the eventual "agents" liability, it could at least limit their own (human) responsibility for the 'agents' behaviour.[85]

---

[85] Francisco Andrade, Paulo Novais, Jose Machado, and Jose Neves, "Contracting Agents: Legal Personality and Representation", *Artificial Intelligence and Law*, Vol. 15 (2007), 357–373.

In addition to the registration of AI on a distributed ledger such as block-chain, that ledger could also display the assets of the AI, with the result that any potential counterparty would know exactly how creditworthy the AI is. The international regulatory framework for banks (the current iteration of which is known as "Basel III") requires that banks hold a minimum amount of regulatory capital which can be called upon in the case of an emergency.[86] Similar requirements might be imposed on AI in order for it to be permitted to take advantages of the various benefits that personality brings.[87] If an AI's assets or credit rating drop below a certain level, then it could be frozen automatically out of certain legal and economic rights.

4.3.2 What Are the Limits?

The same bundle of rights accorded to companies in different legal systems need not be directly transposed on to AI. We are unlikely to want AI to hold various "civil" rights which are deeply entwined into shared concepts of *human* society, such as the right to vote, or to marry.[88]

There is also no need for AI legal rights to be absolute or indefeasible. Most human rights—including even the right to life—can be restricted or overridden in appropriate circumstances: police can shoot a dangerous assailant where necessary. Any AI rights must sit alongside other legal rights and norms, which occasionally clash and are subject to regulatory or judicial adjudication. This realisation also goes to answering the simplistic objection to AI personality which assumes that we will thereby create some kind of master race capable of always defeating the rights of humans. Balancing AI rights against those of existing legal persons will be a complex exercise and (as with many of the unanswered questions) is best answered through societal deliberation.

•

[86]"Basel III: International Regulatory Framework for Banks", *Website of the Bank for International Settlements*, https://www.bis.org/bcbs/basel3.htm, accessed 1 June 2018.

[87]Giovanni Sartor, "Agents in Cyberlaw", *Proceedings of the Workshop on the Law of Electronic Agents (LEA 2002)*, 3–12.

[88]Samir Chopra and Laurence White, "Artificial Agents—Personhood in Law and Philosophy", *Proceedings of the 16th European Conference on Artificial Intelligence* (Amsterdam: IOS Press, 2004), 635–639.

4.4   Would Anyone Own AI?

It might be thought that an entity cannot be both a person and property. This is incorrect; though today it may be repugnant to think of any human person as being owned by another, we experience no cognitive dissonance in viewing a company as both a legal person and the property of its shareholders.

At present, most corporate structures, no matter how complicated, end with humans as the ultimate beneficial owners.[89] Just as no one "owns" a human, could we have a situation where no one "owns" AI? In theory, this would be possible, but we would need to decide as a society whether or not this would be desirable. For example, Koops et al. predict a three-tiered progression in terms of AI personality: in the short term, they predict "interpretation and extension of existing laws". In the middle term, they predict "limited [AI] personhood with strict liability",[90] involving "introduc[tion of] strict liability for electronic agents if their unpredictable actions are felt to be too risky for business or consumers".[91] In the long term, they consider that we may develop "full AI personhood with 'posthuman rights'".[92] The latter, they suggest, should only arise if and when machines develop self-consciousness.

4.5   Could a Robot Commit a Crime?

Chapter 3 addressed *human* criminal responsibility for the acts of AI. Granting AI legal personality, the subject of the present chapter, opens the door to AI having criminal responsibility for its own acts.

Even though a company has no "soul to damn or body to kick",[93] criminal liability for corporations has existed since the Middle Ages,[94]

---

[89] This is not always the case: a company could be ultimately owned by a trust, whose beneficiaries are non-human, such as a charity supporting animals.

[90] Koops, Hildebrandt, and Jaquet-Chiffell, "Bridging the Accountability Gap: Rights for New Entities in the Information Society?" *Minnesota Journal of Law, Science & Technology*, Vol. 11, No. 2 (2010), 497–561, 554.

[91] Ibid., 555.

[92] Ibid., 557.

[93] John C. Coffee, Jr., "'No Soul to Damn: No Body to Kick': An Unscandalised Inquiry into the Problem of Corporate Punishment", *Michigan Law Review*, Vol. 79 (1981), 386.

[94] Markus D. Dubber, "The Comparative History and Theory of Corporate Criminal Liability", 10 July 2012, http://dx.doi.org/10.2139/ssrn.2114300, accessed 1 June 2018. The situation worldwide is by no means uniform: Italy was a fairly recent adopter of corporate criminal liability, via legislative decree no. 231/2001. German law has no criminal liability for enterprises.

and a similar concept might perhaps be extended to AI if and when it is granted legal personality. On the one hand, this might fill the "retribution gap" identified by John Danaher, namely the psychological expectation that there will be a criminally responsible agent to be punished for causing harm.[95] However, a major difficulty remains in that it is hard to reconcile AI criminality with the general requirement in criminal law that a guilty party must "intend" to commit the criminal act.

4.5.1 Case Study: Random Darknet Shopper

In Switzerland, an artistic collective created a piece of software called Random Darknet Shopper which was enabled once a week to access the deep web, a hidden portion of the Internet, and purchase an item at random.[96] Random Darknet Shopper purchased items including a pair of fake diesel jeans, a baseball cap with a hidden camera, 200 Chesterfield cigarettes, a set of fire-brigade issued master keys and 10 ecstasy pills.[97]

The ecstasy purchase came to the attention of the local St Gallen Police Force, which seized the physical computer hardware from which the Random Darknet Shopper was run, as well as the various items it had purchased. Interestingly, both the human designers *and* the AI system were formally charged with the crime of making an illegal purchase of a controlled substance. Three months later, the charges were dropped and all property was returned to the artistic collective (apart from the ecstasy, which was destroyed).[98]

---

[95] See Chapter 3 at s. 3.3.1.

[96] "Homepage", *Website of !Mediengruppe Bitnik*, https://www.bitnik.org/, accessed 11 June 2017. As *!Mediengruppe Bitnik* have not publicised the sourcecode for their programme, we are not aware of whether the Random Darknet Shopper would indeed have fulfilled our definition of AI but we will assume for the purposes of this discussion that it did.

[97] Mike Power, "What Happens When a Software Bot Goes on a Darknet Shopping Spree?", *The Guardian*, 5 December 2014, https://www.theguardian.com/technology/2014/dec/05/software-bot-darknet-shopping-spree-random-shopper, accessed 1 June 2018.

[98] "Random Darknet Shopper Free", *Website of !Mediengruppe Bitnik*, 14 April 2015, https://www.Bitnik.Org/R/2015-04-15-random-darknet-shopper-free/, accessed 1 June 2018; Christopher Markou, "We Could Soon Face a Robot Crimewave … the Law Needs to be Ready", *The Conversation*, 11 April 2017, http://www.cam.ac.uk/research/discussion/opinion-we-could-soon-face-a-robot-crimewave-the-law-needs-to-be-ready, accessed 1 June 2018.

4.5.2 *Locating Intent*

The key question for AI criminal liability is whether it has the relevant mens rea (guilty mind). There are two elements: *first*, ascertaining the AI's decision-making process on a factual basis, and *secondly*, the social and policy question of how this should be treated under criminal law. Because mens rea criteria apply only to humans at present, they are tailored to (what we perceive to be) human thought processes. AI does not function in the same way, and in seeking to apply anthropomorphic concepts to it we risk incoherence and confusion.

It is possible to distinguish between situations in which AI has made a mistake as to a fact and those in which AI has applied the "wrong" rule to a known fact. When a factory robot thinks that a human operator's head is a component in the manufacturing process and decides to crush it—killing the human—this is akin to a mistake of fact.[99] However, where an AI supplants human instructions in some unexpected way—for example the toaster burning a house down to cook all the bread—then this might seem closer to a concept of a criminally guilty mind. Similarly, if AI was to develop the capability to deliberately disobey clear human instructions, then this might also be considered criminal.

Even if AI's "mental state" could be measured and ascertained, we still need to ask whether it would be appropriate from a social and psychological perspective to apply criminal law tenets to a non-human entity. On one view, the notion of mens rea is something which is by its very nature only appropriate to humans. If correct, this seems to militate against AI ever having criminal intent in the sense currently recognised. A system of law might define a new culpable "mental" state applicable to AI, but then labelling it mens rea may no longer be appropriate.[100]

The designation of an act as criminal is usually linked to some form of penalty. If AI was held criminally responsible, there is a further question as to how AI might be punished. The final part of Chapter 8 explores further sanctions which might be used against AI.

---

[99] This appears to have been the situation in the death of Wanda Holbrook, a worker at an auto-parts factory in Michigan. James Temperton, "When Robots Kill: Deaths by Machines Are Nothing New but AI Is About to Change Everything", *Wired*, 17 March 2017, http://www.wired.co.uk/article/robot-death-wanda-holbrook-lawsuit, accessed 1 June 2018.

[100] See, for example, Lawrence B. Solum, "Legal Personhood for Artificial Intelligences", *North Carolina Law Review*, Vol. 70 (1992), 1231–1287, 1239–1240.

5   CONCLUSIONS ON LEGAL PERSONALITY FOR AI

Chopra and White write in the introduction to their book on AI personality: "The artificial agent is here to stay; our task is to accommodate it in a manner that does justice to our interests and its abilities".[101]

It may be hard to imagine a world without separate legal personality for companies, but Paul G. Mahoney has pointed out in a historical study of the institution: "[h]ad property and contract law been permitted to keep evolving in the field of business operations, the set of asset partitioning rules that reside in the law of property, contract and tort (rather than the law of partnership and corporations) might be much larger".[102] Seen in this way, legal personality is neither special nor inevitable in any area; it is simply one tool available to humans to help achieve our extra-legal aims.

Even if separate legal personality for AI is accepted in theory, there remain various difficult unanswered questions as to how it should be structured. The following chapters suggest how we can build institutions capable of resolving these issues.

The issue is controversial and is beset by misunderstandings. This chapter has sought to argue that AI personality should not be discounted out of hand through a knee-jerk, emotion-driven response. Importantly, granting legal personality to an entity does not mean treating it as a human and should not always allow humans to disclaim all responsibility for AI's actions. As Marshal S. Willick said, with prescience but perhaps a little too much optimism, in 1983:

> Eventually, an intelligent computer will end up before the courts. Computers will be acknowledged as persons in the interest of maintaining justice in a society of equals under the law. We should not be afraid that that day may come soon.[103]

---

[101] Samir Chopra and Laurence White, *A Legal Theory for Autonomous Artificial Agents* (Ann Arbor: The University of Michigan Press, 2011), 3.

[102] Paul G. Mahoney, "Contract or Concession—An Essay on the History of Corporate Law", *Georgia Law Review*, Vol. 34 (1999), 873–894, 878.

[103] Marshal S. Willick, "Artificial Intelligence: Some Legal Approaches and Implications", *AI Magazine*, Vol. 4, No. 2 (1983), 5–16, at 14.

CHAPTER 6

Building a Regulator

1   WHY WE MUST DESIGN INSTITUTIONS BEFORE WE CAN WRITE LAWS

Like many other discussions of how AI should be regulated, this book began by reciting Isaac Asimov's Laws of Robotics.[1] Over and above their deliberate gaps, vagueness and oversimplification, they have one overriding problem: Asimov was starting in the wrong place by writing laws. The question to ask first is: "Who should write them?"

In contrast to the preceding chapters, this one will step back from granular legal issues of responsibility and rights and address instead the more general questions of how we ought to design, implement and enforce new rules tailored to AI.

1.1   Philosophy of Institutional Design

Why start with the design of the system? Among legal philosophers, there are two popular schools of thought as to what is needed to make

---

[1] See the opening paragraph of Chapter 1.

© The Author(s) 2019   207
J. Turner, *Robot Rules*, https://doi.org/10.1007/978-3-319-96235-1_6

laws binding authority on their subjects: Positivism and Natural Law.[2] Legal philosopher John Gardner describes Positivism as the view that "[i]n any legal system, whether a given norm is legally valid, and hence whether it forms part of the law of that system, depends on its sources, not its merits".[3] Natural Law theorists on the other hand believe certain values inhere in nature or human reason, and that these ought to be reflected in the legal system.[4] For Natural Lawyers, a law is binding authority *only* if it is good or just.[5]

Though the two approaches may not be incompatible,[6] they lead to a difference in emphasis: Natural Lawyers focus more on ensuring the laws reflect a particular moral code and Positivists on creating institutions whose laws will be acceptable to their subjects.[7]

All this is significant to AI because if Natural Lawyers are correct, then there is only one set of rules which can be right in given circumstances. Any legal scholarship becomes a search for eternal truths. Natural Lawyers would begin and end, like Asimov did, by writing rules.

---

[2] There are other related legal theories, such as Ronald Dworkin's "interpretivism" (as to which see Ronald Dworkin, *Taking Rights Seriously* (Cambridge, MA: Harvard University Press, 1977); *Law's Empire* (Cambridge, MA: Harvard University Press, 1986); *Justice in Robes* (Cambridge, MA: Harvard University Press, 2006); and *Justice for Hedgehogs* (Cambridge, MA: Harvard University Press, 2011). However, Dworkin's writings largely relate to adjudication of disputes rather than the process of lawmaking and legal validity, which are the concerns here. Various types of legal realism are particularly popular in the USA. For a classic exposition, see Karl Llewellyn, *The Bramble Bush: On Our Law and Its Study* (New York: Oceana Publications, 1930). However, the latter theory is more of a rejection of debates as to validity rather than an attempt to engage with them. Accordingly, it will not be discussed further here.

[3] John Gardner, "'Legal Positivism': 5½ Myths", *The American Journal of Jurisprudence*, Vol. 46 (2001), 199–227, 199. See also Joseph Raz, *Ethics in the Public Domain* (Oxford: Clarendon Press, 1994).

[4] See, for example, John Finnis, *Natural Law and Natural Rights* (Oxford: Clarendon Press, 1981). In the "5½ Myths" paper (op. cit.), John Gardner contends that his version of Positivism would be acceptable to natural lawyers also.

[5] It is perhaps unsurprising that some of the most prominent advocates of Natural Law, including both thirteenth-century Saint Thomas Aquinas and modern legal philosopher John Finnis are men of religion for whom there is a unitary God-given structure of values.

[6] John Gardner, *Law as a Leap of Faith: Essays on Law in General* (Oxford: Oxford University Press, 2012), Chapter 6.

[7] To put matters another way, Positivists concentrate on input legitimacy and Natural Lawyers on output legitimacy.

Positivism has the advantage of not needing to take a position as to whether there is one single morally correct system of values.[8] Moreover, the lack of consensus on many moral issues—whether in relation to AI or otherwise—means that even if one did somehow arrive at the optimal set of rules, securing their adoption and enforcement would likely be impossible without some mechanism for ensuring that the rules are accepted and respected by their subjects.

1.2   AI Needs Principles Formulated by Public Bodies, Not Private Companies

In the absence of concerted governmental efforts to regulate AI, private companies have begun to act unilaterally. In October 2017, DeepMind, the world-leading AI company acquired by Google in 2014 and now owned by the parent company Alphabet, launched a new ethics board "to help technologists put ethics into practice, and to help society anticipate and direct the impact of AI so that it works for the benefit of all".[9] Similarly, in 2016 the Partnership on AI to Benefit People and Society was formed by six major tech companies—Amazon, Apple, DeepMind, Google, Facebook, IBM and Microsoft—"to study and formulate best practices on AI technologies".[10]

Interestingly, the words "to benefit people and society" were subsequently dropped from the majority of the Partnership's branding on its website—it now styles itself merely as the "Partnership on

---

[8]For the view of law as fictions, see Yuval Noah Harari, *Sapiens: A Brief History of Humankind* (London: Random House, 2015).

[9]Verity Harding and Sean Legassick, "Why We Launched DeepMind Ethics & Society", Website of Deepmind, 3 October 2017, https://deepmind.com/blog/why-we-launched-deepmind-ethics-society/, accessed 1 June 2018.

[10]"Homepage", Website of the Partnership on AI, https://www.partnershiponai.org/, accessed 1 June 2018. Microsoft has taken a slightly different approach, of eschewing external oversight for a committee which appears to be composed only from Microsoft insiders. Microsoft describes its AI and Ethics in Engineering and Research Committee in a 2018 publication as "a new internal organization that includes senior leaders from across Microsoft's engineering, research, consulting and legal organizations who focus on proactive formulation of internal policies and on how to respond to specific issues as they arise". Microsoft, *The Future Computed: Artificial Intelligence and Its Role in Society* (Redmond, Washington, DC: Microsoft Corporation, 2018), 76–77, https://msblob.blob.core.windows.net/ncmedia/2018/01/The-Future_Computed_1.26.18.pdf, accessed 1 June 2018.

AI"—although at the time of writing, the Partnership still describes those aims as the organisation's "mission". Though DeepMind states it is prepared to hear "uncomfortable" criticism from its advisors, rules formulated by corporate ethics boards will always lack the legitimacy that a government can provide.[11]

On the one hand, it might be argued that there is no need for government regulation because responsible industry figures can be relied upon to regulate themselves.[12] Proponents of industry-led regulation might say that because the companies understand far better the risks and capabilities of technology, they are best placed to set standards. However, allowing companies to regulate themselves without any government oversight may be dangerous.

It is the purpose of governments to act for the common good of everyone in society.[13] Of course, some governments are swayed by powerful lobbies or corrupt individuals, but these represent divergences from the core concept of how governments are supposed to operate. By contrast, companies are usually required by corporate law to maximise value for their owners. This is not to say that companies will always chase profit no matter what the consequences. Most jurisdictions permit companies to act for wider social goals should they decide to do so *in addition* to profit-making and accord a company's officers' wide discretion to act in the company's best interests. Clearly, corporate social responsibility and ethical considerations can and do form part of companies' business plans. However, considerations of doing good are often secondary to or at the very least in tension with the requirement to create value for shareholders.[14]

[11] Natasha Lomas, "DeepMind Now Has an AI Ethics Research Unit: We Have a Few Questions for It...", *TechCrunch*, https://techcrunch.com/2017/10/04/deepmind-now-has-an-ai-ethics-research-unit-we-have-a-few-questions-for-it/, accessed 1 June 2018.

[12] For an argument along these lines in terms of a slightly different issue: antitrust regulation of Internet platforms, see Maurits Dolmans, Jacob Turner, and Ricardo Zimbron, "Pandora's Box of Online Ills: We Should Turn to Technology and Market-Driven Solutions Before Imposing Regulation or Using Competition Law", *Concurrences*, N°3-2017.

[13] This proposition has been widely recognised since at least the time of Aristotle. See, for instance, Pierre Pellegrin, "Aristotle's Politics", in *The Oxford Handbook of Aristotle*, edited by Christopher Shields (Oxford: Oxford University Press, 2012), 558–585.

[14] See, for example, Thomas Donaldson and Lee E. Preston, "The Stakeholder Theory of the Corporation: Concepts, Evidence, and Implications", *The Academy of Management Review*, Vol. 20, No. 1 (January 1995), 65–91; David Hawkins, *Corporate Social Responsibility: Balancing Tomorrow's Sustainability and Today's Profitability* (Hampshire, UK and New York, NY: Springer, 2006).

Under most legal systems, profit-making entities are accountable to their owners, who can challenge the actions of directors.[15] In one infamous example, the automobile industry pioneer Henry Ford declared that "[m]y ambition is to employ still more men, to spread the benefits of this industrial system to the greatest possible number, to help them build up their lives and their homes". The Michigan Supreme Court upheld a complaint against him by his co-owners, the Dodge brothers, saying Ford's aims were improper: "A business corporation is organized and carried on primarily for the profit of the stockholders".[16]

1.3 Impartiality and Regulatory Capture

In 1954, the tobacco industry published the notorious "Frank Statement to Cigarette Smokers" in hundreds of US newspapers. In the face of growing but not-yet-conclusive evidence that smoking was harmful, the industry announced:

> We are establishing a joint industry group consisting initially of the undersigned. This group will be known as [the] Tobacco Industry Research Committee. In charge of the research activities of the Committee will be a scientist of unimpeachable integrity and national repute. In addition there will be an Advisory Board of scientists disinterested in the cigarette industry. A group of distinguished men from medicine, science, and education will be invited to serve on this Board.[17]

Researchers have since linked the success of the tobacco industry's campaign for self-regulation to millions of extra deaths from smoking and its side effects.[18]

Some technology companies have been keen to emphasise that their AI oversight bodies include independent experts, and are not merely public relations tools. The ethics board of DeepMind features prominent

---

[15] Christian Leuz, Dhananjay Nanda, and Peter Wysocki, "Earnings Management and Investor Protection: An International Comparison", *Journal of Financial Economics*, Vol. 69, No. 3 (2003), 505–527.

[16] *Dodge v. Ford Motor Co.*, 170 N.W. 668 (Mich. 1919).

[17] The full text is available at: http://archive.tobacco.org/History/540104frank.html, accessed 1 June 2018.

[18] Kelly D. Brownell and Kenneth E. Warner, "The Perils of Ignoring History: Big Tobacco Played Dirty and Millions Died: How Similar Is Big Food?", *The Milbank Quarterly*, Vol. 87, No. 1 (March 2009), 259–294.

commentators, and the Partnership's coalition now includes non-governmental not-for-profit organisations such as the American Civil Liberties Union, Human Rights Watch and the United Nations Children's Emergency Fund (UNICEF).[19]

These initiatives may sound promising, but there is a risk that if governments do not act swiftly to create their own AI agencies, a significant proportion of thought-leaders in the field will become aligned to one corporate interest or another. Though experts appointed to tech companies' boards will in most cases aim to maintain their independence, the fact of their association inevitably raises the risk that either they will be influenced to some extent by the interests of the company in question, *or* they will be seen to be so influenced. Either way, the public trust in their impartiality is liable to be compromised.

In some countries, trust in traditional figures of authority is already diminishing. As UK Government Minister Michael Gove put it: "people in this country have had enough of experts".[20] This may have been a dangerous over-generalisation, encouraging a self-fulfilling prophecy of anti-intellectualism. That said, if experts are seen to compromise their impartiality, then it does seem likely that people will take their pronouncements less seriously.

To the extent that private companies are currently driving the agenda in AI regulation, any governmental body which eventually enters this arena risks the phenomenon of "regulatory capture": a situation where a regulator is heavily influenced by private interests. As industry self-regulation becomes more developed, it will be increasingly difficult for governments to start afresh by designing new institutions. Instead, governments are likely to endorse the systems of regulation already adopted by industry, not least because the industry itself will by that point have been shaped by its own internal regulations. Those systems will probably favour the corporate interests, causing the government's system to be hamstrung from its inception.

1.4   Too Many Rules, and Too Few

A further problem with industry self-regulation is that it lacks the force of binding law. If ethical standards are only voluntary, companies

[19] "Partners", Website of the Partnership on AI to Benefit People and Society, https://www.partnershiponai.org/partners/, accessed 1 June 2018.

[20] Henry Mance, "Britain Has Had Enough of Experts, Says Gove", *Financial Times*, 3 June 2016, https://www.ft.com/content/3be49734-29cb-11e6-83e4-abc22d5d108c, accessed 1 June 2018.

may decide which rules to obey, giving some organisations advantages over others. For example, none of the major Chinese AI companies, including Alibaba, Tencent and Baidu, have announced that they will join the Partnership.[21]

Without one unifying framework, multiple private ethics boards could lead to there being too many sets of rules. Hobbes observed that without a central authoritative lawgiver, life would be "nasty, brutish and short".[22] It would be chaotic and dangerous if every major company had its own code for AI, just as it would be if every private citizen could set his or her own legal statutes. Only governments have the power and mandate to secure a fair system that commands this kind of adherence across the board.

2   RULES FOR AI SHOULD BE MADE ON A CROSS-INDUSTRY BASIS

To date, the majority of legal debate on AI has been on two sectors: weapons[23] and cars.[24] The public, legal scholars and policy-makers have

---

[21] We return below to the problem of seeking international standards of regulation at s. 5 below.

[22] Thomas Hobbes, *Leviathan: Or, the Matter, Forme, & Power of a Common-Wealth Ecclesiasticall and Civil* (London: Andrew Crooke, 1651), 62.

[23] See, for example, Armin Krishnan, *Killer Robots: Legality and Ethicality of Autonomous Weapons* (Farnham: Ashgate, 2009); Michael N. Schmitt, "Autonomous Weapon Systems and International Humanitarian Law: A Reply to the Critics", *Harvard National Security Journal Feature* (2013); Kenneth Anderson and Matthew C. Waxman, "Law and Ethics for Autonomous Weapon Systems: Why a Ban Won't Work and How the Laws of War Can", *Stanford University, The Hoover Institution (Jean Perkins Task Force on National Security and Law Essay Series), 2013* (2013); Benjamin Wittes and Gabriella Blum, *The Future of Violence: Robots and Germs, Hackers and Drones: Confronting a New Age of Threat* (New York: Basic Books, 2015); Rebecca Crootof, "The Varied Law of Autonomous Weapon Systems", in *Autonomous Systems: Issues for Defence Policymakers*, edited by Andrew P. Williams and Paul D. Scharre (Brussels: NATO Allied Command, 2015). Daniel Wilson writes, in his semi-satirical book, *How to Survive a Robot Uprising:* "If popular culture has taught us anything, it is that someday mankind must face and destroy the growing robot menace. In print and on the big screen we have been deluged with scenarios of robot malfunction, misuse and outright rebellion". Daniel Wilson, *How to Survive a Robot Uprising: Tips on Defending Yourself Against the Coming Rebellion* (London: Bloomsbury, 2005), 10.

[24] See, for example, Alex Glassbrook, *The Law of Driverless Cars: An Introduction* (Minehead, Somerset, UK: Law Brief Publishing, 2017); *Autonomous Driving: Technical, Legal and Social Aspects*, edited by Markus Maurer, J. Christian Gerdes, Barbara Lenz, and Hermann Winner (New York: SpringerLink, 2017).

focussed on these areas at the expense of others. More importantly, though, it is misguided to approach *the entirety* of the regulation of AI solely on an industry-by-industry basis.

2.1   The Shift from Narrow to General AI

When seeking to create regulatory principles, it is not correct to think of narrow AI (which is adept at just one task) and general AI (which can fulfil an unlimited range of tasks) as being hermetically sealed from each other. Instead, there is a spectrum along which we are gradually moving.

As noted in Chapter 1, various writers have ruminated on how soon we might reach the end point on this spectrum: superintelligence[25] and some have raised powerful objections to the idea of superhuman AI ever being created.[26] The observation that there is a continuum between narrow AI and general AI does not require one to take any position as to how soon (if ever) the singularity or superintelligence might appear. Rather, the spectrum analogy is merely a prediction that advancement in AI technology will involve iterative steps whereby programs individually and collectively become increasingly capable of mastering a range of techniques and tasks. This approach accords with the intelligence tests proposed by Ben Goertzel[27] and José Hernández-Orallo,[28] which focus on the creation of cognitive synergies measurable on a sliding scale, rather than a binary question of whether an entity is or is not intelligent.

Early and less-advanced AI systems were able to achieve the narrowest of tasks within a specific rule-based environment. TD-gammon, a backgammon playing program developed by IBM in 1992, learned entirely by reinforcement and eventually achieved a superhuman level of play.[29]

---

[25] See Chapter 1 at s. 5.

[26] Kevin Kelly, "The Myth of Superhuman AI", *Wired*, 24 April 2017, https://www.wired.com/2017/04/the-myth-of-a-superhuman-ai/, accessed 1 June 2018.

[27] Ben Goertzel, "Cognitive Synergy: A Universal Principle for Feasible General Intelligence," *2009 8th IEEE International Conference on Cognitive Informatics* (Kowloon, Hong Kong, 2009), 464–468. https://doi.org/10.1109/coginf.2009.5250694.

[28] José Hernández-Orallo, *The Measure of All Minds: Evaluating Natural and Artificial Intelligence* (Cambridge: Cambridge University Press, 2017).

[29] Gerald Tesauro, "Temporal Difference Learning and TD-Gammon", *Communications of the ACM*, Vol. 38, No. 3 (1995), 58–68.

DeepMind's DeepQ sits yet further along the spectrum. Researchers demonstrated that DeepQ could play seven different Atari computer games, beating most human players at six and beating master human players at three.[30] DeepQ was not allowed to view the games' source codes so as to manipulate them from the inside. It was limited solely to what a human player sees.[31] DeepQ learned to play each game from scratch, using deep reinforcement learning—a series of neural layers connecting an input to an output through three hidden levels of reasoning[32] as well as a new technique which the DeepMind researchers termed "experience replay".[33]

DeepQ was limited in that it needed to be reset between each game with the effect that its memory of how to play the previous one was

---

[30] Volodymyr Mnih, Koray Kavukcuoglu, David Silver, Alex Graves, Ioannis Antonoglou, Daan Wierstra, and Martin Riedmiller, "Playing Atari with Deep Reinforcement Learning", arXiv:1312.5602v1 [cs.LG], 19 December 2013, https://arxiv.org/pdf/1312.5602v1.pdf, accessed 1 June 2018; see also Volodymyr Mnih, Koray Kavukcuoglu, David Silver, Andrei A. Rusu, Joel Veness, Marc G. Bellemare, Alex Graves, Martin Riedmiller, Andreas K. Fidjeland, Georg Ostrovski, Stig Petersen, Charles Beattie, Amir Sadik, Ioannis Antonoglou, Helen King, Dharshan Kumaran, Daan Wierstra, Shane Legg, and Demis Hassabis, "Human-Level Control Through Deep Reinforcement Learning", *Nature*, Vol. 518 (26 February 2015), 529–533, https://deepmind.com/research/publications/playing-atari-deep-reinforcement-learning/, accessed 1 June 2018.

[31] Ibid., 2.

[32] Ibid., 6. "The input to the neural network consists is an $84 \times 84 \times 4$ image produced by φ. The first hidden layer convolves 16 8×8 filters with stride 4 with the input image and applies a rectifier nonlinearity... The second hidden layer convolves 32 4×4 filters with stride 2, again followed by a rectifier nonlinearity. The final hidden layer is fully-connected and consists of 256 rectifier units. The output layer is a fully connected linear layer with a single output for each valid action".

[33] Ibid., 4–5. The DeepMind researchers explain "experience replay" as follows: "In contrast to TD-Gammon and similar online approaches, we utilize a technique known as experience replay... where we store the agent's experiences at each time-step, $et=(s_t, a_t, r^t, s_{t+1})$ in a data-set $D=e_1, ..., e_N$ pooled over many episodes into a replay memory. During the inner loop of the algorithm, we apply Q-learning updates, or minibatch updates, to samples of experience, $e \sim D$, drawn at random from the pool of stored samples. After performing experience replay, the agent selects and executes an action according to a greedy policy. Since using histories of arbitrary length as inputs to a neural network can be difficult, our Q-function instead works on fixed length representation of histories produced by a function φ".

wiped. By contrast, the human mind's versatility is one of its greatest assets. We can derive inferences from one activity and are able to apply them to the fulfilment of another. Original experiences of certain phenomena from early childhood onwards[34] create lasting mental pathways and heuristics, such that when we are faced with a relevantly similar but non-identical situation, we have a reasonable idea of what to do.[35]

A team from DeepMind and Imperial College London published a paper in 2017 entitled "Overcoming catastrophic forgetting in neural networks", which demonstrated how an AI system could learn to play several games and, crucially, derive lessons from each individual game which could then be applied to the others:

> The ability to learn tasks in a sequential fashion is crucial to the development of artificial intelligence. Until now neural networks have not been capable of this and it has been widely thought that catastrophic forgetting is an inevitable feature of connectionist models. We show that it is possible to overcome this limitation and train networks that can maintain expertise on tasks that they have not experienced for a long time. Our approach remembers old tasks by selectively slowing down learning on the weights important for those tasks.[36]

[34] Steven Piantadosi and Richard Aslin, "Compositional Reasoning in Early Childhood", *PloS one*, Vol. 11, No. 9 (2016), e0147734.

[35] As to the various mental shortcuts, or "heuristics" used by humans in this exercise, see Daniel Kahneman, *Thinking Fast and Slow* (London: Allen Lane, 2011), 55–57.

[36] James Kirkpatrick, Razvan Pascanu, Neil Rabinowitz, Joel Veness, Guillaume Desjardins, Andrei A. Rusu, Kieran Milan, John Quan, Tiago Ramalho, Agnieszka Grabska-Barwinska, Demis Hassabis, Claudia Clopath, Dharshan Kumaran, and Raia Hadsell, "Overcoming Catastrophic Forgetting in Neural Networks", *Proceedings of the National Academy of Sciences of the United States of America*, Vol. 114, No. 13 (2017), James Kirkpatrick, 3521–3526, https://doi.org/10.1073/pnas.1611835114; See also R.M. French and N. Chater, "Using Noise to Compute Error Surfaces in Connectionist Networks: A Novel Means of Reducing Catastrophic Forgetting", *Neural Computing*, Vol. 14, No. 7 (2002), 1755–1769; and K. Milan et al., "The Forget-Me-Not Process", in *Advances in Neural Information Processing Systems 29*, edited by D.D. Lee, M. Sugiyama, U.V. Luxburg, I. Guyon, and R. Garnett (Red Hook, NY: Curran Assoc., 2016).

Other research projects have focused on AI's ability to plan and "imagine" possible consequences of its actions under conditions of uncertainty—another step in the progression from narrower to more general AI.[37]

Leading technology companies are now focusing dedicated projects on multipurpose AI.[38] Indeed, to accomplish many everyday tasks requires not just one discrete acumen, but rather multiple skills. Apple co-founder Steve Wozniak alluded to this point when he suggested in 2007 that we would never develop a robot with the numerous different capabilities needed to enter an unfamiliar home and make a coffee.[39] Kazuo Yano, the Corporate Chief Engineer of the Research and Development Group at technology conglomerate Hitachi, has said:

> Many new technologies are developed for many specific purposes... For example, mobile phones originated from phones specialized for car use. In many cases, a landmark change occurs in which the specialized technology is transformed into a multi-purpose technology... We thus decided to focus our efforts on multi-purpose AIs from the beginning, based on our

[37]Weber and Racaniere, et al., "Imagination-Augmented Agents for Deep Reinforcement Learning", arXiv:1707.06203v1 [cs.LG], 19 July 2017, https://arxiv.org/pdf/1707.06203.pdf, accessed 1 June 2018.

[38]See, for example, Darrell Etherington, "Microsoft Creates an AI Research Lab to Challenge Google and DeepMind", TechCrunch, 12 July 2017, https://techcrunch.com/2017/07/12/microsoft-creates-an-ai-research-lab-to-challenge-google-and-deepmind/, accessed 1 June 2018; Shelly Fan, "Google Chases General Intelligence with New AI That Has a Memory", *SingularityHub*, 29 March 2017, https://singularityhub.com/2017/03/29/google-chases-general-intelligence-with-new-ai-that-has-a-memory/, accessed 1 June 2018.

[39]"Think of the steps that a human being has to do to make a cup of coffee and you have covered basically 10, 20 years of your lifetime just to learn it. So for a computer to do it the same way, it has to go through the same learning, walking to a house using some kind of optical with a vision system, stepping around and opening the door properly, going down the wrong way, going back, finding the kitchen, detecting what might be a coffee machine. You can't program these things, you have to learn it, and you have to watch how other people make coffee. ... This is a kind of logic that the human brain does just to make a cup of coffee. We will never ever have artificial intelligence. Your pet, for example, your pet is smarter than any computer". Steve Wozniak, interviewed by Peter Moon, "Three Minutes with Steve Wozniak", *PC World*, 19 July 2007. See also Luke Muehlhauser, "What Is AGI?", *MIRI*, https://intelligence.org/2013/08/11/what-is-agi/, accessed 1 June 2018.

forecast that AIs would need to gain such versatility very soon... Hitachi has a vast variety of connections with industries and customers all over the world, including electric utilities, manufacturers, distributors, finance companies, railway companies, transportation companies, and water supply companies.[40]

2.2   The Need for General Principles

Even if it is accepted that AI is becoming more multi-purpose, some might argue that in each individual industry the relevant rules should continue to apply, with no need for any additional layer of general regulation. There are several reasons why this approach would be problematic.

First, although most industries have their own technical rules and standards, some principles of the legal system apply generally to all human subjects. The shift from narrow to general AI means that attempting to keep each use of AI compartmentalised will become increasingly difficult. Civil wrongs, contracts and criminal law apply equally to a fireman as to a banker, in addition to individual sectoral regulation. Such general rules aid consistency and predictability for all participants in a legal system. Just as it would be confusing and detract from the rule of law to have a different law for each human profession, it would be equally so to attempt to do the same for each application of AI.

Secondly, AI raises various novel questions which apply across different industries. The Trolley Problem (discussed in Chapter 2 at Section 3.1) might apply equally to AI vehicles travelling by air as by land. Should a passenger's life be valued differently depending on whether they are in a car or a plane? The answer may still be "yes", but if each industry approaches such questions separately, then there is a risk that time and energy will be wasted in repeating the same exercises. However, at present governments are approaching AI cars and AI drones as completely separate issues. For instance, in 2015, the UK Government published a policy paper entitled "The Pathway to Driverless Cars: detailed review of regulations for

---

[40] Interview with Dr. Kazuo Yano, "Enterprises of the Future Will Need Multi-purpose AIs", Hitachi Website, http://www.hitachi.co.jp/products/it/it-pf/mag/special/2016_02th_e/interview_ky_02.pdf, accessed 1 June 2018.

automated vehicle technologies",[41] which did not mention any overlap with drone technology.[42]

Thirdly, differentiated sectoral regulation gives rise to boundary disputes or "edge problems", where arguments abound as to whether a particular practice or asset should be treated as falling in one regime or another. Edge problems are particularly common in tax disputes, where authorities argue that something is classified in a higher tax band and taxpayers say the opposite. In one incident, the makers of a popular snack, the "Jaffa cake", challenged their designation by the UK tax authorities as "cakes" which attracted value-added tax and not as "biscuits", which were exempt. Eventually, the makers of Jaffa cakes prevailed: the court was persuaded by evidence including that on going stale

---

[41] UK Department of Transport, "The Pathway to Driverless Cars: Detailed Review of Regulations for Automated Vehicle Technologies", UK Government Website, February 2015, https://www.gov.uk/government/uploads/system/uploads/attachment_data/file/401565/pathway-driverless-cars-main.pdf, accessed 1 June 2018.

[42] When in 2017 the UK's House of Lords Science and Technology Select Committee published a report entitled "Connected and Autonomous Vehicles: The Future?", it concentrated solely on land-based vehicles. House of Lords, Science and Technology Select Committee, "Connected and Autonomous Vehicles: The Future?", 2nd Report of Session 2016–17, *HL Paper 115* (15 March 2017). The Report expressly says at para. 23: "We have not considered remote control vehicles (RCV) or drones (unmanned aerial vehicles) in this report". The US Department of Transport published its Federal Automated Vehicles Policy in September 2016. The US Department of Transport published its Federal Automated Vehicles Policy in September 2016. Unlike the UK's paper, the Federal Policy mentioned the regulation of drones—but only in a two-page appendix. As the Department of Transport observed, the US Federal Aviation Authority's, "… Challenges Seem Closest to Those That [National Highway Traffic Safety Administration] Faces in Dealing with [Highly Automated Vehicles]", 94, https://www.transportation.gov/sites/dot.gov/files/docs/AV%20policy%20guidance%20PDF.pdf, accessed 1 June 2018. Likewise, the UK's Automated and Electric Vehicles Act 2018, one of the world's first pieces of legislation to address insurance for autonomous vehicles, was restricted to liability for motor vehicles which "are or might be used on roads or in other public places". No thought was given to extending its provisions to drones even though the same issues of liability and insurance arise. "Automated and Electric Vehicles Act 2018", UK Parliament Website, https://services.parliament.uk/bills/2017-19/automatedandelectricvehicles.html, accessed 20 August 2018.

a Jaffa cake becomes hard (like a cake) rather than soft (like a biscuit).[43] The matter may sound flippant, but millions of pounds were at stake, and the Government spent significant amounts of public money on the litigation.[44]

Differential tax treatment for different assets and practices is justified on the basis that a government may wish to encourage and discourage various activities using economic incentives. All things being equal, the more complex a regulatory system is, the more time and energy that companies will expend in trying to comply or to secure themselves the most favourable treatment. Likewise, governments will spend more resources on arguing against companies as to how complex systems of overlapping regulation should be enforced. Differential regulation only makes sense to the extent it can be justified by other extraneous considerations, rather than being the default position. The public and private cost of disputing edge problems therefore presents a further powerful incentive for a regulatory regime which is as clear and consistent as possible across industries.

To be clear, it is not suggested here that every aspect of AI should be governed by entirely new regulations or a new regulator—this would be unworkable. Owing to their expertise and established rule-making infrastructures, individual sectoral regulators will continue to be a major source of governance in their own fields, from aviation to agriculture. Sectoral regulation is a necessary, but not a sufficient source of rules for AI. The key point is that individual regulators ought to be supplemented by governance structures which allow for overarching principles to be applied across different industries. This model might take the form of a pyramid: widest at the bottom where there will remain a multiplicity of individual industry regulators setting detailed rules, with each level of governance above that being responsible for a smaller and more refined set of principles. Rather than burdening companies with excessive rules, it is suggested here that building a coherent regulatory structure will

---

[43] Indeed, so important was this case that the UK Government now lists explicit guidance on this issue on its website: "Excepted Items: Confectionery: The Bounds of confectionery, Sweets, Chocolates, Chocolate Biscuits, Cakes and Biscuits: The Borderline Between Cakes and Biscuits", *UK Government Website*, https://www.gov.uk/hmrc-internal-manuals/vat-food/vfood6260, accessed 1 June 2018.

[44] "Why Jaffa Cakes Are Cakes, Not Biscuits", *Kerseys Solicitors*, 22 September 2014, http://www.kerseys.co.uk/blog/jaffa-cakes-cakes-biscuits/, accessed 1 June 2018.

enable them to operate in an efficient and predictable environment. A tentative suggestion for the top layer of guiding principles for AI is set out in Chapter 8.

3   New Laws for AI Should Be Made by Legislation, Not Judges

How should new laws for AI be created? There are several ways in which laws can be written, altered and adapted, some of which are more suitable for AI than others. In order to explain why this is so, it is first necessary to describe the two main categories of legal systems.[45]

3.1   Civil Law Systems

Civil law systems focus on legislation. In a paradigm civil law system, all rules are contained in a comprehensive written code. The main role for judges is to apply and interpret the law, but not usually to change it. Indeed, Article 5 of the French Civil Code prohibits judges from "pronounc[ing] judgment by way of general and regulatory dispositions", theoretically discouraging judges from making law.

3.2   Common Law Systems

In a common law system,[46] judges are entitled to make law as well as apply it. Judges exercise this role by deciding individual disputes brought before them by two or more opposing parties. Later courts are then bound by the earlier decision unless it is made by a court lower in the hierarchy, in which case it can be overturned.[47] Judicial legal development takes place through judges drawing analogies between sufficiently similar circumstances. The first time a new situation comes before the

---

[45] For a list categorising countries' legal systems, see the Central Intelligence Agency World Factbook, Field Listing: Legal Systems, https://www.cia.gov/library/publications/the-world-factbook/fields/2100.html, accessed 1 June 2018.

[46] Common law developed in the UK, and variants are found in countries including Australia, Canada, Ireland, India, Singapore and the USA.

[47] See further Cross and Harris, *Precedent* 6; Neil Duxbury, *The Nature and Authority of Precedent* (Cambridge, UK: Cambridge University Press, 2008), 103; and *Jowitt's Dictionary of English Law*, edited by Daniel Greenberg (4th edn. London: Sweet & Maxwell 2015), Entry on Precedent.

courts is often described as a "test case". Other courts will apply the same principles, allowing change to take place in incremental steps.

US judge Oliver Wendell Holmes Jr. summed up the common law approach by saying: "The life of the law has not been logic; it has been experience... The law embodies the story of a nation's development through many centuries, and it cannot be dealt with as if it contained only the axioms and corollaries of a book of mathematics".[48]

3.3   Writing Rules for AI

When judges change or apply rules in a common law system, this happens ex post facto—after the dispute has arisen. Although in some circumstances it is possible to seek interim relief from courts in the expectation of potential future harm, the damage has often already occurred by the time a matter comes before a judge.

Those who consider that no new laws are needed for AI often argue that the common law is well suited to drawing analogies between new and existing phenomena, for example by repurposing the law applicable to animals.[49] Legal writer, politician and humorist A.P. Herbert parodied this tendency in the common law in the following satirical judgment, where an (imaginary) bench of the English Court of Appeal held in favour of a claimant who was injured by a motor car when crossing the road:

> [The defendant's] motor car should in law be regarded as a wild beast; and the boast of its makers that it contains the concentrated power of forty-five horses makes the comparison just. If a man were to bring upon the public street forty-five horses tethered together, and were to gallop them at their full speed past a frequented crossroad, no lack of agility, judgment, or presence of mind in the pedestrian would be counted such negligence as to excuse his injury.[50]

---

[48] Oliver Wendell Holmes, *The Common Law* (Boston, MA: Little, Brown and Company, 1881), 1.

[49] Kenneth Graham, "Of Frightened Horses and Autonomous Vehicles: Tort Law and Its Assimilation of Innovations", *Santa Clara Law Review*, Vol. 52 (2012), 101–131. See also the views of Mark Deem: "What I think is important... is that we do it through case law...the law has this ability to be able to fill the gaps, and we should embrace that", in "The Law and Artificial Intelligence", *Unreliable Evidence*, BBC Radio 4, first broadcast 10 January 2015, http://www.bbc.co.uk/programmes/b04wwgz9, accessed 1 June 2018.

[50] A.P. Herbert, *Uncommon Law: Being 66 Misleading Cases Revised and Collected in One Volume* (London: Eyre Methuen, 1969), 127.

In written evidence provided for a UK House of Commons Science and Technology Committee Report on *Robotics and artificial intelligence*,[51] the Law Society (the regulatory professional body for part of the UK legal profession) commented:

> One of the disadvantages of leaving it to the Courts to develop solutions through case law is that the common law only develops by applying legal principles after the event when something untoward has already happened. This can be very expensive and stressful for all those affected. Moreover, whether and how the law develops depends on which cases are pursued, whether they are pursued all the way to trial and appeal, and what arguments the parties' lawyers choose to pursue. The statutory approach ensures that there is a framework in place that everyone can understand.[52]

The Law Society's assessment is correct. It is often said that "hard cases make bad law", referring to a tendency that when faced with compelling facts such as a victim of a tragic misfortune who might otherwise go uncompensated, a judge may strain to bend legal rules in order that justice be done to the individual litigants, but at a detriment to overall coherence in the system.

Judges usually decide under significant time pressure with limited information as to the wider consequences of their decisions, whereas a legislator often has the freedom to deliberate for many years and undertake significant research when preparing rules.[53]

Attempting to create rules for AI through contested cases is vulnerable to misaligned incentives. In litigation, each party's lawyers are compelled—both economically and as a matter of professional conduct—to

---

[51] UK House of Commons Science and Technology Committee Report on *Robotics and Artificial Intelligence*, Fifth Report of Session 2016–2017, Published on 12 October 2016, HC 145, https://www.publications.parliament.uk/pa/cm201617/cmselect/cmsctech/145/145.pdf, accessed 1 June 2018.

[52] Written Evidence submitted to the UK House of Commons Science and Technology Committee by the Law Society (ROB0037), http://data.parliament.uk/writtenevidence/committeeevidence.svc/evidencedocument/science-and-technology-committee/robotics-and-artificial-intelligence/written/32616.html, accessed 1 June 2018.

[53] For similar criticisms of case law as a means of creating rules for AI, see Matthew U. Scherer, "Regulating Artificial Intelligence Systems: Risks, Challenges, Competencies and Strategies", *Harvard Journal of Law & Technology*, Vol. 29, No. 2 (Spring 2016), 354–398, 388–392.

achieve the best outcome possible for their client alone.[54] There is no guarantee that either sides' objectives will align with societal goals. Although in a test case judges might be able to lay down certain principles which could apply in future situations, the problem is that those principles are likely to have been shaped by the arguments in the case that was in front of the judges on the day, and not the wider concerns which a legislature could take into account.

A legislature is usually formed from society as a whole, whereas case law systems are presided over predominantly by judges who, in most systems, are not elected and will represent only a small (usually privileged) stratum of the population. This is not to say that judges all suffer from incorrigible elitism, but leaving important societal decisions *solely* to the judiciary risks creating a democratic deficit. Indeed, it is in the light of concerns such as these that judges will sometimes decline to rule on a certain issue, where it touches on matters beyond their institutional or constitutional competence.[55]

Finally, many cases of harm do not reach the stage of judicial determination at all. *First*, AI companies are likely to want to settle disputes outside of the courtroom, so as to avoid any damaging publicity and disclosure associated with a drawn-out legal battle. They may well be willing to pay substantial settlements out of court so as to preserve secrecy. Indeed, it is notable that most of the private claims arising from the few known self-driving car fatalities to date appear to have been settled swiftly out of court, and presumably accompanied by robust non-disclosure agreements to keep their outcomes secret.[56] *Secondly*, the costs and uncertainty of any litigation are likely to encourage parties to at least

---

[54] Jeremy Waldron, "The Core of the Case Against Judicial Review", *The Yale Law Journal* (2006), 1346–1406, 1363.

[55] See, for example, a speech by UK Supreme Court Justice setting out the limits of the judicial function, Lord Sumption, "The Limits of the Law", 27th Sultan Azlan Shah Lecture, Kuala Lumpur, 20 November 2013, https://www.supremecourt.uk/docs/speech-131120.pdf, accessed 1 June 2018. See also the decision of the majority in *R (Nicklinson) v. Ministry of Justice* [2014] UKSC 38, where the Supreme Court declined to find that a terminally ill person had a right to be administered euthanasia, in the absence of any imprimatur from Parliament to that effect.

[56] Jack Stilgoe and Alan Winfield, "Self-Driving Car Companies Should Not Be Allowed to Investigate Their Own Crashes", *The Guardian*, 13 April 2018, https://www.theguardian.com/science/political-science/2018/apr/13/self-driving-car-companies-should-not-be-allowed-to-investigate-their-own-crashes, accessed 1 June 2018.

consider settling outside court. *Thirdly*, at least where there is some form of prior agreement in place between the victim and the potentially liable party (e.g. the manufacturer of an AI car and the owner), then that agreement may provide for a secret and binding arbitration as regards any civil liability. The combination of these trends is likely to stymie further the development of new laws for AI via judicial decisions.

In conclusion, judge-made law could be helpful to smooth out the rough edges of any new legislation, but it would be both risky and inefficient for society to delegate the big decisions on regulating AI entirely to the judiciary.

4   CURRENT TRENDS IN GOVERNMENT AI REGULATION

Government AI policies generally fall into at least one of the following three categories: promoting the growth of a local AI industry; ethics and regulation for AI; and managing the problem of unemployment caused by AI. Sometimes, these categories may be in tension, at other times, they can be mutually supportive. The focus in this section is on regulatory initiatives rather than economic or technological ones, though as will be seen the three are often interlinked. The brief survey below is not intended to be a comprehensive examination of all laws and government initiatives concerning AI regulation; matters are developing fast and any such information would soon go out of date. Instead, our aim is to capture some general regulatory approaches with a view to establishing the direction of travel for several of the major jurisdictions involved in the AI industry.

4.1   UK

At present, governmental bodies such as the UK's House of Lords Select Committee on AI[57] and the All-Party Parliamentary Group on AI[58] seem to be in danger of attempting both too much and too little. They are attempting too much because their mandate tends to include economic questions such as AI's impact on employment. This is an important issue,

---

[57] "Homepage", Website of the House of Lords Select Committee on A.I., http://www.parliament.uk/ai-committee, accessed 1 June 2018.

[58] "Homepage", Website of the All-Party Parliamentary Group on A.I., http://www.appg-ai.org/, accessed 1 June 2018.

but it is distinct from the question of what new legal rules we might use to regulate AI.[59]

Conversely, UK Government initiatives are in danger of doing too little because there has been no concerted effort to develop comprehensive standards to govern AI. In a 2016 report, the UK Parliament's Science and Technology Committee concluded:

> ... initiatives [for the regulation of AI] are being developed at the company level... at an industry-wide level ... and at the European level .... It is not clear, however, if any cross-fertilisation of ideas, or learning, is taking place across these layers of governance or between the public and private sectors. As the Chief Executive of Nesta [a charitable foundation focussed on innovation] has argued, 'it's currently no-one's job to work out what needs to be done'.[60]

In a speech to the World Economic Forum at Davos in 2018, Prime Minister Theresa May emphasised the importance of AI to the UK's economy and signalled its willingness to participate in international regulation.

> ... in a global digital age we need the norms and rules we establish to be shared by all.

> This includes establishing the rules and standards that can make the most of Artificial Intelligence in a responsible way, such as by ensuring that algorithms don't perpetuate the human biases of their developers.

> So we want our new world-leading Centre for Data Ethics and Innovation to work closely with international partners to build a common understanding of how to ensure the safe, ethical and innovative deployment of Artificial Intelligence.... the UK will also be joining the World Economic

---

[59]Another area of focus for discussions on AI and law which is outside of the problems addressed by this book is the impact of AI on the legal industry itself, for example as a replacement for lawyers and judges. As to which, see the website of the International Association for Artificial Intelligence and Law, "Homepage", http://www.iaail.org/, accessed 30 December 2017.

[60]"House of Commons Science and Technology Committee, Robotics and Artificial Intelligence", Fifth Report of Session 2016–17, 13 September 2016, para. 64.

Forum's new council on Artificial Intelligence to help shape global govern-ance and applications of this new technology.[61]

Despite these fine words, specific policy developments remain elu-sive. Rowan Manthorpe, writing in the influential technology magazine *Wired*, argued that "May's Davos speech exposed the emptiness in the UK's AI strategy". He continued: "…there are only bland pronounce-ments about the promise of innovation, that brush aside difficult ques-tions, elide compromises, and obscure the trade-offs made in the name of the national good".[62] Another journalist, Rebecca Hill, has wondered whether the vaunted Centre for Data Ethics and Innovation will turn out to be "[a]nother toothless wonder".[63] Likewise, AI policy expert Michael Veale has voiced concerns that this body "will descend into one of many talking shops, producing a series of one-off reports looking at single abstract issues".[64] So long as it continues to lack a clear mandate, leadership or programme of action these concerns will remain.

The title of a report published in April 2018 by the House of Lords AI Committee asked whether the UK's approach to AI was "ready, willing and able?". It concluded that "[t]here is an opportunity for the UK to shape the development and use of AI worldwide, and we recommend that the Government work with Government-sponsored AI organisations in other leading AI countries to convene a global summit to establish international norms for the design, development, regulation and deployment of artificial intelligence".[65] In the light of the enormous upheavals caused by Brexit, both internally and in terms of the UK's international relations, it remains

[61] Theresa May, "Address to World Economic Forum", 25 January 2018, https://www.weforum.org/agenda/2018/01/theresa-may-davos-address/, accessed 1 June 2018.

[62] Rowan Manthorpe, "May's Davos Speech Exposed the Emptiness in the UK's AI Strategy", *Wired*, 28 January 2018, http://www.wired.co.uk/article/theresa-may-davos-artificial-intelligence-centre-for-data-ethics-and-innovation, accessed 1 June 2018.

[63] Rebecca Hill, "Another Toothless Wonder? Why the UK.gov's Data Ethics Centre Needs Clout", *The Register*, 24 November 2017, https://www.theregister.co.uk/2017/11/24/another_toothless_wonder_why_the_ukgovs_data_ethics_centre_needs_some_clout/, accessed 1 June 2018.

[64] Ibid.

[65] House of Lords Select Committee on Artificial Intelligence, *AI in the UK: Ready, Willing and Able? Report of Session 2017–19* HL Paper 100, https://publications.parliament.uk/pa/ld201719/ldselect/ldai/100/100.pdf, accessed 1 June 2018.

to be seen whether the UK Government will have the resources, commitment or indeed the international clout to make good on this proposal.

4.2   *France*

In March 2018, France's President Emmanuel Macron announced a major new AI strategy for his country, in a speech[66] and an interview with *Wired*.[67] Macron emphasised his aim for France and Europe generally to become leaders in the development of AI. He noted in this regard the crucial importance of rules saying:

> My goal is to recreate a European sovereignty in AI... especially on regulation. You will have sovereignty battles to regulate, with countries trying to defend their collective choices. You will have a trade and innovation fight precisely as you have in different sectors. But I don't believe that it will go to the extreme extents Elon Musk talks about [in terms of a third world war, for AI superiority], because I think if you want to progress, there is a huge advantage in an open innovation model.[68]

President Macron's announcement was followed in March 2018 by the Villani Report,[69] a major study commissioned by the French Prime Minister authored by mathematician and Member of Parliament Cédric Villani. The Villani Report was wide-ranging in focus, covering economic initiatives to grow the industry in France and Europe as well as its potential impact on employment. Part 5 of the Report was dedicated to the ethics of AI. In particular, Villani proposed: "the creation of a digital technology and AI ethics committee that is open to society". He recommended that "[s]uch a body could be modelled on the CCNE (Comité consultatif national d'éthique - National

---

[66]The speech is available at: https://www.pscp.tv/w/1RDGldoaePmGL, accessed 1 June 2018.

[67]Nicholas Thompson, "Emmanuel Macron Talks to Wired About France's AI Strategy", *Wired*, 31 March 2018, https://www.wired.com/story/emmanuel-macron-talks-to-wired-about-frances-ai-strategy/, accessed 1 June 2018.

[68]Ibid.

[69]Cédric Villani, "For a Meaningful Artificial Intelligence: Towards a French and European Strategy", March 2018, https://www.aiforhumanity.fr/pdfs/MissionVillani_Report_ENG-VF.pdf, accessed 1 June 2018.

Consultative Ethics Committee), created in 1983 for health and life sciences".

These are certainly an encouraging steps in terms of governmental action, but it is as yet unclear how Macron's grand strategy will be implemented, or indeed if Villani's more detailed proposals will be adopted more widely.

4.3 EU

The EU has launched several initiatives aimed at the development of a comprehensive AI strategy, which include its regulation.[70] The two key documents in this regard are the European Parliament's Resolution of February 2017 on Civil Laws for Robotics and the General Data Protection Regulation (GDPR). Important provisions of both are addressed in some detail in the following two chapters. The February 2017 Resolution contained much interesting content, but it did not create binding law. Instead, it was a recommendation to the Commission for future action. The GDPR by contrast was not aimed specifically at AI, but its provisions seem likely to have some fairly drastic effects on the industry—potentially even beyond what its drafters might have intended.[71]

Taking forward the European Parliament's call to bring forward binding legislation, in March 2018 the European Commission issued a call for a high-level expert group on artificial intelligence, which according to the Commission "will serve as the steering group for the European AI Alliance's work, interact with other initiatives, help stimulate a multi-stakeholder dialogue, gather participants' views and reflect them in its analysis and reports".[72] The work of the high-level expert group will include to "[p]ropose to the Commission AI ethics guidelines, covering issues such as fairness, safety, transparency, the future of work, democracy and more broadly the impact on the application of the Charter of

---

[70] Anne Bajart, "Artificial Intelligence Activities", *European Commission Directorate-General for Communications Networks, Content and Technology*, https://ec.europa.eu/growth/tools-databases/dem/monitor/sites/default/files/6%20Overview%20of%20current%20action%20Connect.pdf, accessed 1 June 2018.

[71] See Chapter 8 at Sections 2.4 and 4.2.

[72] European Commission, "Call for a High-Level Expert Group on Artificial Intelligence", Website of the European Commission, https://ec.europa.eu/digital-single-market/en/news/call-high-level-expert-group-artificial-intelligence, accessed 1 June 2018.

Fundamental Rights, including privacy and personal data protection, dignity, consumer protection and non-discrimination".

In April 2018, 25 EU countries signed a joint declaration of cooperation on AI, the terms of which included a commitment to: "[e]xchange views on ethical and legal frameworks related to AI in order to ensure responsible AI deployment".[73] Despite these encouraging signs and worthy intentions, the EU's regulatory agenda remains at an incipient stage.

4.4   USA

In its final months, the Obama administration produced a major report on the Future of Artificial Intelligence, along with an accompanying strategy document.[74] Though these focussed primarily on the economic impact of AI, they also covered (briefly) topics such as "AI and Regulation", and "Fairness, Safety and Governance in AI".[75] In late 2016, a large group of US Universities, sponsored by the National Science Foundation, published "A Roadmap for US Robotics: From Internet to Robotics", a 109-page document edited by Ryan Calo.[76] This included calls for further work on AI ethics, safety and liability. However, the subsequent Trump administration appears to have abandoned the topic as a major priority.[77] Though

---

[73]"EU Member States Sign Up to Cooperate on Artificial Intelligence", Website of the European Commission, 10 April 2018, https://ec.europa.eu/digital-single-market/en/news/eu-member-states-sign-cooperate-artificial-intelligence, accessed 1 June 2018.

[74]"The Administration's Report on the Future of Artificial Intelligence", Website of the Obama White House, 12 October 2016, https://obamawhitehouse.archives.gov/blog/2016/10/12/administrations-report-future-artificial-intelligence, accessed 1 June 2018. For the reports themselves, see https://obamawhitehouse.archives.gov/sites/default/files/whitehouse_files/microsites/ostp/NSTC/preparing_for_the_future_of_ai.pdf; and https://obamawhitehouse.archives.gov/sites/default/files/whitehouse_files/microsites/ostp/NSTC/national_ai_rd_strategic_plan.pdf, accessed 1 June 2018.

[75]"Preparing for the Future of Artificial Intelligence", *Executive Office of the President National Science and Technology Council Committee on Technology*, 17–18 and 30–32 October 2016, https://obamawhitehouse.archives.gov/sites/default/files/whitehouse_files/microsites/ostp/NSTC/preparing_for_the_future_of_ai.pdf, accessed 1 June 2018.

[76]"A Roadmap for US Robotics: From Internet to Robotics", 31 October 2016, http://jacobsschool.ucsd.edu/contextualrobotics/docs/rm3-final-rs.pdf, accessed 1 June 2018.

[77]Cade Metz, "As China Marches Forward on A.I., the White House Is Silent", *New York Times*, 12 February 2018, https://www.nytimes.com/2018/02/12/technology/china-trump-artificial-intelligence.html, accessed 1 June 2018.

extensive private sector AI development is taking place, as well as considerable government investment (especially via the Department of Defense), at the time of writing the US Federal Government does not appear to be pursuing major national or international regulatory initiatives in AI.

4.5   Japan

Industry in Japan has for many years had a particular focus on automation and robotics.[78] The Japanese Government has generated various strategy and policy papers with a view to maintaining this position. For instance, in its 5th Science and Technology Basic Plan (2016–2020), the Japanese Government declared its aim to "guide and mobilize action in science, technology, and innovation to achieve a prosperous, sustainable, and inclusive future that is, within the context of ever-growing digitalization and connectivity, empowered by the advancement of AI".[79]

In line with these goals, the Japanese Government's Cabinet Office convened an Advisory Board on Artificial Intelligence and Human Society in May 2016 under the initiative of the Minister of State for Science and Technology Policy "with the aim to assess different societal issues that could possibly be raised by the development and deployment of AI and to discuss its implication for society".[80] The Advisory Board produced a report in March 2017, which recommended further work on issues including ethics, law, economics, education, social impacts and R&D.[81]

The Japanese Government's proactive approach, driven by its national industrial strategy and aided by the strong public discourse on AI, is an excellent example of how governments can foster discussion nationally and internationally. The challenge for Japan will be to sustain this early momentum, something which will be assisted if other countries follow its approach.

[78] "The Japanese Robot market was worth approximately 630 billion (JPY) (approximately 4.8 billion (EUR)), in 2015". Fumio Shimpo, "The Principal Japanese AI and Robot Strategy and Research Toward Establishing Basic Principles", *Journal of Law and Information Systems*, Vol. 3 (May 2018).

[79] "Report on Artificial Intelligence and Human Society", Japan Advisory Board on Artificial Intelligence and Human Society, 24 March 2017, Preface, http://www8.cao.go.jp/cstp/tyousakai/ai/summary/aisociety_en.pdf, accessed 1 June 2018.

[80] Ibid.

[81] Advisory Board on Artificial Intelligence and Human Society, "Report on Artificial Intelligence and Human Society, Unofficial Translation", http://www8.cao.go.jp/cstp/tyousakai/ai/summary/aisociety_en.pdf, accessed 1 June 2018.

4.6   China

In July 2017, China's State Council issued "A Next Generation Artificial Intelligence Development Plan",[82] a document described by two experienced analysts of Chinese digital technology as "[o]ne of the most significant developments in the artificial intelligence (AI) world"[83] that year.

Although its main focus was on fostering economic growth through AI technology, the Plan also provided that "[b]y 2025 China will have seen the initial establishment of AI laws and regulations, ethical norms and policy systems, and the formation of AI security assessment and control capabilities". Jeffrey Ding of the Future of Humanity Institute at Oxford University has commented of this statement: "[n]o further specifics were given, which fits in with what some have called opaque nature of Chinese discussion about the limits of ethical AI research".[84]

In November 2017, Tencent Research, an institute within one of China's largest technology companies, and the China Academy of Information and Communications Technology (CAICT) produced a book of 482 pages, the title of which translates roughly to: "A National Strategic Initiative for Artificial Intelligence". Topics covered include law, governance and the morality of machines.

In a report entitled "Deciphering China's AI Dream",[85] Ding hypothesises that "AI may be the first technology domain in which China

[82] Available in English translation from the New America Institute: "A Next Generation Artificial Intelligence Development Plan", China State Council, translated by Rogier Creemers, Leiden Asia Centre; Graham Webster, Yale Law School Paul Tsai China Center; Paul Triolo, Eurasia Group; and Elsa Kania (Washington, DC: New America, 2017), https://na-production.s3.amazonaws.com/documents/translation-fulltext-8.1.17.pdf, accessed 1 June 2018.

[83] Paul Triolo and Jimmy Goodrich, "From Riding a Wave to Full Steam Ahead As China's Government Mobilizes for AI Leadership, Some Challenges Will Be Tougher Than Others", *New America*, 28 February 2018, https://www.newamerica.org/cybersecurity-initiative/digichina/blog/riding-wave-full-steam-ahead/, accessed 1 June 2018.

[84] Jeffrey Ding, "Deciphering China's AI Dream", in *Governance of AI Program, Future of Humanity Institute* (Oxford: Future of Humanity Institute, March 2018), 30, https://www.fhi.ox.ac.uk/wp-content/uploads/Deciphering_Chinas_AI-Dream.pdf, accessed 1 June 2018.

[85] Jeffrey Ding, "Deciphering China's AI Dream", in *Governance of AI Program, Future of Humanity Institute* (Oxford: Future of Humanity Institute, March 2018), https://www.fhi.ox.ac.uk/wp-content/uploads/Deciphering_Chinas_AI-Dream.pdf, accessed 1 June 2018.

successfully becomes the international standardsetter".[86] The report points out that the National Strategic Initiative for Artificial Intelligence book identified Chinese leadership on AI ethics and safety as a way for China to seize the strategic high ground. Tencent Research and CAICT wrote, "China should also actively construct the guidelines of AI ethics, play a leading role in promoting inclusive and beneficial development of AI. In addition, we should actively explore ways to go from being a follower to being a leader in areas such as AI legislation and regulation, education and personnel training, and responding to issues with AI".[87] Ding observes further:

> One important indicator of China's ambitions in shaping AI standards is the case of the International Organization for Standardization... Joint Technical Committee (JTC), one of the largest and most prolific technical committees in the international standardization, which recently formed a special committee on AI [SC 42]. The chair of this new committee is Wael Diab, a senior director at [Chinese multinational company] Huawei, and the committee's first meeting will be held in April 2018 in Beijing, China - both the chair position and first meeting were hotly contested affairs that ultimately went China's way.[88]

In furtherance of its policies, China established a national AI standardisation group and a national AI expert advisory group in January 2018.[89] At these groups' launch event, a division of China's Ministry of Industry and Information Technology released a 98-page White Paper on AI standardisation.[90] The White Paper noted that AI raised challenges in terms of legal liability, ethics and safety, and stated:

---

[86] Ibid., 31.

[87] Ibid., unofficial translation by Jeffrey Ding.

[88] Ibid.

[89] Paul Triolo and Jimmy Goodrich, "From Riding a Wave to Full Steam Ahead As China's Government Mobilizes for AI Leadership, Some Challenges Will Be Tougher Than Others", *New America*, 28 February 2018, https://www.newamerica.org/cybersecurity-initiative/digichina/blog/riding-wave-full-steam-ahead/, accessed 1 June 2018.

[90] "White Paper on Standardization in AI", National Standardization Management Committee, Second Ministry of Industry, 18 January 2018, http://www.sgic.gov.cn/upload/f1ca3511-05f2-43a0-8235-eeb0934db8c7/20180122/5371516606048992.pdf, accessed 9 April 2018. Contributors to the white paper included: the China Electronics Standardization Institute, Institute of Automation, Chinese Academy of Sciences, Beijing Institute of Technology, Tsinghua University, Peking University, Renmin University, as well as private companies Huawei, Tencent, Alibaba, Baidu, Intel (China) and Panasonic (formerly Matsushita Electric) (China) Co., Ltd.

...considering that the current regulations on artificial intelligence management in various countries in the world are not uniform and relevant
standards are still in a blank state, participants in the same AI technology
may come from different countries which have not signed a shared contract for artificial intelligence. To this end, China should strengthen international cooperation and promote the formulation of a set of universal
regulatory principles and standards to ensure the safety of artificial intelligence technology.[91]

China's aim to become a leader in the regulation of AI may be one of
the motivations behind its call in April 2018 "to negotiate and conclude
a succinct protocol to ban the use of fully autonomous weapon systems",
made to United Nations Group of Governmental Experts on lethal
autonomous weapons systems.[92] In so doing, China adopted for the first
time a different approach regarding autonomous weapons to the USA.
The Campaign to Stop Killer Robots announced that China had joined
25 other nations in calling for such a ban.[93]

Paul Triolo and Jimmy Goodrich of the *New America Institute*, a
think tank, state that "[a]s in many other areas, Chinese government
leadership on AI at least nominally comes from the top. Xi has identified AI and other key technologies as critical to his goal of transforming
China from a 'large cyber power' to a 'strong cyber power' (also translated as 'cyber superpower')".[94] This approach seems to be born out by
the White Paper. In its key recommendations, the authors suggested:

[91] Ibid., para. 3.3.1.

[92] Elsa Kania, "China's Strategic Ambiguity and Shifting Approach to Lethal Autonomous
Weapons Systems", *Lawfare Blog*, 17 April 2018, https://www.lawfareblog.com/chi-
nas-strategic-ambiguity-and-shifting-approach-lethal-autonomous-weapons-systems,
accessed 1 June 2018. Kania notes that China's announcement may not be all that it seems,
especially given that China appears simultaneously to be developing its own autonomous
weapon systems, whilst calling for a potential future ban. The original recording of the
Chinese delegation's statement is available on the UN Digital Recordings Portal website, at:
https://conf.unog.ch/digitalrecordings/index.html?guid=public/61.0500/E91311E5-
E287-4286-92C6-D47864662A2C_10h14&position=1197, accessed 1 June 2018.

[93] "Convergence on Retaining Human Control of Weapons Systems", *Campaign to Stop
Killer Robots*, 13 April 2018, https://www.stopkillerrobots.org/2018/04/convergence/,
accessed 1 June 2018.

[94] Paul Triolo and Jimmy Goodrich, "From Riding a Wave to Full Steam Ahead As
China's Government Mobilizes for AI Leadership, Some Challenges Will Be Tougher Than
Others", *New America*, 28 February 2018, https://www.newamerica.org/cybersecuri-
ty-initiative/digichina/blog/riding-wave-full-steam-ahead/, accessed 1 June 2018.

Development of key, urgently needed standards such as reference frameworks, algorithm models, and technology platforms; promotion of international standardization work on artificial intelligence, gathering domestic resources for research and development, participating in the development of international standards, and improving international discourse power.

Reference to China's development of "international discourse power" (国际话语权) concerning AI is particularly significant.[95] The postmodernist term "discourse", popularised by sociologist Michel Foucault, generally refers to "systems of thoughts composed of ideas, attitudes, courses of action, beliefs, and practices that systematically construct the subjects and the worlds of which they speak".[96] It is an example of "soft power": the projection of influence through social, cultural and economic means.[97] International discourse power was adopted as an official national policy aim in 2011.[98] As Chinese analyst Jin Cai explained, "[t]o control the narrative, then, is the first step to controlling the situation".[99]

[95] WangHun Jen, "Contextualising China's Call for Discourse Power in International Politics", *China: An International Journal*, Vol. 13, No. 3 (2015), 172–189. Project MUSE, muse.jhu.edu/article/604043, accessed 9 April 2018. See also Jin Cai, "5 Challenges in China's Campaign for International Influence", *The Diplomat*, 26 June 2017, https://thediplomat.com/2017/06/5-challenges-in-chinas-campaign-for-international-influence/, accessed 1 June 2018.

[96] See, for example, Michel Foucault, *Archeology of Knowledge*, translated by A.M. Sheridan Smith (New York: Pantheon Books, 1972). The definition quoted is from Iara Lesser, "Discursive Struggles Within Social Welfare: Restaging Teen Motherhood", *The British Journal of Social Work*, Vol. 36, No. 2 (1 February 2006), 283–298.

[97] See Joseph S. Nye, Jr., "Soft Power", *Foreign Policy*, No. 80, Twentieth Anniversary (Autumn 1990), 153–171.

[98] "Decision of the Central Committee of the Communist Party of China on Deepening Cultural System Reforms to Promote Major Development and Prosperity of Socialist Culture", Xinhua News Agency, Beijing, 25 October 2011, http://www.gov.cn/jrzg/2011-10/25/content_1978202.htm, accessed 1 June 2018.

[99] Jin Cai, "5 Challenges in China's Campaign for International Influence", *The Diplomat*, 26 June 2017, https://thediplomat.com/2017/06/5-challenges-in-chinas-campaign-for-international-influence/, accessed 1 June 2018.

4.7   Conclusions on Current Trends in Government AI Regulation

National AI policies are bound up with countries' current positions in the global order, as well as where they are hoping to be in the future. Japan sees the development of AI regulation as part of its industrial strategy. For China, the issue is both one of economics and one of international politics. China's efforts to develop a world-leading home-grown industry in AI are connected to but not the same as its efforts to influence the international discourse on AI; even if the first aim does not succeed, the second might. Recent indications suggest that China may now seek a leading role in shaping worldwide AI regulation, much as the USA did in numerous fields over the twentieth century. The US Government appears, temporarily at least, to have stepped back generally from a global rule-making role. Though the EU is now beginning to make moves towards the development of its own comprehensive AI regulatory strategy, it may find itself competing with China and Japan to be the main driver of any worldwide standards.

In the nineteenth century, the major European powers competed for influence over physical territory, in the "Great Game" for Afghanistan, and the "Struggle for Africa". In the twentieth century, the USA and USSR competed against each other over technology in the "Space Race". The twenty-first century may be characterised by a similar competition for power over AI—not just for developing the technology, but also in terms of writing the regulations.[100]

The following sections explore how international regulations could be designed and implemented for the benefit of all, notwithstanding divergent national interests.

[100]Julian E. Barnes and Josh Chin, "The New Arms Race in AI", *The Wall Street Journal*, 2 March 2018, https://www.wsj.com/articles/the-new-arms-race-in-ai-1520009261, accessed 1 June 2018. See also John R. Allen and Amir Husain, "The Next Space Race Is Artificial Intelligence: And the United States Is Losing", *Foreign Policy*, 3 November 2017, http://foreignpolicy.com/2017/11/03/the-next-space-race-is-artificial-intelligence-and-america-is-losing-to-china/, accessed 1 June 2018.

5   INTERNATIONAL REGULATION

5.1   An International Regulatory Agency for AI

The section above on current trends in government regulation described numerous proposals for national or even regional AI regulators.[101] These bodies will play a vital role in shaping certain aspects of regulation to local demands, but ultimately both suggestions are too narrow. In addition to national and regional institutions, all countries stand to benefit from having a *global* regulator.

5.2   Arbitrary Nature of National Boundaries

Late on the night of 10 August 1945, two young US military officers, Dean Rusk and Charles Bonesteel, drew one of the most important lines of the twentieth century. In the final stages of World War II, the Allied Powers were deciding how Japan's colonies ought to be divided between them following its likely defeat. For Rusk and Bonesteel, the task was to propose a division which would protect the interests of the USA, but also be acceptable to the USSR.[102] They decided to draw a horizontal line, tracking the "38th Parallel"—a circle of latitude measured based on its distance from the Earth's equator. One country ceased to exist, and in its place, two new countries were born: North Korea and South Korea. North Korea, which fell originally under Soviet control, is a brutal, secretive and repressive dictatorship beset by extreme poverty. South Korea is one of the world's most economically developed and socially liberal countries. Though at the time of writing North Korea and South

---

[101] See above at s. 4. See also UK House of Commons Science and Technology Committee, *Robotics and Artificial Intelligence*, Fifth Report of Session 2016–17, 13 September 2016, para. 64; Mathew Lawrence, Carys Roberts, and Loren King, "Inequality and Ethics in the Digital Age", IPPR Commission on Economic Justice Managing Automation Employment, December 2017, 37–39; Ryan Calo, "The Case for a Federal Robotics Commission", Brookings Institution, 15 September 2014, 3, https://www.brookings.edu/research/the-case-for-a-federal-robotics-commission/, accessed 1 June 2018; and Matthew U. Scherer, "Regulating Artificial Intelligence Systems: Risks, Challenges, Competencies and Strategies", *Harvard Journal of Law & Technology*, Vol. 29, No. 2 (Spring 2016), 354–398, 393–398.

[102] "Why Is the Border Between the Koreas Sometimes Called the '38th Parallel'?", *The Economist*, 5 November 2013.

Korea may be moving towards an historic reconciliation, this potential *rapprochement* only serves to accentuate the absurdity of the original fissure. It is hard to think of greater differences between two nations resulting from so arbitrary a decision.

Though some borders follow physical divisions, such as a mountain range or a river, all such frontiers are ultimately human inventions. They can shift through war, gift or even sale.[103] National systems of law are particularly effective when the subjects and objects of regulation have a tangible form which can be located in one place or another. The model begins to break down when the subject is not constrained by physical or political boundaries.

5.3   Cost of Uncertainty

If an AI entity operates in several jurisdictions, its designers will need to ensure that it is compatible with the rules in each of them. Where standards differ then barriers to trade and additional costs will arise, as AI that conforms to one country's standards is barred from another. Because we lack rules which address the novel legal issues raised by AI, there is an opportunity to design a comprehensive set of principles which could be applied worldwide. This would save individual legislatures the costs and difficulty of regulating on their own, and it would save AI designers the costs of seeking to comply with multiple different codes.[104] In turn, consumers and taxpayers would benefit from lower costs and more diverse AI products.

Unlike other products where individual countries' regulations are shaped by many years of cultural, economic and political differences, for AI we have a blank slate. A single code would be far more efficient than waiting for individual countries to each develop their own ones. If we fail to prepare international standards, then regulation for AI will likely become Balkanised, with each territory setting its own mutually incompatible rules. The sunk costs and entrenched cultural differences in the

---

[103] One of the most famous examples is the "Louisiana Purchase" whereby the USA acquired the territory (later the state) of Louisiana from France in 1803, for 50 million Francs.

[104] For an example of the types of costs and difficulties caused by regulatory compliance see Stacey English and Susannah Hammond, *Cost of Compliance 2017* (London: Thompson Reuters, 2017).

regulation of AI may well render any future consolidation of standards impossible.

5.4 Avoiding Arbitrage

It is common for companies to restructure or shift their corporate location from one territory to another so as to achieve tax or regulatory advantages. Companies are thereby able to provide their goods or services in one territory, yet avoid its taxes and at least some of its regulations. This practice is known as arbitrage.

There have been various—largely unsuccessful—initiatives to harmonise tax laws across the world in order to diminish the opportunities for this practice.[105] Part of the reason why it is so difficult to do so successfully is that countries have strong incentives to cut their taxes in order to attract businesses to register there, leading to a "race to the bottom". Similarly, there may be an economic advantage for some countries to seek to incentivise less scrupulous AI developers to establish in their jurisdiction by adopting minimal regulations. When operating a technology as powerful and potentially dangerous as AI, this would be a worrisome trend. An international system of regulation could avoid at least some of these differences by stipulating a single standard which would apply wherever the AI may be.

6 Why Would Countries Agree to a Global Code?

6.1 Balancing Nationalism and Internationalism

Despite their fictional and arbitrary nature, the continuing psychological importance of nation states cannot be denied. Predictions that national boundaries would melt away have proved unfounded; the early twenty-first century has in fact seen resurgence in nationalism.[106]

Critics of international regulation are likely to argue that antagonistic national interests will prevent countries coming together to govern AI.

---

[105] Vanessa Houlder, "OECD Unveils Global Crackdown on Tax Arbitrage by Multinationals", *Financial Times*, 19 July 2013, https://www.ft.com/content/183c2e26-f03c-11e2-b28d-00144feabdc0, accessed 1 June 2018.

[106] "Whither Nationalism? Nationalism Is Not Fading Away: But It Is Not Clear Where It Is Heading", *The Economist*, 19 December 2017.

Well-publicised splits and deadlocks in international bodies such as the UN Security Council would seem to support this pessimistic appraisal.

Even so, there are a number of less prominent examples of international regulation functioning efficiently and garnering wide support despite the many differences which separate nations otherwise.[107] The solution to reconciling the urge for national self-determination with the need for international rules is to combine best practices.

6.2   Case Study: ICANN

The rather banal-sounding Internet Corporation for Assigned Names and Numbers (ICANN) means little to most people, yet billions each day make use of facilities which it maintains. ICANN is the organisation which administers, maintains and updates key infrastructure behind the internet. This includes assigning domain names and IP addresses. These "unique identifiers" are aligned with a standard set of protocol parameters which ensure computers can communicate on an agreed basis.[108]

ICANN started with a single individual: John Postel, an academic who established its forerunner at the University of Southern California to administer the assignment of Internet addresses under a contract with Defense Advanced Research Project Agency, part of the US Department of Defense.[109] Despite its origins as a national military project, the Clinton administration committed to the privatisation of domain name systems management in a manner that would increase competition and facilitate international participation in its management.[110] Following a wide consultation which received over 430 comments from members of governments, the private sector and civil society around the world, in February 1998 the US Government announced it would transfer the management of domain names to a new non-profit corporation based in

---

[107] Ibid.

[108] ICANN, "The IANA Functions", December 2015, https://www.icann.org/en/system/files/files/iana-functions-18dec15-en.pdf, accessed 1 June 2018.

[109] ICANN, "History of ICANN", https://www.icann.org/en/history/icann-usg, accessed 1 June 2018.

[110] Ibid. See also Clinton White House, "Framework for Global Electronic Commerce", 1 July 1997, https://clintonwhitehouse4.archives.gov/WH/New/Commerce/read.html, accessed 1 June 2018.

the USA but with global representation.[111] Later that year ICANN was created to fulfil this commitment.[112]

Since gaining independence ICANN has introduced numerous changes which are crucial to the Internet as we now know it. These include: the accreditation of private sector registrars to create and maintain domain names from 1999 (now over 3000)[113] and the expansion of top-level domain names, including from 2012 in Chinese, Russian and Arabic scripts. Today ICANN's mission remains to "organize the voices of volunteers worldwide dedicated to keeping the Internet secure, stable and interoperable", as well as to promote competition and develop internet policy.[114] ICANN explains its internal organisation as follows:

> At the heart of ICANN's policy-making is what is called a "multistakeholder model". This decentralized governance model places individuals, industry, non-commercial interests and government on an equal level. Unlike more traditional, top-down governance models, where governments make policy decisions, the multistakeholder approach used by ICANN allows for community-based consensus-driven policy-making. The idea is that Internet governance should mimic the structure of the Internet itself – borderless and open to all.[115]

ICANN's "At-Large" governance structure incorporates more than 165 local organisations, including professional societies (engineers, lawyers etc.), academic and research organisations, community networks, consumer advocacy groups, and civil society. These are grouped into five

[111] "National Telecommunications Information Administration, Responses to Request for Comments", https://www.ntia.doc.gov/legacy/ntiahome/domainname/130dftmail/, accessed 1 June 2018.

[112] "Articles of Incorporation of Internet Corporation for Assigned Names and Numbers", *ICANN*, as revised 21 November 1998, https://www.icann.org/resources/pages/articles-2012-02-25-en, accessed 1 June 2018.

[113] ICANN-Accredited Registrars, https://www.icann.org/registrar-reports/accredited-list.html, accessed 18 January 2018.

[114] ICANN, "Beginner's Guide to At-Large Structures", June 2014, https://www.icann.org/sites/default/files/assets/alses-beginners-guide-02jun14-en.pdf, accessed 1 June 2018, 3.

[115] Ibid., 4.

regions: Africa, Asia, Europe, Latin America and North America, thereby fostering global discussions.[116]

On 6 January 2017, the final formal agreement between ICANN and the US Department of Commerce expired, thereby ending the US Government's authority to approve key changes for the internet. Lawrence Strickling, the US Assistant Secretary of Commerce for Communications and Information, 2009–2017 commented: "The successful transition of the IANA stewardship proves that the multistakeholder model can work".[117]

6.3   Self-Interest and Altruism

In September 2017, President Trump addressed the General Assembly of the United Nations. He began by reiterating his campaign doctrine of "America First":

> As president of the United States, I will always put America first. Just like you, as the leaders of your countries, will always and should always put your countries first. All responsible leaders have an obligation to serve their own citizens, and the nation state remains the best vehicle for elevating the human condition.[118]

This appears to be a statement *par excellence* of the view in foreign policy that each nation should act solely in its own interests.[119] Yet President Trump continued:

> But making a better life for our people also requires us to with work together in close harmony and unity, to create a more safe and peaceful future for all people.

---

[116] Ibid., 7–8.

[117] "History of ICANN", ICANN, https://www.icann.org/en/history/icann-usg, accessed 1 June 2018.

[118] "Remarks by President Trump to the 72nd Session of the United Nations General Assembly", Website of the White House, 19 September 2017, https://www.whitehouse.gov/briefings-statements/remarks-president-trump-72nd-session-united-nations-general-assembly/, accessed 1 June 2018.

[119] For discussion, see Arthur A. Stein, *Why Nations Cooperate: Circumstance and Choice in International Relations* (Ithaca and London: Cornell University Press, 1990).

This caveat is crucial and demonstrates how even the most powerful nation in the world, led by one of its most nativist leaders, still acknowledges the importance of international coordination on certain global issues.

A 2013 paper produced by the UN System Task Team—a large group of UN bodies—described a "global commons", namely "the High Seas, the Atmosphere, the Antarctica and the Outer Space", noting that "These resource domains are guided by the principle of the common heritage of mankind".[120] Though it is not a physical resource, AI also qualifies as an equivalent global issue with potential to affect the whole of humanity.

Some nations may already recognise the enormous potential of AI and its potential to serve the world as a whole, if its power can be harnessed. Such countries are more likely to support an international rules-based system as a matter of altruistic principle.[121] There are also pragmatic reasons why even the most self-regarding state might wish to see international regulation for AI, in addition to the economic incentives identified above. Game theory explains why self-interested rational actors might cooperate and indeed establish rules on which basis future cooperation may take place.[122]

For less economically developed nations, one major barrier to international regulation of certain industries, such as climate change, is a feeling that more developed countries made unfettered use of technologies with harmful side effects to grow rich in previous decades, and that it is now unfair to seek to impose constraints which could slow growth for those nations now attempting to catch up.[123] Because AI technology remains

---

[120]OHCHR, OHRLLS, UNDESA, UNEP, UNFPA, "Global Governance and Governance of the Global Commons in the Global Partnership for Development Beyond 2015: Thematic Think Piece", January 2013, http://www.un.org/en/development/desa/policy/untaskteam_undf/thinkpieces/24_thinkpiece_global_governance.pdf, accessed 1 June 2018.

[121]Arthur A. Stein, *Why Nations Cooperate: Circumstance and Choice in International Relations* (Ithaca and London: Cornell University Press, 1990), 7–10.

[122]Thomas C. Schelling, *The Strategy of Conflict* (Cambridge, MA: Harvard University Press, 1960). See also Glenn H. Snyder, "'Prisoner's Dilemma' and Chicken' Models in International Politics", *International Studies Quarterly*, Vol. 15 (March 1971), 66–103.

[123]Tucker Davey, "Developing Countries Can't Afford Climate Change", Future of Life Institute, 5 August 2016, https://futureoflife.org/2016/08/05/developing-countries-cant-afford-climate-change/, accessed 1 June 2018.

relatively new even for developed countries, there are fewer structural disparities than in other industries. In consequence, there is an opportunity to forestall arguments against regulation based on historic injustice, by instituting international principles now rather than at a later juncture when the field is more mature.

Although the immediate prospect of superintelligence may be low this does not mean that the chances of developing AI which humans are subsequently unable to control can be discounted altogether. Moreover, even if the existential risks to all of humanity appear minimal at present, there are very many less powerful and advanced AI technologies that could cause serious harm, short of the singularity. As such our best chance of protecting against these is to pool resources and expertise in developing the technology within agreed parameters. An untrammelled international arms race in AI could lead to some countries to develop it in an irresponsible manner, prioritising achievement of immediate goals over safety.

Given the arbitrary nature of international borders, there is no reason why AI's impacts should be self-contained within the country in which it originates. Instead, much like a wildfire, tsunami or virus, AI's impacts will cross man-made boundaries with impunity. The danger of a country being cross-infected ought to encourage its leaders to promote international standards as a matter of national self-preservation as much as anything else.

6.4   Case Study: Space Law

On 17 December 1903, over a windy beach in North Carolina, USA, Orville Wright piloted the first powered aeroplane flight. Fewer than 60 years later, the USSR propelled the first human into the earth's orbit. During the Cold War—which was at its most intense in the early 1960s—space technology gave rise to a number of concerns, the most immediate of which was the possibility of nuclear and conventional weapons being used from space.

As well as the security element, space technology was significant also to the scientific and cultural competition between the West and the Soviets. Each sought to prove that it was the world's dominant civilisation through feats including putting the first man in space or the first man on the moon.

Following the launch of the USSR's Sputnik 1 rocket in 1957—the first artificial satellite—the Western powers made a series of proposals to ban the use of outer space for military purposes.[124] The USA and the USSR were the main participants in discussions given that they were the powers most advanced in this field.[125] However, at an early stage the discussion was also internationalised to include the views of nations without space technology: the UN General Assembly passed a unanimous resolution entitled the "Declaration of Legal Principles Governing the Activities of States in the Exploration and Use of Outer Space" in October 1963, calling upon all states to refrain from introducing weapons of mass destruction into outer space.[126] This was despite the lack of any provisions within the treaty for verification of whether states were adhering to its terms.

After successive draft treaties were submitted by the USA and USSR, their positions gradually aligned. The text of the Treaty on Principles Governing the Activities of States in the Exploration and Use of Outer Space, Including the Moon and Other Celestial Bodies (the Outer Space Treaty), was agreed on 19 December 1966. The Outer Space Treaty was put to a vote of the General Assembly, which approved it by unanimous acclamation. It entered into force in October 1967.

To date, the Outer Space Treaty has been ratified by 62 States, which include all those with space exploration capacities. Article I provides: "… exploration and use of outer space, including the moon and other celestial bodies, shall be carried out for the benefit and in the interests of all countries, irrespective of their degree of economic or scientific development, and shall be the province of all mankind". Key provisions include an undertaking not to place in orbit around the Earth, install on the moon or any other celestial body, or otherwise station in outer space, any weapons of mass destruction; and limiting the use of celestial bodies exclusively to peaceful purposes.

---

[124] Helpfully, there was precedent for this: the Antarctic Treaty was agreed in 1957 by the 12 countries whose scientists had been active in that area, which included several of the major world powers: the US, France, the UK and Russia. "The Antarctic Treaty", Website of the Antarctic Treaty Secretariat, http://www.ats.aq/e/ats.htm, accessed 1 June 2018.

[125] US Department of State, "Treaty on Principles Governing the Activities of States in the Exploration and Use of Outer Space, Including the Moon and Other Celestial Bodies: Narrative", Website of the US Department of State, https://www.state.gov/t/isn/5181.htm, accessed 1 June 2018.

[126] General Assembly resolution 1962 (XVIII) of 13 December 1963.

Other clauses emphasise the need for "co-operation and mutual assistance", as well as the importance of "appropriate international consultations before proceeding with any [potentially harmful] activity or experiment".[127]

These norms have been successful both in terms of adherence to the various prohibitions of militarisation and also in regard to fostering a continuing spirit of international cooperation concerning outer-space activities. The International Space Station was launched in 1998 and operates as a joint project between five space agencies.[128] It is highly unlikely that this achievement would have been possible were it not for the Outer Space Treaty. The development of space law contains a number of lessons for AI.

First, it stands as a rebuke to those who suggest that states will be unwilling or unable to agree to international regulation of AI as a matter of principle; indeed when the Outer Space Treaty was agreed at the height of the Cold War, the use of space was far more intertwined with national and international security, as well as prestige and pride, than AI is at present.

Secondly, the process of negotiation and affirmation of the principles eventually enshrined in the Outer Space Treaty was carried out both between the nations which were at the time most advanced in the relevant technology, but also on an inclusive basis, thus ensuring that the principles agreed had legitimacy not just as between scientifically-advanced nations, but the entire international community.

Thirdly, the framers of the Outer Space Treaty adopted an incremental approach, starting with broad propositions that could be agreed by all nations, whilst leaving some gaps to be filled by later instruments. The Outer Space Treaty articulates a small number of high-level principles and prohibitions. It was followed by four other major international treaties on the topic.[129]

---

[127] Art. IX, Treaty on Principles Governing the Activities of States in the Exploration and Use of Outer Space, Including the Moon and Other Celestial Bodies 1967.

[128] These are: NASA (the US), ESA (various European states), CSA (Canada), JAXA (Japan), and Roscosmos (Russia). "International Space Station", Website of NASA, https://www.nasa.gov/mission_pages/station/main/index.html, accessed 1 June 2018.

[129] The 1967 Agreement on the Rescue of Astronauts, the Return of Astronauts and the Return of Objects Launched into Outer Space provides for aiding the crews of spacecraft in the event of accident or emergency landing. It also establishes a procedure for returning to a launching authority a space object found beyond the territorial limits of that authority. The 1971 Convention on International Liability for Damage Caused by Space Objects provides

Fourthly, an international regulatory body, the United Nations Office for Outer Space Affairs, contributes to the sharing of information as between nations as well as capacity-building for less developed states, enabling them to benefit also from development in the field. In so doing it works closely alongside individual countries' and regions' own space agencies.[130]

7   Applying International Law to AI: The Toolbox

7.1   Traditional Structure of Public International Law

Laws operate at different levels. Civil and common law systems refer to regulatory choices *within* countries. There is a separate body of law which operates to regulate relations *between* countries: public international law.[131]

The traditional sources of international law are set out in Article 38(1) of the Statute of the International Court of Justice: treaties; international custom, as evidence of a general practice accepted as law[132]; the

---

that launching States are liable for damage caused by their space objects on the Earth's surface or to aircraft in flight and/or to space objects of another State or to persons or property on board those objects. The 1974 Convention on Registration of Objects Launched into Outer Space provides that launching States shall maintain registries of space objects and furnish specified information on each space object launched for inclusion in a central United Nations register. The 1979 Agreement Governing the Activities of States on the Moon and Other Celestial Bodies elaborates, in more specific terms, the principles relating to the Moon and other celestial bodies set out in the 1966 Treaty. See "General Assembly Resolutions and Treaties Pertaining to the Peaceful Uses to Outer Space", *United Nations Website*, http://www.un.org/events/unispace3/bginfo/gares.htm, accessed 1 June 2018.

[130]"Role and Responsibilities", Website of UNOOSA, http://www.unoosa.org/oosa/en/aboutus/roles-responsibilities.html, accessed 1 June 2018.

[131]Public International Law is often distinguished from Private International Law. The latter stipulates how the national legal systems of countries interact, particularly with regard to rules on when courts of different countries will accept jurisdiction over a dispute, as well as the recognition and enforcement of foreign judgments. See for the English approach, Collins, ed., *Dicey, Morris and Collins on the Conflict of Laws* (15th edn. London: Sweet & Maxwell, 2016). Private International Law is not discussed further in this work.

[132]Customary International Law is particularly difficult to define with precision, given that it is formed by a combination of state practice and *opinio juris*: whether states actually regard themselves as bound by law to do or not do the action in question. See, for example, Jörg Kammerhofer, "Uncertainty in the Formal Sources of International Law: Customary International Law and Some of Its Problems", *European Journal of International Law*, Vol. 15, No. 3, 523–553; Frederic L. Kirgis, Jr., "Custom on a Sliding Scale", *The American Journal of International Law*, Vol. 81, No. 1 (January 1987), 146–151.

general principles of law recognised by civilised nations[133]; certain judicial decisions[134]; and the teachings of well-respected legal academics.[135] Additionally, certain resolutions of the United Nations Security Council are accepted as being legally binding.[136] Though public international law has applied historically to govern relations between sovereign states, its subjects now include private individuals, companies and international bodies, and non-governmental organisations.[137]

Aside from a small category of "peremptory" (or fundamental) laws, such as the prohibition on slavery, much public international law is voluntary to start with, but binding once agreed.[138] For instance, a country can decide whether or not to accede to a treaty, and even if it does it can usually do so with reservations such that certain provisions of that treaty do not apply.[139] The main reason is that individual countries have been viewed traditionally as independent sovereigns which are able to act unconstrained as regards their internal affairs.[140]

---

[133] Again, there is significant uncertainty as to what principles are generally recognised as such. See *The Barcelona Traction Case*, ICJ Reports (1970), 3.

[134] Unlike common law systems, there is no *stare decisis* rule in international law. Rebecca M. Wallace, *International Law* (London: Sweet & Maxwell, 1986, 22).

[135] See The Statute of the International Court of Justice, art. 38(1).

[136] See art. 25 of the UN Charter. See also sources cited at entry entitled "Are UN Resolutions Binding", Website of Dag Hammarskjold Library, http://ask.un.org/faq/15010, accessed 3 June 2017; Philippe Sands, Pierre Klein, and D.W. Bowett, *Bowett's Law of International Institutions* (6th edn. London: Sweet & Maxwell, 2009).

[137] Math Noortmann, August Reinisch, and Cedric Ryngaert, eds. *Non-state Actors in International Law* (Oxford: Hart Publishing, 2015).

[138] Legal theorist Hans Kelsen identified the proposition *"Pacta sunt servanda"*: agreements are to be honoured, as the most important norm in international law. Hans Kelsen, "Théorie générale du droit international public. Problèmes choisis", Collected Courses of The Hague Academy of International Law 42 (Boston: Brill Nijhoff, 1932), IV, 13. Discussed in Francois Rigaux, "Hans Kelsen on International Law", *European Journal of International Law*, Vol. 9, No. 2 (1998), 325–343.

[139] Ryan Goodman, "Human Rights Treaties, Invalid Reservations, and State Consent", *American Journal of International Law*, Vol. 96 (2002), 531–560; Alan Boyle and Christine Chinkin, *The Making of International Law* (Oxford: Oxford University Press, 2007).

[140] This is referred to as the "Westphalian Model", following the Peace of Westphalia of 1648, ending the 30 Years War by affirming the individual control of States over their internal affairs.

Any international system of regulations for AI is likely to require, at least at a high level, some form of treaty agreement so as to create a basic structural framework from which other norms can easily develop, for instance by creating the international regulator. Beyond the traditional forms of international law outlined above, the following sections set out a number of additional methods and techniques which might be used to build an effective system of regulation for AI.

7.2 Subsidiarity

The Catholic Church has, for over a millennium, balanced a system of centralised law-making focussed predominantly on the Vatican, with an incredibly wide jurisdiction stretching across much of the world. The Church developed a principle known as "subsidiarity", whereby decisions are taken as closely as possible to the smallest possible unit of administration whilst maintaining the coherence and efficiency of the system as a whole. As Catholic theologian and legal academic Russell Hittinger explains, "the principle does not require 'lowest possible level' but rather the 'proper level'".[141]

The EU has also adopted subsidiarity, requiring decisions to be taken as closely as possible to the citizen and that constant checks are made to verify that action at EU level is justified in light of the possibilities available at national, regional or local level.[142] In particular the EU offers a structured approach to this principle: action at the EU level is only justified if (a) the objectives of the proposed action cannot be sufficiently achieved by individual Member States (i.e. necessity); and (b) the action can by reason of its scale or effects, be implemented more successfully by the EU (i.e. added value).[143]

---

[141] Russell Hittinger, "Social Pluralism and Subsidiarity in Catholic Social Doctrine", *Annales Theologici*, Vol. 16 (2002), 385–408, 396.

[142] "Subsidiarity", EUR-Lex (official website for EU law), http://eur-lex.europa.eu/summary/glossary/subsidiarity.html, accessed 9 December 2017. It is now enshrined as a principle within art. 5(3) of the Treaty on the Functioning of the EU.

[143] "The Principle of Subsidiarity", *European Parliament*, January 2018, http://www.europarl.europa.eu/ftu/pdf/en/FTU_1.2.2.pdf, accessed 1 June 2018.

An AI regulator should utilise subsidiarity as a guiding principle when deciding whether and to what extent to lay down international rules. As is the case in the EU, it would be sensible if the actions of a global AI regulator could be challenged and overturned if they are found to breach the subsidiarity requirement.[144]

7.3   Varying Intensity of Regulation

It is a widely held misconception there is a binary choice between laws which are "hard" and binding, or "soft" and merely persuasive.[145] In fact, there is a range of options which international organisations can use to maintain the efficacy of regulation whilst respecting national sovereignty. The EU has a particularly nuanced menu, which includes the following[146]:

Regulations

A 'regulation' is a binding legislative act. It must be applied in its entirety across the EU. For example, when the EU wanted to make sure that there are common safeguards on goods imported from outside the EU, the Council adopted a regulation.[147]

Directives

A 'directive' is a legislative act that sets out a goal that all EU countries must achieve. However, it is up to the individual countries to devise their own laws on how to reach these goals. One example is the EU consumer rights directive, which strengthens rights for consumers across the EU....[148]

---

[144] See art. 263 of the Treaty on the Functioning of the European Union.

[145] Christine M. Chinkin, "The Challenge of Soft Law: Development and Change in International Law", *International and Comparative Law Quarterly* (1989), 850–866.

[146] Regulations, Directives and Other Acts", Website of the European Union, https://europa.eu/european-union/eu-law/legal-acts_en, accessed 1 June 2018.

[147] Regulation (EU) 2015/478 of the European Parliament and of the Council of 11 March 2015 on common rules for imports.

[148] Directive 2011/83/EU of the European Parliament and of the Council of 25 October 2011 on consumer rights.

Decisions

A 'decision' is binding on those to whom it is addressed (e.g. an EU country or an individual company) and is directly applicable. For example, the Commission issued a decision on the EU participating in the work of various counter-terrorism organisations...[149]

Recommendations

A 'recommendation' is not binding. When the Commission issued a recommendation that EU countries' law authorities improve their use of videoconferencing to help judicial services work better across borders, this did not have any legal consequences.[150] A recommendation allows the institutions to make their views known and to suggest a line of action without imposing any legal obligation on those to whom it is addressed.

In addition to the above, another mechanism for creating softer international law is the promulgation of "guidance" or "guidelines", which provide how an institution considers a certain rule or result ought to be achieved or implemented, albeit without formally requiring subjects to comply.[151]

[149]Joint Decision of the European Commission and the High Representative of the Union for Foreign Affairs and Security Policy on the participation of the European Union in various organisations for cooperation to prevent and counter terrorism; JOIN/2015/032. A "decision" by the European Commission might perhaps be seen as executive rather than legislative. Indeed the website of the European Commission characterises it as the EU's "politically independent executive arm". However, this belies the role that the European Commission plays in generating legislation. Moreover, executive acts—particularly when they have an effect of creating or communicating policies—have the effect of creating or confirming law. As such, the Commission's decisions can properly be characterised as a form of law-making, even if they do not emanate from the purely legislative arm (i.e. the Parliament). See "European Commission", Website of the EU, https://europa.eu/european-union/about-eu/institutions-bodies/european-commission_en, accessed 1 June 2018.

[150]Council Recommendations—'Promoting the use of and sharing of best practices on cross-border videoconferencing in the area of justice in the Member States and at EU level'. *OJ C 250*, 31 July 2015, 1–5.

[151]See, for example, various EU guidelines and guidance on sanctions: Maya Lester QC and Michael O'Kane, "Guidelines", *European Sanctions Blog*, https://europeansanctions.com/eu-guidelines/, accessed 1 June 2018.

International regulation for AI ought to be formed by a combination of the above options. Regulations are the bluntest instrument because they allow nations no discretion whatsoever as to their implementation. Therefore their use ought to be restricted to only the most fundamental principles, for which national derogation of any kind would be impossible.

Rules (such as the EU's Directives) which are binding only as to the result to be achieved offer a good compromise between the desirability of international rules and the prevalent instinct towards nations being able to choose their own methods and structures. The other selectively binding or non-binding options can be used as appropriate. Complete harmonisation of all laws relating to AI need not occur immediately. One model may be to begin with non-binding recommendations in certain areas, with a view to gradually increasing the extent to which they are mandatory over a number of years or even decades.[152]

7.4   Model Laws

Model laws allow an organisation to create a piece of legislation which its constituent members can adopt entirely, partially, or not at all. The advantage of a model law is that it has the detail of a full mandatory regulation but it does not require adherence.

Particularly in technical areas of regulation, it may be too costly and time consuming for certain less wealthy nations to devote the resources to design such laws themselves independently. Model laws allow for nations to pool and share expertise, creating a common good which reflects each of their input. After model laws are enacted, countries can then draw on each others' experiences as an aid to implementation and interpretation.

There can be an economic advantage in terms of increased trade between countries which harmonise their laws. Model laws are thus especially useful in fields which feature interstate commerce. One example of a particularly successful model law is the United Nations Commission on International Trade Law's Model Law on International Commercial

---

[152]Willem Riphagen, "From Soft Law to Jus Cogens and Back", *Victoria University of Wellington Law Review*, Vol. 17 (1987), 81.

Arbitration, versions of which have since 1985 been adopted by many nations.[153]

Model laws are popular in some federal countries where each individual state has discretion to set its own laws, yet there are demonstrable advantages to those laws being similar or the same. To this end, the US National Conference of Commissioners on Uniform State Laws[154] was formed in 1892 for the purpose of promoting "uniformity in state laws on all subjects where uniformity is deemed desirable and practicable". To date, the Commissioners have approved more than two hundred uniform laws, some of which have been adopted across all states.[155]

A global AI regulator could be a source of model laws, drawing on expertise from nations around the world. This option may well be more attractive to nations which are wary of giving up their freedom to legislate more generally. As with the use of non-binding recommendations and principles, model laws could form the first step towards greater harmonisation, depending on their uptake and effectiveness.

7.5   An International Academy for AI Law and Regulation

One major objection to the idea of international regulation for AI is that it might come to be dominated by personnel from the most powerful and developed nations on the basis that these countries have more resources to train the relevant experts in computer science, law and other fields. If a supposedly international body was controlled by specialists from just a handful of countries then its legitimacy would be severely diminished. Without sufficiently trained personnel some nations may not be able to develop or articulate their own viewpoints and may therefore be more inclined to simply follow regional leaders or bloc groupings to which they are aligned.

A lack of properly trained personnel distributed around the world would lead also to a diminution in the effectiveness of any AI laws.

---

[153] See "UNICTRAL Model Law on International Commercial Arbitration", http://www.uncitral.org/uncitral/en/uncitral_texts/arbitration/1985Model_arbitration.html, accessed 1 June 2018.

[154] Deanna Barmakian and Terri Saint-Amour, "Uniform Laws and Model Acts", *Harvard Law School Library*, https://guides.library.harvard.edu/unifmodelacts, accessed 1 June 2018.

[155] The Uniform Commercial Code is an example of a widely-adopted uniform law.

Passing global regulations is one thing; implementing them is another. Careful coordination and interaction will be needed between any global body and the national or regional mechanisms which are to enforce its directives. Absent local personnel who understand and are aligned with the goals of the global regulatory structure, the actual enforcement of any regulation in many areas will be impossible.

A partial solution is to create an International Academy for AI Law and Regulation, which would aim to contribute to the development and dissemination of knowledge and expertise in international AI law. There are certainly social benefits to having classes at one centralised location, through which participants from around the world could meet each other in person and thereby foster a sense of shared objectives and international camaraderie. However, it would also be possible to disseminate courses via an online platform, as has been achieved with several recent highly popular online courses run by universities including Harvard and MIT.

There is precedent for such a body: in 1988, the International Maritime Organisation, the body which oversees much of the implementation of the global Law of the Sea created the International Maritime Law Institute (IMLI) in Malta. The IMLI website explains: "The Institute provides suitably qualified candidates, particularly from developing countries, with high-level facilities for advanced training, study and research in international maritime law. It also focuses on legislative drafting techniques designed to assist participants in the process of incorporating international treaty rules into domestic law".[156]

8   IMPLEMENTATION AND ENFORCEMENT OF AI LAWS

8.1   Coordination with National Regulators

Different structures are available for an international institution. At one extreme, under a complete "top down" model the AI regulator would have its own staff, who might establish local offices and operate without recourse to or discussion with any national governments. The benefit of this would be a high degree of uniformity of application and enforcement of the relevant norms. However, such an intrusive model would

---

[156]"About the IMLI", Website of the IMLI, http://www.imli.org/about-us/imo-international-maritime-law-institute, accessed 1 June 2018.

doubtless be objectionable to governments and indeed many citizens as an interference with their sovereignty. Disobedience and resentment may result.

A far better model would be for the AI regulator to work in conjunction with established—or yet to be established—national authorities. These need not be new bodies, but states may find it useful to create new agencies at a local level. The EU's regime for financial regulation does not refer to specific authorities in terms of setting national requirements, but refers rather to the "competent authority", which is appointed by each Member State, and can be split between different national bodies if the Member State so desires.[157]

The designated national regulators of AI should each be required to have a minimum set of powers and competencies. For instance under the EU's regime for financial markets, the competent authority in each Member State must have powers including to: "(a) have access to any document or other data in any form which the competent authority considers could be relevant for the performance of its duties and receive or take a copy of it; (b) require or demand the provision of information from any person and if necessary to summon and question a person with a view to obtaining information; (c) carry out on-site inspections or investigations; ... (e) require the freezing or the sequestration of assets, or both; (f) require the temporary prohibition of professional activity; ... [and] (q) issue public notices...".[158]

To the extent that the local regulators lack the ability to achieve any of the minimum requirements, the global AI regulator might facilitate local institution and capacity-building programmes so as to train personnel on the ground, or for example to provide loans of the software and hardware needed to achieve the task. Training of personnel at the AI Academy might also foster such local growth.

A national AI regulatory body might be required to have the above powers, as well as others specific to AI, such as the ability to demand to view the source code of an AI system, and perhaps to be able to insist on amendments to programs which breach its requirements. As well as these more punitive measures, national AI bodies could also provide facilitative services such as the "sandboxing" of new technologies, namely the

[157] Directive 2014/65/EU of the European Parliament and of the Council of 15 May 2014 on Markets in Financial Instruments, art. 67.
[158] Ibid.

ability to test them in safe environments, as well as the licensing and certification of individuals and AI systems for compliance with relevant standards. Section 3.4 of Chapter 7 elaborates further on the methodology of regulatory sandboxes.

In an example of the type of international co-operation which could be applied to AI, in August 2018 the UK Financial Conduct Authority (FCA) and 11 other organisations created the Global Financial Innovation Network (GFIN). The FCA explained that GFIN "... will seek to provide a more efficient way for innovative firms to interact with regulators, helping them navigate between countries as they look to scale new ideas. It will also create a new framework for co-operation between financial services regulators on innovation-related topics, sharing different experiences and approaches". Notably, the GFIN's initial members included not just national financial regulators (such as the Australian Securities & Investments Commission and Hong Kong Monetary Authority) but also an NGO: the Consultative Group to Assist the Poor.[159]

8.2   *Monitoring and Inspections*

In order to ensure consistency of implementation and enforcement, the national model could be supplemented by a regime of periodic monitoring and inspections, either by the global regulator itself or by bodies operating at a regional level. The principle of subsidiarity ought to be used to decide which is most appropriate, though all things being equal it would usually be best for a country to be inspected and rated by its peers rather than at the global level. Regional organisations which might play such a role might include, for instance, the African Union, the EU and the Association of Southeast Asian Nations.

Various international bodies already utilise periodic inspections. The International Atomic Energy Agency monitors civilian and military use of nuclear power in this manner. As with any monitoring of AI development, nuclear power is a highly technical field which requires a significant degree of training and expertise. Inspectors from any AI regulator would

---

[159] "Global Financial Innovation Network", FCA Website, 7 August 2018, updated 9 August 2018, https://www.fca.org.uk/publications/consultation-papers/global-financial-innovation-network, accessed 16 August 2018. There is further discussion of the functioning and nature of regulatory sandboxes in Chapter 7 at 7.3.4.

likewise need to be expert in their field. In order to facilitate trust in their independence and legitimacy, it would be advisable for such personnel to be drawn from all around the world and to operate in teams which are diverse in terms of national origin. Unlike the physical inspection of individuals and sites required for controlling doping in sports, nuclear regulation and chemical weapons, it is possible that AI inspections could occur by remote access or even via distributed ledgers. These features may render an international monitoring system of AI less prone to obstruction by recalcitrant regimes than is the case for other technologies.

8.3   Sanctions for Non-compliance

After achieving the initial agreement to be bound to any international system of norms, sanctions for non-compliance are among the most difficult aspects of a regulatory scheme to design and enforce. Indeed some international accords do not contain any form of sanction-based enforcement mechanism at all: the Paris Climate Agreement of 2015 creates a mechanism to secure compliance, but states expressly that it shall be "non-adversarial and non-punitive".[160] Even where sanctions are available in theory, political considerations may render them impossible to implement. The UN Security Council is one of the most prominent bodies under international law empowered to order sanctions, and yet frequently it is unable to do so owing to structural deadlock—not least because of the power of the Permanent Five members (the USA, Russia, France, the UK and China) to veto any resolution with which they disagree.

Furthermore, some states have refused to become party to treaties such as the Rome Statute of the International Criminal Court out of a concern that its enforcement mechanisms might be used to target their citizens for political purposes, rather than those for which the institution was originally established.[161] It will be important to ensure that

---

[160] Art. 15(2) of the Paris Climate Agreement 2015 provides: "The [compliance] mechanism referred to in paragraph 1 of this Article shall consist of a committee that shall be expert-based and facilitative in nature and function in a manner that is transparent, non-adversarial and non-punitive".

[161] See, for example, the views of Israel on the International Criminal Court: "Israel and the International Criminal Court", *Office of the Legal Adviser to the Ministry of Foreign Affairs*, 30 June 2002, http://www.mfa.gov.il/MFA/MFA-Archive/2002/Pages/Israel%20and%20the%20International%20Criminal%20Court.aspx, accessed 1 June 2018.

any regulatory body for AI is not debased by political machinations, and instead remains true to its role as a regulatory and standard-setting body. One possible way to reduce the risk of politicisation of an AI regulator is to require that the membership of any body that has the power to recommend sanctions is properly qualified, and not filled by purely political appointees answerable to national governments.[162]

Rather than having to resort to direct sanctions, it would be preferable that parties to an international agreement on AI adhere to its provisions through self-interest in maintaining the integrity of the system as a whole. However, there will be occasions where states choose not to comply, in which case a system of sanctions may be necessary as a last resort. Instead of economic penalties, it would be preferable in the first instance to develop sanctions which are self-contained within the structure of the global regulator itself. These might include matters such as suspending certain membership or voting rights from a member which is in persistent violation. If well-designed, the desire of states to be part of the international standard-setting body would be sufficiently strong as to provide its own incentive to comply. Failure to do so could see the relevant state lose its place at the table.

8.4   Case Study: The EU's Sanctioning Method for Member States

The EU has a form of self-contained sanctions, which were invoked for the first time against Poland in late 2017, in response to judicial reforms in that country which were deemed to breach minimum standards required of EU Member States to safeguard the rule of law.[163] In order for such sanctions to take effect, they must pass through a number of stages, at each of which the country in default of its obligations is encouraged to enter into dialogue with a view to rectifying the situation.

---

[162] One successful example of a mechanism to ensure a balance of quality as well as geographic diversity is the EU's "Article 255 Committee", which since 2010 has assessed nominees for judicial appointment to the EU's Courts. For discussion see Tomas Dumbrovsky, Bilyana Petkova, and Marijn Van der Sluis, "Judicial Appointments: The Article 255 TFEU Advisory Panel and Selection Procedures in the Member States", *Common Market Law Review*, Vol. 51 (2014), 455–482.

[163] European Commission, "Press Release—Rule of Law: European Commission Acts to Defend Judicial Independence in Poland", Website of the European Commission, 20 December 2017, http://europa.eu/rapid/press-release_IP-17-5367_en.htm, accessed 1 June 2018.

The first stage of this process was the European Commission proposing to another EU body, the Council of Ministers that sanctions be imposed. This triggered a three-month period for Poland to comply with the Council's requests.[164]

The EU Member States considered that Poland's actions constituted a "clear risk of a serious breach by a Member State of the values referred to in Article 2", namely: "respect for human dignity, freedom, democracy, equality, [and] the rule of law and respect for human rights, including the rights of persons belonging to minorities". Rather than fining Poland or seeking personal sanctions against its Ministers, the EU voted to begin the process of invoking Article 7 of the Treaty on the European Union, which provides that Member States "may decide to suspend certain of the rights deriving from the application of the Treaties to the Member State in question, including the voting rights of the representative of the government of that Member State in the Council".[165]

The EU's sanctioning method represents a somewhat useful precedent because: (a) it stipulates a limited number of high-level principles; and (b) there is a relatively high threshold required for their breach to have legal effects ("clear risk of *serious* breach"). The EU's sanctions are not perfect however: Article 7.3 of the Treaty requires that Member States vote unanimously to move to the final stage of enforcement: something which is unlikely to be achieved save in the most extreme of cases (the above Polish example being a situation where further sanctions will likely be vetoed by the country's regional allies). A better system might allow punishments based merely on some kind of super-majority.

8.5   Case Study: OECD Guidelines for Multinational Enterprises

The Organisation for Economic Cooperation and Development (OECD) is a forum whereby the governments of 30 democracies collaborate to address the economic, social and environmental challenges of globalisation.[166] Originally articulated in 1976, the OECD's Guidelines

---

[164] Ibid.

[165] Art. 7, Treaty on the European Union.

[166] The OECD member countries are: Australia, Austria, Belgium, Canada, the Czech Republic, Denmark, Finland, France, Germany, Greece, Hungary, Iceland, Ireland, Italy, Japan, Korea, Luxembourg, Mexico, the Netherlands, New Zealand, Norway, Poland, Portugal, the Slovak Republic, Spain, Sweden, Switzerland, Turkey, the United Kingdom and the United States. The Commission of the European Communities takes part in the work of the OECD.

for Multinational Enterprises (the Guidelines) have undergone a number of revisions, most notably the addition of a human rights chapter in 2011.[167] The Guidelines are a series of recommendations addressed by governments to multinational enterprises (in other words, a type of soft law). They provide voluntary principles and standards for responsible business conduct consistent with applicable laws.

The Guidelines are designed to apply to multinational enterprises (i.e. companies, organisations or groups) which operate in a number of jurisdictions and to secure a minimum degree of compliance with international best practices, especially in developing countries where the methods of protection and enforcement of such standards are otherwise weak.[168]

The principal means of enforcement of the Guidelines is a series of National Contact Points (NCPs), which are required by the OECD to be established in each state party. The role of the NCPs includes furthering the effectiveness of the Guidelines by undertaking promotional activities and handling enquiries. Governments have discretion as to how the NCP should be formed, for example whether it is part of the Executive, or independent of it. NCPs must, however, be "functionally equivalent" to each other, and to this end must all function in "a visible, accessible, transparent, and accountable manner".[169]

As well as their educative function, a major feature of NCPs is to facilitate the resolution of complaints made against multinational enterprises for alleged breaches of the Guidelines. In the event that the NCP considers that there is a case to answer for a complained breach, it will attempt

---

[167] The 2011 revision of the Guidelines added a chapter on Human Rights aligned with the language of the UN Guiding Principles on Business and Human Rights. The Guidelines also make reference to relevant provisions of the ILO Tripartite Declaration of Principles concerning Multinational Enterprises and Social Policy as well as the Rio Declaration. See OECD Secretariat, *Implementing the OECD Guidelines for Multinational Enterprises: The National Contact Points from 2000 to 2015* (2016), 11, http://mneguidelines.oecd.org/OECD-report-15-years-National-Contact-Points.pdf, accessed 1 June 2018.

[168] Ibid., 12. NGOs accounted for 80 specific instances or 48% of all complaints from 2011 to 2016, followed by trade unions which account for a quarter of all complaints since 2011. Individuals have filed 33 complaints from 2011 to 2016 accounting for 19% of all complaints in this time period. Approximately a third of all closed specific instances were not accepted for further consideration at the initial assessment stage. A non-acceptance rate of between 30 and 40% has been relatively stable since 2000.

[169] Ibid., Glossary.

to establish dialogue between the complainant and the multinational enterprise with a view to resolving the matter to the satisfaction of both parties. If this is not possible and the breach is proved, then the NCP can issue a declaration of non-compliance against the party in breach. As at 2016 over 360 complaints had been handled by NCPs, addressing impacts from business operations in over 100 countries and territories.[170]

Despite lacking a specific punishment mechanism, the "naming and shaming" approach as well as the facilitation of dialogue between parties has been largely successful. Reasons for compliance despite the lack of punishment include the avoidance of bad publicity.[171] Governments also consider the fulfilment of the Guidelines with regard to economic decisions, including as to public procurement, and in terms of providing diplomatic support to a company's operations abroad. The OECD records that:

> Between 2011 and 2015, approximately half of all specific instances which were accepted for further examination by NCPs resulted in an agreement between the parties. Agreements reached through NCP processes were often paired with other types of outcomes such as follow-up plans and have led to significant results, including changes to company policies, remediation of adverse impacts, and strengthened relationships between parties. Of all specific instances accepted for further examination between 2011-2015, approximately 36% resulted in an internal policy change by the company in question, contributing to potential prevention of adverse impacts in the future.[172]

In addition to the indirect economic and reputational risk for companies which breach the Guidelines, they may come to influence substantive law in some of the countries where they are implemented, particularly where the local laws require compliance with international best practices.[173]

In summary, the Guidelines are an example of how a non-binding and non-punitive system of rules and norms can achieve a high degree of compliance and effectiveness through gradualist, behaviour-shaping activities, whilst at the same time respecting national differences.

---

[170] Ibid., 12.

[171] Ibid.

[172] Ibid., 13.

[173] *Vilca v. Xstrata* [2016] EWHC 2757 (QB) at [22], [25].

9   CONCLUSIONS ON BUILDING A REGULATOR

According to ancient Mesopotamian legend, repeated in the Old Testament, at one time "the whole earth was of one language, and of one speech".[174] At Babel, the people decided to build a tower so high that it would reach the heavens. God saw this tower, and realised the extraordinary power that mankind was able to exercise through acting together:

> And the Lord said, Behold, the people is one, and they have all one language; and this they begin to do: and now nothing will be restrained from them, which they have imagined to do.[175]

God's solution to this challenge was to "confound their language, that they may not understand one another's speech". People still had the physical tools to rebuild the tower, but without a shared language they lacked a common purpose. The legend of Babel is usually recounted as a cautionary tale of mankind's overweening vanity, but it illustrates also the achievements humanity can make if we can overcome ethno-nationalism and instead learn to collaborate across cultures and borders.

Nations have not yet reached definitive positions on how AI should be governed. The clay of public opinion remains unformed. We have a unique opportunity to create laws and principles to govern AI from a shared basis, a new common language. If each country adopts its own rules for AI—or worse, none at all—we stand to bring upon ourselves the curse of Babel once again.

---

[174] Genesis 11:1, King James Bible.
[175] Ibid.

CHAPTER 7

Controlling the Creators

1   CREATORS AND CREATIONS

Questions of responsibility address legal liability for harm after it has occurred. The final two chapters of this book consider how we can prevent undesirable consequences arising in the first place.[1] In so doing, we will engage with the third major issue raised by AI: "How should ethical standards be applied to the new technology?"

This chapter and Chapter 8 distinguish between those rules which apply to "creators" and those which apply to "creations". The term creators refers to the humans who (at present) design, programme, operate, collaborate with and otherwise interact with AI. Creations means the AI itself.

Technology journalist John Markoff writes in *Machines of Loving Grace*: "[t]he best way to answer the hard questions about control in a world full of smart machines is by understanding the values of those who are actually building these systems".[2] This is true up to a point, but the answers to "hard questions" will also be shaped by what is technologically possible. Although human input is needed to write both sets of rules, the distinction relates to the addressees of standards rather than

---

[1] The distinction is sometimes referred to as regulation *ex ante* (before the event) and *ex post* (after the event).

[2] John Markoff, *Machines of Loving Grace: The Quest for Common Ground Between Humans and Robots* (New York: ECCO, 2015).

© The Author(s) 2019
J. Turner, *Robot Rules*, https://doi.org/10.1007/978-3-319-96235-1_7

their source: rules for creators tell humans what to do; rules for creations do this for AI.

Rules for creators are a set of design ethics. They have an *indirect* effect in that the potential benefit or harm that they are seeking to promote or restrain happens via another entity: the AI. As a rule of thumb, rules for creators will be expressed in the following form: "When designing, operating or interacting with AI, you should...". By contrast, the usual formulation of rules for creations will be simply: "You (the AI) should...".

The remainder of this chapter will focus on rules for creators. Building from the bottom, *first* it will discuss how to build appropriate institutions for setting ethical rules; *secondly*, it will assess various codes proposed to date; and *thirdly*, it will consider how rules for creators might be implemented and enforced.

Most of the bodies writing moral codes for AI engineers[3] have started—like Asimov did in writing his Laws of Robotics—at the second stage, with little consideration for the first and third. If any effective regulation for AI is to be possible, these other elements will be just as important as the rules' substance, if not more so.

2   A Moral Regulator: The Quest for Legitimacy

Some choices in industrial and technical regulation are important but arbitrary. For example, people are more interested in being able to listen to their favourite radio stations than in choosing the delineation between radio frequencies allocated for civilian use and those allocated to public services such as the police or military. By contrast, most members of the public will have an opinion on the ethics and legality of "moral" matters such as euthanasia, abortion or gay marriage. Chapter 2 showed how AI is now engaging in these important moral choices.

It is suggested below that far-reaching decisions on moral questions should not be left to a technocratic elite. When designing ethical standards for AI, the first task is to ensure that ethical regulations take into account input from an appropriate range of sources.

---

[3]The term "AI engineers" is generally adopted in this book in preference to "programmers", in order to avoid giving the impression that each AI decision is programmed or set by the human(s) in question, and because the term engineer connotes a wider class of activities than traditional programming.

2.1   *"...of the People, by the People, for the People"*

At Gettysburg, Abraham Lincoln stated his intention to create "government of the people, by the people, for the people". In their discussion of how to legislate for new technology, Morag Goodwin and Roger Brownsword use this quotation to illustrate the point that: "...procedure lies at the heart of what we understand political legitimacy to be".[4]

The UK's decision to leave the EU—"Brexit"—is an example of what can happen if a legal system is perceived to lack legitimacy by all/or some of its subjects. British citizens benefited from a range of social and economic protections under EU laws,[5] but this did not stop 52% of voters rejecting these so-called "foreign" and "undemocratic" laws in order to "take back control"[6] of their laws in a 2016 referendum.

One major source of frustration for "Leave" voters seems to have been a notion that the EU institutions lacked legitimacy to make laws binding on the UK.[7] This visceral feeling of distance and disenfranchisement from the EU serves as a cautionary tale of how a benign system of laws might be rejected by even a relatively prosperous and well-educated population if those laws do not garner sufficient public support.[8]

Public attitudes towards AI are at a crossroads: a consumer survey in Germany, the USA and Japan suggested that most people are comfortable with robots being part of their daily life.[9] Likewise, an IPSOS Mori

---

[4] Morag Goodwin and Roger Brownsword, *Law and the Technologies of the Twenty-First Century: Text and Materials* (Cambridge: Cambridge University Press, 2012), 246.

[5] See, for example, Directive 2003/88/EC of the European Parliament and of the Council of 4 November 2003 concerning certain aspects of the organisation of working time, or Council Directive 85/374/EEC of 25 July 1985 on the approximation of the laws, regulations and administrative provisions of the Member States concerning liability for defective products.

[6] "Take Back Control" was the slogan of the Vote Leave campaign.

[7] See, for example, the website of the Vote Leave campaign: "In the EU, decisions are made by three key bodies; the European Commission (which is unelected), the Council of Ministers (where the UK is outvoted) and the European Parliament. This system is deliberately designed to concentrate power into the hands of a small number of unelected people and undermines democratic government". *Briefing, Taking Back Control from Brussels*, http://www.voteleavetakecontrol.org/briefing_control.html, accessed 1 June 2018.

[8] For a critical but ultimately hopeful vision of Europe and its failure to engender a sense of shared identity, see Larry Siedentop, *Democracy in Europe* (London: Allen Lane, 2000).

[9] Hiroyuki Nitto, Daisuke Taniyama, and Hitomi Inagaki, "Social Acceptance and Impact of Robots and Artificial Intelligence—Findings of Survey in Japan, the U.S. and Germany", *Nomura Research Institute Papers*, No. 2011, 1 February 2017, https://www.nri.com/~/

poll in April 2017 found that 29% of UK public considered the risks of machine learning outweighed the benefits, 36% viewed them as balanced, and 29% considered machine learning to be more risky than beneficial (the remaining 7% said they didn't know).[10] It appears that whilst many have misgivings, people are as yet relatively open to AI having a greater role. It is possible that public opinion could tip either way, in favour or against AI.[11]

Chapter 5 discussed at Section 3.6.4 the potential development of a dichotomy between Technophiles who embrace AI wholeheartedly and neo-Luddites who are fearful of or even hostile towards AI based on a combination of economic and social concerns. Unless there is a process of public consultation, the void in public discourse is liable to be filled by pressure groups advocating a dogmatic view that a new technology be banned from some or perhaps even all applications. The following sections suggest how governments might avoid such a situation.

2.2   Case Study: GM Crops and Food Safety

The varying reactions to genetically modified (GM) crops demonstrate the importance of public consultation on new technology.[12] In the early 1970s, scientists developed a technique to transfer DNA from one

media/PDF/global/opinion/papers/2017/np2017211.pdf, accessed 1 June 2018. The definition of "robots" in this survey was somewhat unclear, meaning that participants likely included both simple automation and what this book would define as true artificial intelligence in their responses.

[10] Sarah Castell, Daniel Cameron, Stephen Ginnis, Glenn Gottfried, and Kelly Maguire, "Public Views of Machine Learning: Findings from Public Research and Engagement Conducted on Behalf of the Royal Society", Ipsos MORI, April 2017, https://royalsociety.org/~/media/policy/projects/machine-learning/publications/public-views-of-machine-learning-ipsos-mori.pdf, accessed 1 June 2018.

[11] Vyacheslav Polonski, "People Don't Trust AI—Here's How We Can Change That", *Scientific American*, 10 January 2018, https://www.scientificamerican.com/article/people-dont-trust-ai-heres-how-we-can-change-that/, accessed 1 June 2018.

[12] The House of Lords Science and Technology Committee has commented: "For any new technology to succeed, the trust of consumers is vital. In the food sector gaining that trust is a particular challenge—as recently demonstrated by the public reaction to the introduction of technologies such as genetic modification and irradiation", *House of Lords Science and Technology Committee, First Report of Session 2009–2010: Nanotechnologies and Food*, s. 7.1.

organism to another. The prospects for improved agriculture were clear from the outset: if selected strands of DNA from one organism could be transferred to another, the second organism might then acquire a new trait.[13] In a famous example, DNA from a fish able to survive in very cold water was added to a tomato plant, enabling the latter to better resist frost.[14]

It might be expected that the ability to produce plants which can resist common scourges such as extreme weather, parasites and disease would be celebrated. Instead, there was a significant public backlash in the EU. Although there is little evidence of GM crops being dangerous to humans or environmentally harmful, a fear of unknown risks and a sense of scientists "tampering with nature" led many to refuse to purchase GM crops or even to support campaigns that they be banned altogether.[15] In April 2004, various major biotech companies abandoned GM field trials in England, citing concerns raised by British consumers.[16] In 2015, more than half of the EU Member States banned their farmers from growing GM crops.[17] In fact, this ban made very little practical difference: prior to 2015, only one GM crop had ever been approved and grown in the EU.[18]

It did not have to be this way; the US Department of Agriculture reported in 2017 that 77% of corn grown in the USA was GM.[19] At least part of the difference in attitude between the EU and the USA lies in

[13] "What is genetic modification (GM) of crops and how is it done?" *Website of the Royal Society*, https://royalsociety.org/topics-policy/projects/gm-plants/what-is-gm-and-how-is-it-done, accessed 1 June 2018.

[14] Charles W. Schmidt, "Genetically Modified Foods: Breeding Uncertainty", *Environmental Health Perspectives*, Vol. 113, No. 8 (August 2005), A526–A533.

[15] L. Frewer, J. Lassen, B. Kettlitz, J. Scholderer, V. Beekman, and K.G. Berdalf, "Societal Aspects of Genetically Modified Foods", *Food and Chemical Toxicology*, Vol. 42 (2004), 1181–1193.

[16] Ibid.

[17] Andy Coghlan, *More Than Half of EU Officially Bans Genetically Modified Crops*, 5 October 2015, https://www.newscientist.com/article/dn28283-more-than-half-of-european-union-votes-to-ban-growing-gm-crops/, accessed 1 June 2018.

[18] Ibid. This was a weevil-resistant maize grown in Spain.

[19] "Recent Trends in GE [Genetically-Engineered] Adoption", US Department of Agriculture, 17 July 2017, https://www.ers.usda.gov/data-products/adoption-of-genetically-engineered-crops-in-the-us/recent-trends-in-ge-adoption.aspx, accessed 1 June 2018.

the behaviour of regulators when the technology was in its infancy.[20] Consumer surveys and psychological studies have demonstrated that trust in information sources is an important determinant of the way people respond to information about GM foods.[21] Shortly after GM technology was first developed, the US Government appointed the Food and Drug Administration (FDA) to oversee it. The FDA facilitated discussions among stakeholders including scientists, regulators, farmers and environmentalists, which were followed by field tests in the mid-1980s. Data from those tests were then shared between stakeholders, and further experiments were carried out in order to address concerns raised by the discussants.[22] As behavioural scientists Finucane and Holup observe:

> In contrast [to the US], Europe had no central regulator to greenlight the technology and allay public fears and biotechnology was dealt with as a novel process requiring novel regulatory provisions… European field tests in the early 1990s failed to engage discussions between the public and governmental agencies[23]

It might be objected that consultative rule-making is a Western conceit suited only to liberal democracies. However, even in countries which have a non-democratic system of government, regulators have recognised the importance of public trust. Although the Chinese state exercises a significant degree of control over citizens' lives, a 2015 survey indicated that 71% of Chinese consumers considered food and drug safety to be a problem, following repeated scandals concerning tainted milk, oil, meat and even counterfeit eggs.[24] Many of the opportunities for adulteration arise from increased use of technology in the production

---

[20] Melissa L. Finucane and Joan L. Holup, "Psychosocial and Cultural Factors Affecting the Perceived Risk of Genetically Modified Food: An Overview of the Literature", *Social Science & Medicine*, Vol. 60 (2005), 1603–1612.

[21] L. Frewer, C. Howard, and R. Shepherd, "Public Concerns About General and Specific Applications of Genetic Engineering: Risk, Benefit and Ethics", *Science, Technology, & Human Values*, Vol. 22 (1997), 98–124.

[22] Roger N. Beachy, "Facing Fear of Biotechnology", *Science* (1999), 285, 335.

[23] Melissa L. Finucane and Joan L. Holup, "Psychosocial and Cultural Factors Affecting the Perceived Risk of Genetically Modified Food: An Overview of the Literature", *Social Science & Medicine*, Vol. 60 (2005), 1603–1612, 1608.

[24] Lin Fu, "What China's Food Safety Challenges Mean for Consumers, Regulators, and the Global Economy", *The Brookings Institution*, 21 April 2016.

and processing of food. In a traditional agrarian society, people purchase their food from a known food source, such as the farmer or perhaps through a single intermediary such as a market seller. Abandoning this model in favour of industrialised mass-production of food requires great trust in the integrity of the new system. In direct response to these issues, the Chinese state has in recent years attempted to tighten standards, for instance by passing a Food Safety Law in 2015.[25]

The Chinese example indicates that regardless of the culture or type of society involved, public confidence in rule-making standards for new technology is paramount to successful implementation and adoption.[26] Without this, those people who have a choice whether or not to use the new technology may elect not to do so (as is the case with European consumers and GM foods). Where people do not have a choice (as is the case with Chinese food consumers), then they will be forced to engage with a technology they do not trust and in consequence social cohesion and trust in the institutions of government stands to be eroded.

3   Collaborative Lawmaking

The above examples illustrate how important it is to include subjects in the development of rules for new technology. Even if the legislation would have looked the same without public consultation, the important point is that the lawmakers should be seen to involve citizens and stakeholders. Doing so allows the public, and particularly those groups most affected by any new technology, to feel that they are part of the process and thereby to take greater ownership of any eventual regulations created. This is likely to precipitate a virtuous circle where

---

[25] Ibid.

[26] See also the discussion at s. 4.9 of this chapter of the January 2018 White Paper prepared by a division of China's Ministry of Industry and Information Technology and its observation at para. 3.3 that "[i]n the case of AI technology, issues of safety, ethics and privacy have a direct impact on people's trust in AI technology in their interaction experience with AI tools.": "White Paper on Standardization in AI", *National Standardization Management Committee, Second Ministry of Industry*, 18 January 2018, http://www.sgic.gov.cn/upload/f1ca3511-05f2-43a0-8235-eeb0934db8c7/20180122/5371516606048992.pdf, accessed 1 June 2018.

collaborative regulation leads to greater uptake of the technology, which in turn leads to better feedback and adjustment of the rules.[27]

Rousseau wrote in *The Social Contract* that "the general will, to be really such... must both come from all and apply to all".[28] The fundamental right of citizens to participate in public affairs received international recognition in the twentieth century. It is enshrined in Article 25 of the International Covenant on Civil and Political Rights of 1954, a treaty with 169 State Parties: "Every citizen shall have the right and opportunity... to take part in the conduct of public affairs".[29]

When putting this lofty rhetoric into practice, there is no "one size fits all" solution to involving the public in decision-making on AI. Instead, each country and, if appropriate, region ought to define its position in accordance with local political traditions. Although this book supports deliberative and responsive lawmaking, its aim is to suggest regulation for AI which is compatible with all existing legal and political systems. In order to achieve this balance, a degree of flexibility is necessary. Nonetheless, there are certain important tools and techniques which governments of all kinds can use to achieve the aim of civic participation.

Two of the most important factors to the success of public engagement with regulation will be the provision of information and education concerning the new technology. These prerequisites encourage people to make informed decisions as and when their opinion is sought.[30] As shown later in this chapter, public education is also deeply important to the effective enforcement of norms. Another background condition for effective public participation is the freedom of speech for individuals and groups to voice their opinions and create a marketplace of ideas.[31]

---

[27] Ulrich Beck, "The Reinvention of Politics: Towards a Theory of Reflexive Modernization", in *Reflexive Modernization: Politics, Tradition and Aesthetics in the Modern Social Order*, edited by Ulrich Beck, Anthony Giddens, and Scott Lash (Cambridge: Polity Press, 1994), 1–55.

[28] Jean-Jacques Rousseau, *The Social Contract*, edited and translated by Victor Gourevitch (Cambridge: Cambridge University Press: 1997), Book 2, 4.

[29] Human Rights Committee General Comment No. 25: CCPR/C/21/Rev.1/Add.7, 12 July 1996.

[30] Morag Goodwin and Roger Brownsword, *Law and the Technologies of the Twenty-First Century: Text and Materials* (Cambridge: Cambridge University Press, 2012), 262.

[31] This justification for free speech was set out in the writings of John Stuart Mill and was invoked by Justice Oliver Wendell Holmes in a celebrated dissent in the US Supreme Court Case *Abrams v. United States*, 250 U.S. 616 (1919), at 630.

The public does not speak with one voice. When a regulator has to take a decision between competing regulatory options, one or more parts of the population may well be left dissatisfied with the outcome. The solution for policy-makers is to engender a sense of what political philosopher John Rawls' called "public reason" among subjects. This describes the notion that in a just society, rules which regulate public life should be justifiable or acceptable to all those affected. This does not mean that each individual citizen needs to *agree* with every rule, but they should at least consent to the system. There should be some common ideal which is accepted by all and which forms the basis for the legitimacy of the relevant lawmaking institutions.[32]

An AI regulator should take care to ensure that participation includes so far as possible a representative sample reflecting the entirety of society, adjusted, for example, by features including gender, geographic distribution, socio-economic background, religion and race. If one or more groups are deliberately or accidentally excluded from the consultation process, then policy decisions will lack legitimacy among those parts of the population, and future social fissures may result.[33] Diversity is an issue particularly pertinent to AI, where many have already voiced fears that AI systems are likely to reflect the inherent biases of predominantly white, male engineers.[34]

A range of methods should be used to solicit opinions. For instance, a government or legislature could hold public consultations to gauge public views. Such consultations could be augmented by methodologies popular in the private sector, including targeted focus groups to solicit the opinions of key segments of the population who might otherwise not be reached. The legislature might also invite interest groups and experts to a series of open forums. The UK's All Party Parliamentary Group (APPG) on AI held meetings over the course of 2017 and 2018 where experts were questioned by members of the legislature and the public on various issues. These meetings were streamed live and made available

---

[32] John Rawls, *A Theory of Justice: Revised Edition* (Oxford: Oxford University Press, 1999). See also Jurgen Habermas, "Reconciliation Through the Public Use of Reason: Remarks on John Rawls's Political Liberalism", *The Journal of Philosophy*, Vol. 92, No. 3 (1995), 109–131.

[33] Morag Goodwin and Roger Brownsword, *Law and the Technologies of the Twenty-First Century: Text and Materials* (Cambridge: Cambridge University Press, 2012), 255.

[34] See further Chapter 8 at s. 3.3.1.

online.[35] In the USA, a proposed rule is published on the Federal Register and then opened to public discussion, a procedure known as "notice and comment".[36]

Between February and June 2017, the European Parliament undertook an online consultation on public attitudes to AI, with a particular emphasis on civil law rules. This was open worldwide to anyone who wished to respond, and published in all EU official languages. It included two separate questionnaires, adapted to their audience: a shorter version for the general public and a longer one for specialists.[37] The survey provides empirical support for some of the policy solutions suggested in this book; among its key findings, a large majority of respondents expressed "the need for public regulation in the area" and considered "this regulation should be done at EU and/or International level".[38]

Though it was theoretically a worldwide survey, the number of respondents was very small: there were only 39 responses from organisations and 259 from private individuals. Of the private individuals, 72% were male, and 65% had a master's level or more advanced degree, putting them into a tiny minority of the world's population. More meaningful surveys will need to do a better job of ensuring that a broad spectrum of the world's population participates.

The Open Roboethics Institute (ORI) is perhaps a superior example of how an organisation can take an inclusive "bottom-up" approach to ethical questions. Founded in 2012, it has explored ethical questions on

---

[35] Resources are available on the website of the UK's All Party Parliamentary Group on AI, http://www.appg-ai.org/, accessed 1 June 2018.

[36] "Notice-and-Comment' Rulemaking", *Centre for Effective Government*, https://www.foreffectivegov.org/node/2578, accessed 1 June 2018. For discussion see D.J. Galligan, "Citizens' Rights and Participation in the Regulation of Biotechnology", in *Biotechnologies and International Human Rights*, edited by Francesco Francioni (Oxford: Hart Publishing, 2007).

[37] European Parliament Research Service, "Summary of the public consultation on the future of robotics and artificial intelligence (AI) with an emphasis on civil law rules", October 2017, summary of the public consultation on the future of robotics and artificial intelligence (AI) with an emphasis on civil law rules, accessed 1 June 2018.

[38] Tatjana Evas, "Public Consultation on Robotics and Artificial Intelligence First (Preliminary) Results of Public Consultation", *European Parliament Research Service*, 13 July 2017, http://www.europarl.europa.eu/cmsdata/128665/eprs-presentation-first-results-consultation-robotics.pdf, accessed 1 June 2018.

topics including self-driving vehicles, care robots and lethal autonomous weapons systems, with a particular emphasis on involving stakeholders from different groups.[39] The ORI's methods include questionnaires and surveys which are accompanied by neutral and balanced explanations of the technology and issues involved. Importantly, it uses simple and accessible language as opposed to the dry technical wording sometimes favoured by experts in AI or law; for instance an ORI poll on the role of AI in social care was snappily titled "Would you trust a robot to take care of your grandma?".[40]

MIT's "Moral Machine" simulator is another interesting method of a bottom-up approach to ethical standard setting. This is a website operated by the MIT Media Lab, available at the time of writing in ten languages, which operates as "a platform for gathering a human perspective on moral decisions made by machine intelligence, such as self-driving cars".[41] The website explains: "[w]e show you moral dilemmas, where a driverless car must choose the lesser of two evils... As an outside observer, you judge which outcome you think is more acceptable". In other words, the simulator is a practical example of various iterations of the Trolley Problem described in previous chapters. The Moral Machine project has been successful in gathering responses from a wide range of participants: by the end of 2017, 1.3 million people had responded. It has generated scientific papers of significant interest.[42] Results also revealed certain regional variations in opinions.[43]

[39] "What Is Open Roboethics Institute?", *ORI Website*, http://www.openroboethics. org/about/, accessed 1 June 2018.

[40] "Would You Trust a Robot to Take Care of Your Grandma?", *ORI Website*, http:// www.openroboethics.org/would-you-trust-a-robot-to-take-care-of-your-grandma/, accessed 1 June 2018.

[41] "Homepage", *Moral Machine Website*, http://moralmachine.mit.edu/, accessed 1 June 2018.

[42] Jean-François Bonnefon, Azim Shariff, and Iyad Rahwan, "The Social Dilemma of Autonomous Vehicles", *Science*, Vol. 352, No. 6293 (2016), 1573–1576; Ritesh Noothigattu, Snehalkumar 'Neil' S. Gaikwad, Edmond Awad, Sohan Dsouza, Iyad Rahwan, Pradeep Ravikumar, and Ariel D. Procaccia, "A Voting-Based System for Ethical Decision Making", arXiv:1709.06692v1 [cs.AI], accessed 1 June 2018.

[43] Oliver Smith, "A Huge Global Study On Driverless Car Ethics Found the Elderly Are Expendable", *Forbes*, 21 March 2018, https://www.forbes.com/sites/oliver-smith/2018/03/21/the-results-of-the-biggest-global-study-on-driverless-car-ethics-are-in/#7fbb629f4a9f, accessed 1 June 2018.

This type of crowd-sourced research is only one aspect of the type of consultative exercise needed in developing regulation. Care would still need to be taken to ensure that those taking the tests are representative, and governments ought also avoid succumbing to the tyranny of the majority. Nonetheless, the Moral Machine project represents a valuable example of how public engagement can be encouraged in novel ethical issues to which AI gives rise.

3.1 Multidisciplinary Experts

Regulating AI requires expertise from a variety of fields. Clearly, computer scientists and designers of AI ought to be involved. As noted above, any guidance would need to have a firm grounding in what is technically achievable.

Within the field of computer science, there are many different approaches to developing AI. Though deep learning and the use of neural networks are perhaps the most promising techniques at present, there are several other methods, including whole brain emulation and human–computer interfaces which may also be capable of generating AI (and indeed perhaps even more powerful AI than that can be generated by neural networks). For these reasons, it will be important to ensure that the range of technical AI experts is internally diverse.

Lawyers would be needed to draft the relevant guidance and also to explain how it would interact with existing laws. Though there is no closed list of those areas affected by AI, other professions to be represented in any consultation (and indeed in any regulator) would include ethicists, theologians and philosophers, medical personnel and specialists in robotics and engineering. Discussions are already underway within many of these professional groups as to how to react to AI, but the greater challenge will be to bring about a cross-fertilisation of ideas between them.

3.2 Stakeholders, Interest Groups and NGOs

Consultation should include those with a special interest in the regulation of AI and its particular applications. For instance, when designing any rules pertinent to medicine and care, it would be appropriate to consult with medical organisations, such as professional doctors' bodies, as well as patient representative groups.

Those collating the information should bear in mind that NGOs, interest groups and other stakeholders are likely to have particularly strong and pronounced views. In the light of the GM example above, it will be important for AI regulators to ensure that they are not swayed by a small but noisy minority. In a notorious example, when the Netherlands government attempted a public consultation on GM foods, a coalition of anti-GM NGOs attempted to influence the evidence that could be made available to the public in order to shape the debate in their favour.[44]

3.3   Companies

Companies which produce AI technologies will clearly be an important determinant of regulation, as they are likely to be its most immediate subjects. The largest companies already have highly developed policy teams who are very experienced in liaising with regulators and governments—particularly in the fields of antitrust and, increasingly, in data privacy. Smaller companies often join industry associations such as the Confederation of British Industries, which form powerful lobbies for their members' interests.

One growing source of industry-led regulation for AI is collectives of companies as well as other interest groups, such as the Partnership on AI which, as discussed in Chapter 6, was formed originally by US tech giants[45] Google, DeepMind, IBM, Facebook, Microsoft, Amazon and Apple.[46] Organisations such as the Partnership should certainly play a role in the regulation of AI, but for the reasons given in Chapter 5

---

[44]Joel D'Silva and Geert van Calster, "For Me to Know and You to Find Out? Participatory Mechanisms, the Aarhus Convention and New Technologies", *Studies in Ethics, Law, and Technology*, Vol. 4, No. 2 (2010).

[45]Strictly speaking, DeepMind is UK based, though it is a subsidiary of Alphabet, the US-based parent of Google.

[46]"Homepage", *Website of the Partnership on AI*, https://www.partnershiponai.org/, accessed 1 June 2018. The Partnership's governing board now includes six representatives from for-profit organisations and six from not-for-profit ones. See "Frequently Asked Questions: Who Runs PAI today?". At the time of writing the Executive Director of the Partnership is Terah Lyons, a former Policy Advisor to the U.S. Chief Technology Officer in the White House Office of Science and Technology Policy. Notwithstanding this formal balance between companies and NGOs, it remains to be seen whether the Partnership will present any real challenge to the major technology firms.

(including their focus on shareholder benefit rather than public profit, as well as the voluntary nature of self-regulation), they are inappropriate as the sole source of rules on AI.

A further problem with the Partnership is that it is formed, funded and at least partially controlled by major tech powers (albeit that its Board of Directors now contains an equal split between for-profit and not-for-profit entities). It does not include the many small to medium-sized enterprises which are also developing AI. If the major tech powers are able to play a significant role in shaping AI regulatory policy, then they might tend to do so in a way which is harmful to competition and innovation from smaller rivals.

3.4   Case Study: The FCA FinTech Sandbox

One particularly useful mechanism for governments and technology companies to collaborate is "sandboxing". This describes a process where regulators allow a new technology to be used and tested within a closed or limited environment and in close dialogue with policy-makers. In addition to the technology being assessed, sandboxing also allows regulators to try out new rules and observe their impact on the technology in an environment where wider damage or danger to the public is limited. One notable example of such a sandbox is that used by the UK's Financial Conduct Authority (FCA) for new financial technology (FinTech). The FCA explains its approach as follows:

The sandbox seeks to provide firms with:

- the ability to test products and services in a controlled environment

- reduced time-to-market at potentially lower cost

- support in identifying appropriate consumer protection safeguards to build into new products and services

- better access to finance

The sandbox also offers tools such as restricted authorisation, individual guidance, informal steers, waivers and no enforcement action letters.[47]

---

[47] "Regulatory Sandbox", *FCA Website,* 14 February 2018, https://www.fca.org.uk/firms/regulatory-sandbox, accessed 1 June 2018.

The FCA sandbox is not tailored to AI in particular, but many of its techniques could be applied in this field. For instance, the FCA sandbox allows FinTech companies to test their products on real consumers; for non-consumer facing (or back end) AI, the relevant challenge might be testing how the programme interacts with existing technology or even other AI still under development.

For AI, sandboxes will work particularly well in circumstances where current laws require that a human always be in control of a particular decision or process, such that absent a sandbox, using the AI at all might be illegal. A sandbox could be used to demonstrate the safety and efficiency of the AI system on a small scale, precipitating its wider legalisation for the rest of the jurisdiction (accompanied of course by appropriate safety standards, which the government will have also tested out in the sandbox).

A similar sandbox-type approach has been used in other countries and industrial sectors. The Monetary Authority of Singapore also operates a regulatory sandbox for FinTech companies and products.[48] The Spanish region of Catalonia provides test bed facilities for autonomous vehicles, linking car manufacturers (Seat, Nissan), industry representatives (Ficosa, which produces automobile parts), telecommunication companies, academia and legislators (such as the transportation service and the Mayoralty of Barcelona).[49]

Data collected could be shared between both government and industry, thereby allowing each to benefit from the improved information. Sandboxing has an advantage over traditional industry consultation in that governments should be less reliant on slick presentations prepared by expensive lobbyists and public relations consultants and can instead focus on the empirical results of a trial. Developing technology in virtual reality simulations allows for yet further scope in terms of being able to model the behaviour of AI in an extremely complex system.

Sandboxes are an example of where two of the areas of governmental policy in AI—encouraging growth in the local sector and creating new

---

[48] "FinTech Regulatory Sandbox", *Moneyart Authority of Singapore Website*, 1 September 2017, http://www.mas.gov.sg/Singapore-Financial-Centre/Smart-Financial-Centre/FinTech-Regulatory-Sandbox.aspx, accessed 1 June 2018.

[49] See Geoff Mulgan, "Anticipatory Regulation: 10 Ways Governments Can Better Keep UP with Fast-Changing Industries", *Nesta Website*, 15 May 2017, https://www.nesta.org.uk/blog/anticipatory-regulation-10-ways-governments-can-better-keep-up-with-fast-changing-industries/, accessed 1 June 2018.

regulation—can be mutually supportive. Far from stifling competition, this type of collaborative and iterative regulation can actually encourage it by allowing market entrants lacking in expensive research and design facilities or government relations departments the opportunity to educate regulators on the potential benefits of new products. In its "Lessons Learned" impact assessment publication concerning the sandbox, the FCA said:

> A number of indicators suggest that the sandbox is beginning to have a positive impact in terms of price and quality... As more firms with better products and services enter the market, we expect competitive pressure to improve incumbent firms' consumer propositions[50]

Not only does a sandbox approach promote competition, it can also enable governments to better achieve societal goals in terms of protecting underrepresented areas of the population, which might not otherwise be obtained through a purely market-driven approach. The FCA observed in its impact assessment:

> The sandbox has enabled a variety of tests from firms with innovative business models that look to address the needs of more vulnerable consumers who may be particularly at risk of financial exclusion. The House of Lords Select Committee on Financial Inclusion published a report in March 2017 which cited the FCA sandbox as a positive way of encouraging fintech solutions to aspects of financial exclusion.[51]

Promoting the inclusion of the whole of society is essential to creating a sustainable environment for AI regulation and growth in the longer term. As discussed in Chapter 6 at Section 8.1, the FCA FinTech sandbox is now part of a global collaboration of financial regulators—demonstrating that this type of flexible and responsive governance technique presents multiple lessons for future AI regulation.

---

[50] FCA, *Regulatory Sandbox Lessons Learned Report*, October 2017, para. 4.1, https://www.fca.org.uk/publication/research-and-data/regulatory-sandbox-lessons-learned-report.pdf, accessed 1 June 2018.

[51] Ibid., para. 4.16.

3.5   Industry Standards Bodies

Another type of industry-led regulation comes from standard-setting bodies. At the national level, these include the British Standards Institute,[52] the US National Institute of Standards and Technology[53] and the Japanese Industrial Standards Committee. Some operate internationally, such as the International Organisation for Standardisation (ISO).[54] The Institute of Electrical and Electronics Engineers (IEEE) is a professional body rather than a standards organisation, but in AI (as well as some other fields) it plays a standard-setting role. The Association for Computer Machinery is another professional body which promulgates non-binding standards of best practices in this field.[55]

At both the national and international levels, standards bodies play a crucial role in both setting and updating standards, as well as ensuring interoperability between different products and technologies.[56] Standards bodies are usually formed by a large number of members; as at January 2018, the IEEE website listed over 420,000 across 160 countries.[57] The ISO is the umbrella body for national standards bodies, again incorporating over 160 countries.[58] This diffuse membership means that

---

[52] Some national standards bodies have promulgated their own AI guidance, such as the British Standards Institute's BS 8611:2016 on "Robots and robotic devices - Guide to the ethical design and application of robots and robotic systems". These ought also to be factored into any standard-setting conversation internationally.

[53] "Artificial Intelligence", *Website of the National Institute of Standards and Technology*, https://www.nist.gov//topics/artificial-intelligence, accessed 1 June 2018.

[54] Readers may have noted that the acronym IOS does not match the name of the organisation; this is deliberate: "ISO" is derived from the Greek word isos (equal) and remains the same across all languages. "ISO and Road Vehicles—Great Things Happen When the World Agrees", *ISO*, September 2016, 2, https://www.iso.org/files/live/sites/isoorg/files/archive/pdf/en/iso_and_road-vehicles.pdf, accessed 1 June 2018.

[55] "About the ACM Organisation", *Website of the Association of Computer Machinery*, https://www.acm.org/about-acm/about-the-acm-organization, accessed 2 July 2018.

[56] See, for example, "ISO and Road Vehicles—Great Things Happen When the World Agrees", *ISO*, September 2016, https://www.iso.org/files/live/sites/isoorg/files/archive/pdf/en/iso_and_road-vehicles.pdf, accessed 1 June 2018.

[57] "About IEEE", *Website of IEEE*, https://www.ieee.org/about/about_index.html, accessed 1 June 2018.

[58] "About ISO", *Website of ISO*, https://www.iso.org/about-us.html, accessed 1 June 2018.

they are likely to be less-easily dominated by a small group of power-ful corporate interests than organisations such as the Partnership, which have far fewer members, and less-transparent decision-making processes.

The international standards bodies provide a good example for how worldwide regulation of AI might function, in terms of their wide mem-bership, scope of coverage and technical expertise. It might therefore be argued that industry standard-setting bodies ought to be the sole source of regulation for AI. This would be going too far; the ISO, IEEE and organisations like them are well-suited to the formulation of technical standards which do not have an ethical or societal dimension. Technical standards bodies are adept at dealing with arbitrary or uncontroversial, as opposed to moral choices.

The following case study explores the type of extra elements which an ethical regulator ought to incorporate.

3.6   Case Study: The UK Human Fertilisation and Embryology Authority

The Human Fertilisation and Embryology Authority (HFEA) is the UK's independent regulator of fertility treatment and research using human embryos. The process by which it came into being and now oper-ates contains many lessons for developing a similar regulator for AI.

In the late 1970s and early 1980s, scientists made significant advances in the field of biological reproduction, including in particular fertilisation and embryology. The first child resulting from in vitro fertilisation (pop-ularly known as a "test-tube baby") was born in 1978. Though much of this technology was at a theoretical stage, it seemed likely that the new developments would allow far greater scope for detecting and potentially remedying defects in embryos at an early point. Matters previously in the realm of science fiction, such as animal and human cloning, were no longer out of the question.

In 1982, the UK Government commissioned a report by a Committee of Inquiry into Human Fertilisation and Embryology chaired by Dame Mary Warnock (the Warnock Inquiry). The panel included a judge, consultant obstetricians and gynaecologists, a pro-fessor of theology, professors of psychology and directors of research institutes.[59] Its mandate was: "[t]o consider recent and potential

---

[59] Report of the Committee of Inquiry into Human Fertilisation and Embryology, July 12984, Cmnd. 9314, ii–iii.

developments in medicine and science related to human fertilisation and embryology; to consider what policies and safeguards should be applied, including consideration of the social, ethical and legal implications of these developments; and to make recommendations".[60]

Though views on the matter were diverse and often very strongly held, the Warnock Inquiry Committee concluded:

> What is common (and this too we have discovered from the evidence) is that people generally want *some principles or other* to govern the development and use of the new techniques... But in our pluralistic society it is not to be expected that any one set of principles can be enunciated to be completely accepted by everyone... The law itself, binding on everyone in society, whatever their beliefs, is the embodiment of a common moral position... In recommending legislation, then, we are recommending a kind of society that we can, all of us: praise and admire, even if, in detail, we may individually wish that it were different.[61]

Chapter 13 of the Report issued by the Committee recommended the creation of a new regulator, noting that though it should have significant representation of scientific and medical interests,

> ... this is not exclusively, or even primarily, a medical or scientific body. It is concerned essentially with broader matters and with the protection of the public interest. If the public is to have confidence that this is an independent body, which is not to be unduly influenced by sectional interests, its membership must be wide-ranging and in particular the lay interests should be well represented.[62]

In accordance with the Committee's proposals, the HFEA was created in 1990. Today, clinics and research centres in the field of reproductive technology must be inspected at least every two years by the HFEA to make sure they are continuing to operate safe, legal and quality services and research. Mindful of its public-facing role, the HFEA seeks to educate and inform not just those involved in the embryology industry but

---

[60] Ibid., 4.
[61] Ibid., 2–3.
[62] Ibid., 75–76.

also the general public, for instance by maintaining a website with clear explanations of its role.[63]

3.7   A Minister for AI?

The HFEA is successful model but the same aims can be achieved through different institutional structures. Another option is to create a dedicated ministry on AI within the government.

In late 2017, the UAE created the world's first Minister for AI: His Excellency Omar Bin Sultan Al Olama.[64] Several months later, the UAE augmented the Minister's role by creating a National Council for AI tasked with overseeing AI integration in government departments and the education sector.[65] Al Olama commented: "AI is not negative or positive. It's in between. The future is not going to be a black or white. As with every technology on Earth, it really depends on how we use it and how we implement it… People need to be part of the discussion. It's not one of those things that just a select group of people need to discuss and focus on".[66] He emphasised the need for multiple voices to be brought together, including governments, organisations and citizens—nationally and internationally:

> At this point, it's really about starting conversations — beginning conversations about regulations and figuring out what needs to be implemented in order to get to where we want to be. I hope that we can work with other governments and the private sector to help in our discussions and to

---

[63] "About Us", *Website of the HFEA*, https://www.hfea.gov.uk/about-us/, accessed 1 June 2018.

[64] "Cabinet Members: Minister of State for Artificial Intelligence", *Website of the Government of the UAE*, https://uaecabinet.ae/en/details/cabinet-members/his-excellency-omar-bin-sultan-al-olama, accessed 11 June 2018. See also "UAE Strategy for Artificial Intelligence", *Website of the Government of the UAE*, https://government.ae/en/about-the-uae/strategies-initiatives-and-awards/federal-governments-strategies-and-plans/uae-strategy-for-artificial-intelligence, accessed 1 June 2018.

[65] Anna Zacharias, "UAE Cabinet Forms Artificial Intelligence Council", *The UAE National*, https://www.thenational.ae/uae/uae-cabinet-forms-artificial-intelligence-council-1.710376, accessed 1 June 2018.

[66] Dom Galeon, "An Inside Look at the First Nation with a State Minister for Artificial Intelligence", *Futurism*, https://futurism.com/uae-minister-artificial-intelligence/, accessed 1 June 2018.

really increase global participation in this debate. With regards to AI, one country can't do everything. It's a global effort[67]

In its 2017 findings, the UK's APPG on AI key recommendation was for the Government to create a new UK Minister for AI.[68] Of course, there is a significant gap between rhetoric and action, but the UAE's move is nonetheless significant; it may not be long before other countries follow suit. Once the institution has been created, the next question is what output it might have.

4   Proposed Regulatory Codes

4.1   The Roboethics Roadmap

The European Robotics Research Network produced an important paper in 2006 entitled "The Roboethics Roadmap", which identified a number of ethical issues concerning robotics. Its aim was "to increase the understanding of the problems at stake, and to promote further study and transdisciplinary research".[69] The authors, led by Gianmarco Veruggio, were clear as to the limitations of their project:

> It is not a list of *Questions & Answers*. Actually there are no easy answers and the complex fields require careful consideration.
>
> It cannot be a *Declaration of Principles*. The Euron Roboethics Atelier, and the sideline discussion undertaken, cannot be regarded as the institutional committee of scientists and experts entitled to draw a Declaration of Principles on Roboethics.[70]

The Roboethics Roadmap did not seek to create regulations, but rather laid down a challenge to others to do so. Since then, various organisations have taken up the mantle. What follows is a non-exhaustive selection of some of the most influential proposals to date.

---

[67] Ibid.

[68] APPG on AI, "APPG on AI: Findings 2017", http://www.appg-ai.org/wp-content/uploads/2017/12/appgai_2017_findings.pdf, accessed 1 June 2018.

[69] "EURON Roboethics Roadmap", July 2006, 6, http://www.roboethics.org/atelier2006/docs/ROBOETHICS%20ROADMAP%20Rel2.1.1.pdf, accessed 1 June 2018.

[70] Ibid., 6–7.

4.2   The EPSRC and AHRC "Principles of Robotics"

In September 2010, a multidisciplinary group of UK academics including representatives from technology, industry, the arts, law and social sciences met at the joint Engineering and Physical Sciences Research Council (EPSRC) and Arts and Humanities Research Council (AHRC) Robotics Retreat to design a set of Principles of Robotics.[71] The authors made clear that their principles are "rules for robotics (not robots)", applicable "designers, builders and users of robots", putting them firmly within the scope of the present chapter.

The EPRSC/AHRC Principles are as follows[72]:

Rule	Semi-legal	General audience
1	Robots are multi-use tools. Robots should not be designed solely or primarily to kill or harm humans, except in the interests of national security	Robots should not be designed as weapons, except for national security reasons
2	Humans, not robots, are responsible agents. Robots should be designed and operated as far as is practicable to comply with existing laws, fundamental rights and freedoms, including privacy	Robots should be designed and operated to comply with existing law, including privacy
3	Robots are products. They should be designed using processes which assure their safety and security	Robots are products: as with other products, they should be designed to be safe and secure
4	Robots are manufactured artefacts. They should not be designed in a deceptive way to exploit vulnerable users; instead, their machine nature should be transparent	Robots are manufactured artefacts: the illusion of emotions and intent should not be used to exploit vulnerable users
5	The person with legal responsibility for a robot should be attributed	It should be possible to find out who is responsible for any robot

[71] "Principles of Robotics", *EPRSC Website*, https://www.epsrc.ac.uk/research/ourportfolio/themes/engineering/activities/principlesofrobotics/, 1 June 2018.

[72] Margaret Boden, Joanna Bryson, Darwin Caldwell, Kerstin Dautenhahn, Lilian Edwards, Sarah Kember, Paul Newman, Vivienne Parry, Geoff Pegman, Tom Rodden, Tom Sorrell, Mick Wallis, Blay Whitby, and Alan Winfield, "Principles of Robotics: Regulating Robots in the Real World", *Connection Science*, Vol. 29, No. 2 (2017), 124–129.

The simplicity of the project's output is laudable. But rather like Asimov's Laws, with brevity comes hidden dangers of under-specification and over-generalisation.

For each principle, one part of the rule is aimed at a technical audience, with another more user-friendly version directed towards the general public. Though this approach may be intended to advance public understanding, great care will need to be taken to ensure that in rendering the technical rule more digestible its meaning is not changed. Otherwise, there is a risk that there will be a clash between the two norms, leading to uncertainty as to which is binding.

Rule 2 is an example of where the transposition between the two versions is somewhat incomplete. The technical rule includes the sentence "Humans, not robots, are responsible agents", whereas that for the general audience does not. This raises the question whether that aspect of the rule is intended to be binding for all. Most likely, it was felt by the authors that questions of "responsibility" and "agency", which are complex legal and philosophical terms, were too esoteric for a general audience. However, the authors' failure to even attempt to describe them is problematic. The notion of providing simplified explanations breaks down if the authors omit to mention key parts of their rules in the public version.

4.3 CERNA Ethics of Robot Research

L'Alliance des Sciences et Technologies du Numerique (Allistene) is a major French academic and industry think tank focused on science and technology.[73] Within Allistene, La Commission de Réflexion sur l'Éthique de la Recherche en sciences et technologies du Numérique d'Allistene (CERNA) is a sub-committee dealing with questions of ethics.[74]

---

[73] Its founding members include the Conference of Engineering College and Training Directors, the French Atomic Energy Commission, the French National Centre for Scientific Research, the Conference of University Chairmen, the French National Institute for computer science and applied mathematics and the Institut Télécom. "Foundation of Allistene, the Digital Sciences and Technologies Alliance", *Website of Inria*, https://www.inria.fr/en/news/mediacentre/foundation-of-allistene?mediego_ruuid=4e8613ea-7f23-4d58-adfe-c01885f10420_2, accessed 1 June 2018.

[74] "Cerna", *Website of Allistene*, https://www.allistene.fr/cerna-2/, accessed 1 June 2018.

In 2014, CERNA produced a code of ethics for robotics research.[75] It is available only in French, so what follows is an unofficial English translation. The section on recommendations for researchers concerning entities with autonomy and decision-making capabilities is most relevant for present purposes[76]:

1. **Maintaining control over transfers of decision-making**
   The researcher must consider when the operator or the user should take back control of a process (from a robot) and the roles that the robot can perform (at the expense of the human being), including in what circumstances such transfers of power should be permitted or obligatory. The researcher must also study the possibility for a human to "disengage" the autonomous functions of the robot.

2. **[Decisions] Outside of the knowledge of the operator**
   The researcher must ensure that robot decisions are not made without the knowledge of the operator so as not to create breaks in his understanding of the situation (i.e. so that the operator does not believe that the robot is in a certain state when in fact it is in another state).

3. **Influences [of Robots] on the behaviour of the operator**
   The researcher must be aware of the phenomena of [i] confidence bias, namely the tendency of an operator to rely on robot decisions and [ii] of moral distancing ('Moral Buffer') of the operator in relation to the actions of the robot.

---

[75] "CERNA Éthique de la recherche en robotique": First Report of CERNA, CERNA, http://cerna-ethics-allistene.org/digitalAssets/38/38704_Avis_robotique_livret.pdf, accessed 3 February 2018. The CERNA researchers used a definition of robots which is roughly co-extensive with that adopted in this book.

[76] First, a General S. dealt with matters common to all high-profile emerging technologies, because it was not tailored particularly to AI or robotics this will not be discussed further here. The Cerna principles also cover six recommendations for robots which imitate living entities and engage in emotional and social interactions with humans, as well as medical robots. Both of these topics are too narrow to qualify as general ethical codes and therefore are not discussed further here.

4. **Programme limitations**

The researcher must be careful to evaluate the robot's programmes of perception, interpretation and decision-making, and to clarify the limits of these powers. In particular, programmes that aim to confer moral conduct to the robot ought to be subject to such limitations.

5. **[Robot] Characterisation of a situation**

With respect to robot interpretation software, the researcher must evaluate how far the robot can characterise a situation correctly and discriminate between several apparently similar situations, especially if the decision of action taken by the operator or by the robot itself is based solely on this characterisation. In particular, we must evaluate how uncertainties are taken into account.

6. **Predictability of the human–robot system**

More generally, the researcher must analyse the predictability of the system taken as a whole, taking into account the uncertainties of interpretation and possible failures of the robot and those of the operator, and analyse all the states achievable by this system.

7. **Tracing and explanations**

The researcher must integrate tracing tools as soon as the robot is designed, which should enable the development of at least limited explanations addressed to robotics experts, operators or users.[77]

The CERNA recommendations are directed towards identifying issues, both moral and technical, arising from AI. This limited and modest approach is helpful, in that it seeks to identify potential problems first before charging headlong into an attempt at laying down definitive commands.

4.4   Asilomar 2017 Principles

In 1975, leading DNA researcher Paul Berg convened a conference at Asilomar Beach, California, on the dangers and potential regulation of

---

[77] "CERNA Éthique de la recherche en robotique": First Report of CERNA, *CERNA*, 34–35, http://cerna-ethics-allistene.org/digitalAssets/38/38704_Avis_robotique_livret. pdf, accessed 1 June 2018.

Recombinant DNA technology.[78] Around 140 people participated, including biologists, lawyers and doctors. The participants agreed principles for research, recommendations for the technology's future use, and made declarations concerning prohibited experiments.[79] The Asilomar 1975 Conference later came to be seen as a seminal moment not just in the regulation of DNA technology but also the public engagement with science.[80]

In January 2017, another conference was convened at Asilomar by the Future of Life Institute, a think tank which focusses on "Beneficial AI". Much like the original Asilomar conference, Asilomar 2017 brought together more than 100 AI researchers from academia and industry, as well as specialists in economics, law, ethics and philosophy.[81] The conference participants agreed 23 principles, grouped under three headings[82]:

Research Issues

1. *Research Goal*: The goal of AI research should be to create not undirected intelligence, but beneficial intelligence.

2. *Research Funding*: Investments in AI should be accompanied by funding for research on ensuring its beneficial use, including thorny questions in computer science, economics, law, ethics and social studies, such as:

   - How can we make future AI systems highly robust, so that they do what we want without malfunctioning or getting hacked?

---

[78] The term "Recombinant" refers to the practice of attaching DNA from one organism to DNA of another, with the potential for creating organisms displaying traits from these multiple sources. See Paul Berg, "Asilomar and Recombinant DNA", *Official Website of the Nobel Prize*, https://www.nobelprize.org/nobel_prizes/chemistry/laureates/1980/berg-article.html, accessed 1 June 2018.

[79] Paul Berg, David Baltimore, Sydney Brenner, Richard O. Roblin III, and Maxine F. Singer. "Summary Statement of the Asilomar Conference on Recombinant DNA Molecules", *Proceedings of the National Academy of Sciences* Vol. 72, No. 6 (June 1975), 1981–1984, 1981.

[80] Paul Berg, "Asilomar and Recombinant DNA", *Official Website of the Nobel Prize*, https://www.nobelprize.org/nobel_prizes/chemistry/laureates/1980/berg-article.html, accessed 1 June 2018.

[81] "A principled AI Discussion in Asilomar", *Future of Life Institute*, 17 January 2017, https://futureoflife.org/2017/01/17/principled-ai-discussion-asilomar/, accessed 1 June 2018.

[82] 90% approval from participants was required in order for a principle to be adopted in the final set.

- How can we grow our prosperity through automation whilst maintaining people's resources and purpose?
- How can we update our legal systems to be more fair and efficient, to keep pace with AI and to manage the risks associated with AI?
- What set of values should AI be aligned with, and what legal and ethical status should it have?

3. *Science-Policy Link*: There should be constructive and healthy exchange between AI researchers and policy-makers.
4. *Research Culture*: A culture of cooperation, trust and transparency should be fostered among researchers and developers of AI.
5. *Race Avoidance*: Teams developing AI systems should actively cooperate to avoid corner-cutting on safety standards.

Ethics and Values

6. *Safety*: AI systems should be safe and secure throughout their operational lifetime and verifiably so where applicable and feasible.
7. *Failure Transparency*: If an AI system causes harm, it should be possible to ascertain why.
8. *Judicial Transparency*: Any involvement by an autonomous system in judicial decision-making should provide a satisfactory explanation auditable by a competent human authority.
9. *Responsibility*: Designers and builders of advanced AI systems are stakeholders in the moral implications of their use, misuse and actions, with a responsibility and opportunity to shape those implications.
10. *Value Alignment*: Highly autonomous AI systems should be designed so that their goals and behaviours can be assured to align with human values throughout their operation.
11. *Human Values*: AI systems should be designed and operated so as to be compatible with ideals of human dignity, rights, freedoms and cultural diversity.
12. *Personal Privacy*: People should have the right to access, manage and control the data they generate, given AI systems' power to analyse and utilise that data.
13. *Liberty and Privacy*: The application of AI to personal data must not unreasonably curtail people's real or perceived liberty.

14. *Shared Benefit*: AI technologies should benefit and empower as many people as possible.
15. *Shared Prosperity*: The economic prosperity created by AI should be shared broadly, to benefit all of humanity.
16. *Human Control*: Humans should choose how and whether to delegate decisions to AI systems, to accomplish human-chosen objectives.
17. *Non-subversion*: The power conferred by control of highly advanced AI systems should respect and improve, rather than subvert, the social and civic processes on which the health of society depends.
18. *AI Arms Race*: An arms race in lethal autonomous weapons should be avoided.

Longer-term Issues

19. *Capability Caution*: There being no consensus, we should avoid strong assumptions regarding upper limits on future AI capabilities.
20. *Importance*: Advanced AI could represent a profound change in the history of life on earth and should be planned for and managed with commensurate care and resources.
21. *Risks*: Risks posed by AI systems, especially catastrophic or existential risks, must be subject to planning and mitigation efforts commensurate with their expected impact.
22. *Recursive Self-Improvement*: AI systems designed to recursively self-improve or self-replicate in a manner that could lead to rapidly increasing quality or quantity must be subject to strict safety and control measures.
23. *Common Good*: Superintelligence should only be developed in the service of widely shared ethical ideals and for the benefit of all humanity rather than one state or organisation.[83]

The authors of the Asilomar Principles would probably admit that they need much further detail and specification if they were to form the basis

---

[83] "Asilomar AI Principles", *Future of Life Institute*, https://futureoflife.org/ai-principles/, accessed 1 June 2018.

for any eventual laws. However, the shortcomings in Asilomar was not so much in the content of its proposals, but in the process. The participants were hand-picked from a fairly small group of AI intelligentsia. Moreover, they were predominantly Western-based. Jeffrey Ding notes that "...out of more than 150 attendees, only one was working at a Chinese institution at the time (Andrew Ng, who has now left his role at Baidu)".[84] Another participant expressed surprise at being one of the small minority of non-native English speakers invited.[85]

The patrician approach to AI regulation can certainly generate potentially beneficial output, but risks public rejection if other means of garnering legitimacy are not used alongside. Sociologist Jack Stilgoe and technology ethicist Andrew Maynard have written:

> The new Asilomar principles are a starting point. But they don't dig into what is really at stake. And they lack the sophistication and inclusivity that are critical to responsive and responsible innovation. To be fair, the principles' authors realize this, presenting them as 'aspirational goals'. But within the broader context of a global society that is faced with living with the benefits and the perils of AI, they should be treated as hypotheses – the start of a conversation around responsible innovation rather than the end. They now need to be democratically tested.[86]

4.5   IEEE Ethically Aligned Design

The IEEE Ethically Aligned Design: A Vision for Prioritizing Human Well-being with Autonomous and Intelligent Systems, Version 2 (EAD v2)[87]

---

[84] Jeffrey Ding, "Deciphering China's AI Dream", *Governance of AI Program, Future of Humanity Institute* (Oxford: Future of Humanity Institute, March 2018), 30, https://www.fhi.ox.ac.uk/wp-content/uploads/Deciphering_Chinas_AI-Dream.pdf, accessed 1 June 2018.

[85] Anonymous comment made in discussion with the author, January 2018. Even fewer participants were non-native English speakers working in countries which were not English-speaking.

[86] Jack Stilgoe and Andrew Maynard, "It's Time for Some Messy, Democratic Discussions About the Future of AI", *The Guardian*, 1 February 2017, https://www.the-guardian.com/science/political-science/2017/feb/01/ai-artificial-intelligence-its-time-for-some-messy-democratic-discussions-about-the-future, accessed 1 June 2018.

[87] EAD v2 follows from an initial version ("EAD v1"), published in December 2016, and reflects feedback on that initial document, http://standards.ieee.org/develop/indconn/ec/ead_v1.pdf, accessed 1 June 2018.

was published in December 2017. Its authors describe EAD v2 as "the most comprehensive, crowd-sourced global treatise regarding the ethics of autonomous and intelligent systems available today".[88] The EAD papers are written by committees comprising several hundred multi-disciplinary participants.[89] EAD v2 was opened for public comment, with responses invited by the end of April 2018. A final version is due in 2019.

EAD v2 contains "General Principles" (in bold) and "Candidate Recommendations" for regulation which include the following:

Human Rights
1. Governance frameworks, including standards and regulatory bodies, should be established to oversee processes assuring that the use of Autonomous and Intelligent Systems (A/IS) does not infringe upon human rights, freedoms, dignity and privacy and of traceability to contribute to the building of public trust in A/IS.
2. A way to translate existing and forthcoming legal obligations into informed policy and technical considerations is needed. Such a method should allow for differing cultural norms as well as legal and regulatory frameworks.
3. For the foreseeable future, A/IS should not be granted rights and privileges equal to human rights: A/IS should always be subordinate to human judgment and control.[90]

Prioritise Well-Being
A/IS should prioritise human well-being as an outcome in all system designs, using the best available, and widely accepted, well-being metrics as their reference point.[91]

---

[88] IEEE, EAD v2 website, https://ethicsinaction.ieee.org/, accessed 1 June 2018.

[89] The IEEE Global Initiative on Ethics of Autonomous and Intelligent Systems "Ethically Aligned Design: A Vision for Prioritizing Human Well-being with Autonomous and Intelligent Systems", Version 2. *IEEE*, 2017, 2, http://standards.ieee.org/develop/indconn/ec/autonomous_systems.html, accessed 1 June 2018.

[90] Ibid., 25–26.

[91] Ibid., 28.

Accountability

1. Legislatures/courts should clarify issues of responsibility, culpability, liability and accountability for A/IS where possible during development and deployment (so that manufacturers and users understand their rights and obligations).

2. Designers and developers of A/IS should remain aware of, and take into account when relevant, the diversity of existing cultural norms among the groups of users of these A/IS.

3. Multi-stakeholder ecosystems should be developed to help create norms (which can mature to best practices and laws) where they do not exist because A/IS-oriented technology and their impacts are too new (including representatives of civil society, law enforcement, insurers, manufacturers, engineers, lawyers, etc.).

4. Systems for registration and record-keeping should be created so that it is always possible to find out who is legally responsible for a particular A/IS....[92]

Transparency

Develop new standards that describe measurable, testable levels of transparency, so that systems can be objectively assessed and levels of compliance determined. For designers, such standards will provide a guide for self-assessing transparency during development and suggest mechanisms for improving transparency. (The mechanisms by which transparency is provided will vary significantly, for instance

1. for users of care or domestic robots, a why-did-you-do-that button which, when pressed, causes the robot to explain the action it just took;

2. for validation or certification agencies, the algorithms underlying the A/IS and how they have been verified; and

3. for accident investigators, secure storage of sensor and internal state data, comparable to a flight data recorder or black box.)[93]

EAD v2 is clearly the result of much thoughtful reflection.[94] However, for the reasons given above, international standard-setting bodies remain

---

[92] Ibid., 29–30.

[93] Ibid., 32–33.

[94] In addition to setting standards for human technology designers, the IEEE Global Initiative aims to embed values into autonomous systems and acknowledges the prior need to "identify the norms of the specific community in which the systems are to be deployed and, in particular, norms relevant to the kinds of tasks that they are designed to perform". Ibid., 11.

ill-equipped to form the sole source of standards for AI. Notably, EAD v2 itself suggests at various points that national governments will need to set appropriate regulations to address issues identified therein.[95]

4.6   Microsoft Principles

In an article for *Slate Magazine* published in June 2016,[96] Microsoft CEO Satya Nadella proposed six Principles and Goals:

A.I. must be designed to assist humanity: As we build more autonomous machines, we need to respect human autonomy. Collaborative robots, or co-bots, should do dangerous work like mining, thus creating a safety net and safeguards for human workers.

A.I. must be transparent: We should be aware of how the technology works and what its rules are. We want not just intelligent machines but intelligible machines. Not artificial intelligence but symbiotic intelligence. The tech will know things about humans, but the humans must know about the machines. People should have an understanding of how the technology sees and analyses the world. Ethics and design go hand in hand.

A.I. must maximise efficiencies without destroying the dignity of people: It should preserve cultural commitments, empowering diversity. We need broader, deeper, and more diverse engagement of populations in the design of these systems. The tech industry should not dictate the values and virtues of this future.

A.I. must be designed for intelligent privacy—sophisticated protections that secure personal and group information in ways that earn trust.

A.I. must have algorithmic accountability so that humans can undo unintended harm. We must design these technologies for the expected and the unexpected.

---

[95] See, for example, ibid., 150.

[96] Satya Nadella, "The Partnership of the Future", *Slate*, 28 June 2016, http://www.slate.com/articles/technology/future_tense/2016/06/microsoft_ceo_satya_nadella_humans_and_a_i_can_work_together_to_solve_society.html, accessed 1 June 2018.

A.I. must guard against bias, ensuring proper, and representative research so that the wrong heuristics cannot be used to discriminate.

Technology writer James Vincent argues that "Nadella's goals are as full of ambiguity as Asimov's own Three Laws. But while loopholes in the latter were there to add intrigue to short stories… the vagueness of Nadella's principles reflect the messy business of building robots and AI that deeply affect peoples' lives".[97]

In a 2018 publication: *The Future Computed: Artificial Intelligence and Its Role in Society*, the Microsoft Corporation set out its official manifesto on the societal issues raised by AI. Without explicitly mentioning Nadella's article, the Corporation echoed its content, declaring that there are: "…six principles that we believe should guide the development of AI. Specifically, AI systems should be fair, reliable and safe, private and secure, inclusive, transparent, and accountable".[98]

Interestingly, in the official Microsoft Corporation list, two of Nadella's most far-reaching and altruistic principles, "A.I. must be designed to assist humanity", and "A.I. must maximise efficiencies without destroying the dignity of people", were replaced with more limited, technical aims: that AI be "fair", "inclusive" and "reliable and safe". One is led to wonder to what extent the concerns of Corporation's shareholders might have had an impact on this small but significant shift.

4.7   EU Initiatives

As one of the three legislative bodies of the European Union (alongside the Council and the Commission) and the only one which is directly elected by citizens, the European Parliament plays an important role in scrutinising and enacting EU laws.[99]

---

[97] James Vincent, "Satya Nadella's Rules for AI Are More Boring (and Relevant) Than Asimov's Three Laws", *The Verge*, 29 June 2016, https://www.theverge.com/2016/6/29/12057516/satya-nadella-ai-robot-laws, accessed 1 June 2018.

[98] Microsoft, *The Future Computed: Artificial Intelligence and Its Role in Society* (Redmond, WA: Microsoft Corporation: U.S.A., 2018), 57, https://msblob.blob.core.windows.net/ncmedia/2018/01/The-Future_Computed_1.26.18.pdf, accessed 1 June 2018.

[99] "European Parliament—Overview", *Website of the European Union*, https://europa.eu/european-union/about-eu/institutions-bodies/european-parliament_en, accessed 1 June 2018.

Although the various proposals for AI regulation put forward in the European Parliament's February 2017 resolution are non-binding, crucially it contained the following formula which triggers a lawmaking process[100]:

> [The European Parliament] [r]equests, on the basis of Article 225 TFEU, the Commission to submit, on the basis of Article 114 TFEU, a proposal for a directive on civil law rules on robotics, following the recommendations set out in the Annex hereto[101];

The Annex includes the following proposed rules for creators, a "License for Designers" within a "Code of Ethical Conduct for Research Engineers":

> – You should take into account the European values of dignity, autonomy and self-determination, freedom and justice before, during and after the process of design, development and delivery of such technologies including the need not to harm, injure, deceive or exploit (vulnerable) users.

> – You should introduce trustworthy system design principles across all aspects of a robot's operation, for both hardware and software design, and for any data processing on or off the platform for security purposes.

> – You should introduce privacy by design features so as to ensure that private information is kept secure and only used appropriately.

> – You should integrate obvious opt-out mechanisms (kill switches) that should be consistent with reasonable design objectives.

> – You should ensure that a robot operates in a way that is in accordance with local, national and international ethical and legal principles.

> – You should ensure that the robot's decision-making steps are amenable to reconstruction and traceability.

---

[100] The right of the European Parliament to request that the Commission propose legislation is now found in art. 225 of the Treaty on the Functioning of the European Union (otherwise known as the Lisbon Treaty).

[101] European Parliament Resolution with recommendations to the Commission on Civil Law Rules on Robotics (2015/2103(INL)), art. 65.

– You should ensure that maximal transparency is required in the programming of robotic systems, as well as predictability of robotic behaviour.

– You should analyse the predictability of a human-robot system by considering uncertainty in interpretation and action and possible robotic or human failures.

– You should develop tracing tools at the robot's design stage. These tools will facilitate accounting and explanation of robotic behaviour, even if limited, at the various levels intended for experts, operators and users.

– You should draw up design and evaluation protocols and join with potential users and stakeholders when evaluating the benefits and risks of robotics, including cognitive, psychological and environmental ones.

– You should ensure that robots are identifiable as robots when interacting with humans.

– You should safeguard the safety and health of those interacting and coming in touch with robotics, given that robots as products should be designed using processes which ensure their safety and security. A robotics engineer must preserve human wellbeing while also respecting human rights and may not deploy a robot without safeguarding the safety, efficacy and reversibility of the operation of the system.

– You should obtain a positive opinion from a Research Ethics Committee before testing a robot in a real environment or involving humans in its design and development procedures.[102]

Unusually, the European Parliament has also proposed a separate a license for "users" of robots:

– You are permitted to make use of a robot without risk or fear of physical or psychological harm.

– You should have the right to expect a robot to perform any task for which it has been explicitly designed.

---

[102] Ibid., Annex to the motion for a resolution: detailed recommendations as to the content of the proposal requested.

– You should be aware that any robot may have perceptual, cognitive and actuation limitations.

– You should respect human frailty, both physical and psychological, and the emotional needs of humans.

– You should take the privacy rights of individuals into consideration, including the deactivation of video monitors during intimate procedures.

– You are not permitted to collect, use or disclose personal information without the explicit consent of the data subject.

– You are not permitted to use a robot in any way that contravenes ethical or legal principles and standards.

– You are not permitted to modify any robot to enable it to function as a weapon.[103]

It remains to be seen though whether and to what extent the European Parliament's ambitious proposals will be adopted in legislative proposals by the Commission.

4.8   Japanese Initiatives

A June 2016 Report issued by Japan's Ministry of Internal Affairs and Communications proposed nine principles for developers of AI, which were submitted for international discussion at the G7[104] and OECD:

1) **Principle of collaboration**—Developers should pay attention to the interconnectivity and interoperability of AI systems.

---

[103] Ibid.

[104] G7 refers to the "Group of 7" countries. It consists of Canada, France, Germany, Italy, Japan, the UK and the USA. The EU is also represented at summits. These principles were distributed by Minister Takaichi at the G7 ICT Ministers' Meeting in Takamatsu, Kagawa held on 29–30 April 2016. See: https://www.kagawa-mice.jp/en/g7.html, accessed 1 June 2018; and, for Minister Takaichi's presentation materials, http://www.soumu.go.jp/joho_kokusai/g7ict/english/main_content/ai.pdf, accessed 1 June 2018.

2) Principle of transparency—Developers should pay attention to the verifiability of inputs/outputs of AI systems and the explainability of their judgments.

3) Principle of controllability—Developers should pay attention to the controllability of AI systems.

4) Principle of safety—Developers should take it into consideration that AI systems will not harm the life, body, or property of users or third parties through actuators or other devices.

5) Principle of security—Developers should pay attention to the security of AI systems.

6) Principle of privacy—Developers should take it into consideration that AI systems will not infringe the privacy of users or third parties.

7) Principle of ethics—Developers should respect human dignity and individual autonomy in R&D of AI systems.

8) Principle of user assistance—Developers should take it into consideration that AI systems will support users and make it possible to give them opportunities for choice in appropriate manners.

9) Principle of accountability—Developers should make efforts to fulfill their accountability to stakeholders including users of AI systems.[105]

Japan emphasised that the above principles were intended to be treated as soft law, but with a view to "accelerate the participation of multistakeholders involved in R&D and utilization of AI... at both national and international levels, in the discussions towards establishing 'AI R&D Guidelines' and 'AI Utilization Guidelines'".[106]

Non-governmental groups in Japan have also been active: the Japanese Society for Artificial Intelligence proposed Ethical Guidelines

---

[105]"Towards Promotion of International Discussion on AI Networking", Japan Ministry of Internal Affairs and Communications, http://www.soumu.go.jp/main_content/000499625.pdf (Japanese version), http://www.soumu.go.jp/main_content/000507517.pdf (English version), accessed 1 June 2018.

[106]Ibid.

for an Artificial Intelligence Society in February 2017, aimed at its members.[107] Fumio Shimpo, a member of the Japanese Government's Cabinet Office Advisory Board, has proposed his own Eight Principles of the Laws of Robots.[108]

4.9   Chinese Initiatives

In furtherance of China's Next Generation Artificial Intelligence Development Plan,[109] and as mentioned in Chapter 6, in January 2018 a division of China's Ministry of Industry and Information Technology released a 98-page White Paper on AI Standardization (the White Paper), the contents of which comprise China's most comprehensive analysis to date of the ethical challenges raised by AI.[110]

The White Paper highlights emergent ethical issues in AI including privacy,[111] the Trolley Problem,[112] algorithmic bias,[113] transparency[114]

---

[107] Yutaka Matsuo, Toyoaki Nishida, Koichi Hori, Hideaki Takeda, Satoshi Hase, Makoto Shiono, Hiroshitakashi Hattori, Yusuna Ema, and Katsue Nagakura, "Artificial Intelligence and Ethics", *Artificial Intelligence Journal*, Vol. 31, No. 5 (2016), 635–641; Fumio Shimpo, "The Principal Japanese AI and Robot Strategy and Research toward Establishing Basic Principles", *Journal of Law and Information Systems*, Vol. 3 (May 2018).

[108] Fumio Shimpo, "The Principal Japanese AI and Robot Strategy and Research toward Establishing Basic Principles", *Journal of Law and Information Systems*, Vol. 3 (May 2018).

[109] Available in English translation from the New America Institute: "A Next Generation Artificial Intelligence Development Plan", *China State Council*, Rogier Creemers, Leiden Asia Centre; Graham Webster, Yale Law School Paul Tsai China Center; Paul Triolo, Eurasia Group; and Elsa Kania trans. (Washington, DC: New America, 2017), https://na-production.s3.amazonaws.com/documents/translation-fulltext-8.1.17.pdf, accessed 1 June 2018. See for discussion Chapter 6 at s. 4.6.

[110] National Standardization Management Committee, Second Ministry of Industry, "White Paper on Standardization in AI", translated by Jeffrey Ding, 18 January 2018 (the "White Paper") http://www.sgic.gov.cn/upload/f1ca3511-05f2-43a0-8235-eeb0934d b8c7/20180122/5371516606048992.pdf, accessed 9 April 2018. Contributors to the White Paper included: the China Electronics Standardization Institute, Institute of Automation, Chinese Academy of Sciences, Beijing Institute of Technology, Tsinghua University, Peking University, Renmin University, as well as private companies Huawei, Tencent, Alibaba, Baidu, Intel (China) and Panasonic (formerly Matsushita Electric) (China) Co., Ltd.

[111] Ibid., para. 3.3.3.

[112] Ibid., para. 3.4.

[113] Ibid., para. 3.3.2.

[114] Ibid.

and liability for harm caused by AI.[115] In terms of AI safety, the White Paper explains that:

> Because the achieved goals of artificial intelligence technology are influenced by its initial settings, the goal of artificial intelligence design must be to ensure that the design goals of artificial intelligence are consistent with the interests and ethics of most human beings. So even in facing different environments in the decision-making process, artificial intelligence can make relatively safe decisions.[116]

In the light of these concerns, the White Paper set out the following areas of analysis and further research needed for standardisation in AI:

1) Define the scope of needed artificial intelligence research. Artificial intelligence has turned from laboratory research to practical systems in various fields of application, taking on a fast-paced growth trend. This needs to be defined through a unified terminology, clarifying the core concepts of the connotation, extension and demand of artificial intelligence, and guiding the industry to correctly recognize and understand artificial intelligence technology, making it easier for the widespread use of artificial intelligence technology;

2) Describe the framework of the artificial intelligence system. When faced with the functions and implementation of artificial intelligence systems, users and developers generally regard artificial intelligence systems as a "black box", but it is necessary to enhance the transparency of artificial intelligence systems through technical framework specifications. Due to the wide range of applications of artificial intelligence systems, it may be very difficult to provide a generic artificial intelligence framework. A more realistic approach is to give particular frameworks in particular scopes and problems...;

3) Evaluate the intelligence level of the artificial intelligence system. Differentiating an artificial intelligence system by level of intelligence has always been controversial, and providing a benchmark to measure its intelligence level is a difficult and challenging task...;

---

[115] Ibid.
[116] Ibid., para. 3.3.1.

4) Promote the interoperability of artificial intelligence systems. Artificial intelligence systems and their components have a certain complexity, and different application scenarios involve different systems and components. System-to-system, component-to-component interaction and sharing of information needs to be ensured through interoperability. Artificial intelligence interoperability also involves interoperability between different smart module products to achieve data interoperability, that is, different intelligent products require standardized interfaces...;

5) Conduct assessments of artificial intelligence products. As an industrial product, an artificial intelligence system needs to be evaluated in terms of functions, performance, safety, compatibility, interoperability, etc. in order to ensure the quality and availability of the product and provide safeguards for the industry's sustainable development.... According to standardized procedures and methods, scientific assessment results can be obtained through measurable indicators and quantifiable evaluation systems, and at the same time, coordinate training, promotion, and other means to promote the implementation of standards;

6) Begin standardization of key technologies. For key technologies that have already formed a model and are widely used, they should be standardized in a timely manner to prevent the fragmentation and independence of versions and ensure interoperability and continuity. For example, the user data bound to a deep learning framework should be clearly defined by the neural network's data representation method and compression algorithms, in order to ensure data exchange while not being bound by the platform, and protect the user's rights to the data....;

7) Ensure safety and ethics. Artificial intelligence collects a large amount of personal data, biological data, and data on other characteristics from various devices, applications, and networks. It is not necessarily possible to organize and manage properly and take appropriate privacy protection measures for these data from the very start of system design. Artificial intelligence systems that have a direct impact on human security and human life may pose a threat to humans. Before such artificial intelligence systems are widely used, they must be standardized and evaluated to ensure safety;

8) Standardization of the features of industry application. Apart from common technologies, the implementation of artificial intelligence in specific industries still has individualized needs and technical characteristics....[117]

[117] Ibid., para. 4.5.

The breadth and depth of the Chinese regulatory research agenda are striking. This is all the more so in circumstances where Western commentators sometimes wrongly characterise China's attitude to technology regulation as being purely mercantilist and its policy on privacy protections as non-existent.[118] Notably, the White Paper emphasises also the importance of public trust:

> The issues of safety, ethics and privacy covered in this section are challenges to the development of artificial intelligence. Safety is a prerequisite for sustainable technology. The development of technology poses risks to social trust. How to increase social trust and let the development of technology follow ethical requirements, especially, is an urgent problem to be solved to ensure that privacy will not be violated.[119]

This admission illustrates how one of the world's most secure and powerful governments is nonetheless taking into account the desiderata identified earlier in this chapter concerning the need for legitimacy when regulating for new technology.

As explained in Chapter 6, the proposals in China's standardisation White Paper are part of a coordinated effort by its Government to become a leader in both AI technology and its regulation. China's findings and areas of priority in AI regulations do not differ radically from those suggested elsewhere, but the fact that such proposals are coming from an official, state-sanctioned source is significant.

5   Themes and Trends

In the light of the numerous proposals set out above, it is possible to draw together some broad themes and commonalities. The four most common themes which emerge from this brief survey are the need for some rules as to who is liable if AI causes harm, safety in design of the

---

[118] For instance, Jeffrey Ding notes that there are "common misperceptions of China's relatively lax privacy protections". See Jeffrey Ding, "Deciphering China's AI Dream", *Governance of AI Program, Future of Humanity Institute* (Oxford: Future of Humanity Institute, March 2018), 19, https://www.fhi.ox.ac.uk/wp-content/uploads/Deciphering_Chinas_AI-Dream.pdf, accessed 1 June 2018.

[119] White Paper, para. 3.3.3.

AI, transparency/explainability, and a requirement for AI to operate consistently with established human values.

The overall picture which emerges is one of convergence, suggesting that despite the various bodies' different areas of expertise and focus, the same concerns arise time and again. This is a further reason why a single set of guidelines for the design of AI would be appropriate and achievable.

Rules	Control of "Killer robots"	Safety in Design	Rules for attribution/ liability	Explainability/ Transparency	Benefits shared with all humanity	Act consistently with human rights	Ability to reassert human control	Privacy	Unbiased
EPSRC/AHRC	✓	✓	✓	✓		✓		✓	
CERNA				✓			✓		
Asilomar	✓	✓	✓	✓	✓	✓	✓	✓	
IEEE EAD v2	✓	✓	✓	✓	✓	✓	✓	✓	✓
Satya Nadella/ Microsoft		✓	✓	✓	✓ (Nadella but not Microsoft)	✓		✓	✓
European Parliament Resolution		✓	✓	✓		✓	✓	✓	
Japan Ministry of Communications		✓	✓	✓		✓	✓		
China White Paper		✓	✓	✓		✓	✓	✓	✓

6   LICENSING AND EDUCATION

Once we have arrived at a set of rules, the final question is how they can be implemented and enforced. In addition to the national and international regulatory bodies discussed in the previous chapter, another important aspect is the creation of structures to harmonise and improve the quality of education, training and professional standards for those involved in creating AI.

6.1   Historic Guilds

At least as far back as the late Roman period, skilled artisans and craftsmen formed associations which came to be known as guilds. The guilds imposed controls on the provision of services and the production of various goods, in terms of upholding standards and as cartels, restricting competition in their local area.[120] Stripping away the anti-competitive aspects, which are now largely precluded by antitrust laws, the guilds played an important role in training, quality control and assurance long before these standards were enshrined in national law.[121]

Guilds were not just a set of internal rules: they were a way of life, a self-contained social system with customs, hierarchies and guiding norms. As economists Roberta Dessi and Sheilagh Ogilvie note, "many economists… regard the merchant guild as an exemplar of social capital: these guilds fostered shared norms, transmitted information effectively, punished deviants swiftly, and organized collective action efficiently".[122] Guilds' trade-restricting function may have been curtailed, but their standard-setting role continues today in the form of modern professional associations, sometimes referred to simply as "the professions".

6.2   Modern Professions

Richard and Daniel Susskind suggest that professions today are characterised by the following features:

> (1) they have specialist knowledge; (2) their admission depends on credentials; (3) their activities are regulated; and (4) they are bound by a common set of values.[123]

These four elements are interlocking: the initial (and sometimes ongoing) training inculcates a sense of common professional standards among the cohort of participants. The "common set of values" also provides a

---

[120] "Guild: Trade Association", *Encyclopaedia Britannica*, https://www.britannica.com/topic/guild-trade-association, accessed 1 June 2018.

[121] Avner Greif, Paul Milgrom, and Barry R. Weingast, "Coordination, Commitment, and Enforcement: the Case of the Merchant Guild", *Journal of Political Economy*, Vol. 102 (1994), 745–776.

[122] Roberta Dessi and Sheilagh Ogilvie, "Social Capital and Collusion: The Case of Merchant Guilds" (2004) *CESifo Working Paper No. 1037*. Dessi and Ogilvie do not endorse guilds as an entirely beneficial institution, but they do acknowledge that the social norms which they created.

[123] Richard and Daniel Susskind, *The Future of The Professions* (Oxford: Oxford University Press, 2015).

sense of shared identity for those engaged in the profession. The most well-known example of such a principle is the Hippocratic Oath taken by physicians, first recorded between the third and fifth centuries BC.[124] Though no longer recited in its original form—beginning as that does with an incantation of the names of various Ancient Greek Gods—many of its lessons remain a core part of principles imparted to medical professionals as part of their training: confidentiality, abstaining from corruption and always doing benefit to the sick.[125]

In terms of regulation, detailed rules usually govern day-to-day practice in a modern profession. Finally the disciplinary system acts as the stick of enforcement, providing a signalling factor to other participants, aiming to deter conduct which falls below the specified guideline and again contributing to a sense of shared professional pride in the integrity of the profession. This provides professionals with security in the knowledge that they are only competing for business and collaborating with other individuals who will sustain the same ethical and quality standards. It also benefits the public, who are assured of a certain level of competence, expertise and probity when they deal with a member of a regulated profession.

Some industry regulation is standalone in nature and operates separately to the legal system. In professions considered critical to society or public safety, the internal industry standards are more likely to be backed by the force of law, and practising in the relevant industry without a professional license can be a criminal offence. Professions often covered by such provisions include doctors, airline pilots and lawyers. The following factors justify public interest regulation:

- *Technical complexity:* The most regulated professions are often ones which are particularly impenetrable to the average person. Fields such as the law, medicine or airline piloting are often difficult for a non-specialist to assess. Consequently, the public have little option but to believe the opinion of practitioners, which they are usually unable to second-guess.

---

[124] Ludwig Edelstein, *The Hippocratic Oath: Text, Translation and Interpretation* (Baltimore: Johns Hopkins Press, 1943), 56.

[125] "Hippocratic Oath", *Encyclopaedia Britannica*, https://www.britannica.com/topic/Hippocratic-oath, accessed 1 June 2018, quoting translation from Greek by Francis Adams (1849).

- *Public interaction*: The more engagement that members of the public have with a profession *directly*, the more it is in need of internal regulatory standards. A high level of technical knowledge is more significant relative to the knowledge and training of those who interact most directly with the profession, as well as established regulatory systems for its control. Nuclear physicists have an extremely high degree of technical knowledge, but because they work alongside various other professionals who are able to check and verify their output, imposing professional regulatory standards on those physicists is less pressing. By contrast, a profession where the practitioners deal directly with members of the public without other bodies of professionals acting as checks and balances is much more in need of regulation. Doctors and lawyers are a good example of the latter.
- *Societal importance*: The more fundamental that a given profession is, whether from a commercial or social perspective, the more essential it will be for regulation to be put in place. Thus, musical instrument makers might fulfil the above two criteria, but it would be difficult to describe their role as being of such vital importance that it demands industry regulation. If an instrument maker creates a defective violin, then the violinist (and her audience) might be disappointed, but if a medical professional acts negligently, then the consequences could be fatal.

The development of AI fulfils all of these requirements.

6.3   A Hippocratic Oath for AI Professionals

The increasing importance of AI to society and commerce means that the time is now right for AI engineering to become a regulated profession. In its publication *The Future Computed*, Microsoft Corporation said the following:

> In computer science, will concerns about the impact of AI mean that the study of ethics will become a requirement for computer programmers and researchers? We believe that's a safe bet. Could we see a Hippocratic Oath for coders like we have for doctors? That could make sense. We'll all need to learn together and with a strong commitment to broad societal

responsibility. Ultimately the question is not only what computers can do. It's what computers should do.[126]

Microsoft is not the first tech giant to consider an overarching moral principle ought to be applied to data science. Google's original motto: "Don't be evil" was a modern update to the Hippocratic Oath. Eric Schmidt, one of Google's founders, and co-author Jonathan Rosenberg, have written that this motto:

> ...genuinely expresses a company value and aspiration that is deeply felt by employees. But "Don't be evil" is mainly another way to empower employees... Googlers do regularly check their moral compass when making decisions.[127]

Whether or not the above is true is a matter of some debate,[128] but it is nonetheless significant that one of the major technology giants has consciously limited itself through the adoption of such an overarching principle. Schmidt and Rosenberg described it as "a cultural lodestar that shines over all management layers, product plans and office politics".[129]

Such principles can come back to bite their creators: in April 2018, *The New York Times* reported that various Google developers were protesting against the company's collaboration with the US Department of

---

[126] Microsoft, *The Future Computed: Artificial Intelligence and Its Role in Society* (Redmond, WA: Microsoft Corporation, 2018), 8–9, https://msblob.blob.core.windows.net/ncmedia/2018/01/The-Future_Computed_1.26.18.pdf, accessed 1 June 2018. In March 2018, Oren Etzioni of AI2 responded to Microsoft's book by proposing a draft text for an AI practitioners' Hippocratic Oath. See Oren Etzioni, "A Hippocratic Oath for Artificial Intelligence Practitioners", *TechCrunch*, https://techcrunch.com/2018/03/14/a-hippocratic-oath-for-artificial-intelligence-practitioners/, accessed 1 June 2018.

[127] Eric Schmidt and Jonathan Rosenberg, *How Google Works* (London: Hachette UK, 2014).

[128] Leo Mirani, "What Google Really Means by 'Don't Be Evil'", *Quartz*, 21 October 2014, https://qz.com/284548/what-google-really-means-by-dont-be-evil/, accessed 1 June 2018.

[129] Eric Schmidt and Jonathan Rosenberg, *How Google Works* (London: Hachette UK, 2014).

Defense in using AI technology to scan military drone footage, known by the codename "Project Maven". The developers wrote to CEO Sundar Pichai,[130] citing the company's own motto against it[131]:

> The argument that other firms, like Microsoft and Amazon, are also participating doesn't make this any less risky for Google. Google's unique history, its motto Don't Be Evil, and its direct reach into the lives of billions of users set it apart.[132]

The disgruntled Google employees prevailed. In June 2018, Google announced that it had abandoned Project Maven.[133] Around the same time, Google released a set of ethical principles, which included that it would not design or deploy AI in "[w]eapons or other technologies whose principal purpose or implementation is to cause or directly facilitate injury to people".[134]

A motto, oath or principle is a useful starting point but to achieve the more complex aims of various ethical codes set out above, professional regulation will need to include mechanisms for standard setting, training and enforcement. The final sections of this chapter expand on this idea.

6.4   A Global Professional Body

As with global regulations (discussed in Chapter 6), a single worldwide body regulating AI professionals would encourage the maintenance

---

[130]The text of the letter is available at: https://static01.nyt.com/files/2018/technology/googleletter.pdf, accessed 1 June 2018.

[131]Scott Shane and Daisuke Wakabayashi, "'The Business of War': Google Employees Protest Work for the Pentagon", *The New York Times*, 4 April 2018, https://www.nytimes.com/2018/04/04/technology/google-letter-ceo-pentagon-project.html, accessed 1 June 2018.

[132]Letter from various Google employees to Sundar Pichai, https://static01.nyt.com/files/2018/technology/googleletter.pdf, accessed 1 June 2018.

[133]Hannah Kuchler, "How Workers Forced Google to Drop Its Controversial 'Project Maven'", *Financial Times*, 27 June 2018, https://www.ft.com/content/bd9d57fc-78cf-11e8-bc55-50daf11b720d, accessed 2 July 2018.

[134]Sundar Pichai, "AI at Google: Our Principles", *Google website*, 7 June 2018, https://blog.google/technology/ai/ai-principles/, accessed 2 July 2018.

of standards and avoid creating costly barriers between countries.[135] If professional regulation is undertaken only on a national level, this may lead to significant barriers to the movement of services across borders. For instance, physicians in the USA must obtain a US Medical License in order to practise[136] meaning that foreign doctors with equal qualifications may be frozen out of practising for many years.[137]

An EU law on the recognition of professional qualifications provides recognition for foreign qualifications in certain industries as between EU Member States.[138] However, the system is cumbersome and contains many carve-outs in order to placate local interest groups. Instead of resorting to these byzantine workarounds, it would be far better to begin by having a single standard applicable across all countries.

6.5   AI Auditors

Alternatively or perhaps additionally to the designers and operators being regulated directly, we might create a group of "AI auditors". In the same way as companies and charities in many countries are required to be audited on an annual (or even more frequent) basis by professional financial auditors, organisations using AI might be required to submit their algorithms to professional auditors who could independently assess their compliance with an external set of principles and values. It could be that AI inspectors or auditors will itself become a profession of the future, with its own worldwide standards and disciplinary processes (much like,

---

[135] For a similar proposal, see Joanna J. Bryson, "A Proposal for the Humanoid Agent-Builders League (HAL)", *Proceedings of the AISB 2000 Symposium on Artificial Intelligence, Ethics and (Quasi-)Human Rights*, edited by John Barnden (2000), http://www.cs.bath.ac.uk/~jjb/ftp/HAL00.html, accessed 1 June 2018.

[136] "Homepage", *Website of Federation of State Medical Boards*, http://www.fsmb.org/licensure/spex_plas/, accessed 1 June 2018.

[137] As to the difficulties of foreign doctors, even from those with high quality health systems, practising in the USA, see, for example, "Working in the USA", *Website of the British Medical Association*, https://www.bma.org.uk/advice/career/going-abroad/working-abroad/usa, accessed 1 June 2018.

[138] Directive 2005/36/EC of the European Parliament and Council of 7 September 2005.

for instance, the International Federation of Accountants, which maintains the International Standards on Auditing).

Depending on the dangers involved, AI auditing may not be needed to apply to all instances of AI use, just as some professional regulation applies only where the activity is carried out in a commercial or public setting. For instance, people are entitled to cook food and eat what they cook, including with friends and family. But when a person cooks and sells their food for profit, many governments require independent inspections to be carried out.

6.6   Objections and Responses

6.6.1 "Who Is an AI Professional?"

In order to regulate, we need to know who we are regulating. There are many roles in computer science, including programmers, engineers, analysts, software engineers and data scientists. New ones are constantly being created as the field develops. Further, none of these are terms of art, meaning that an "engineer" in one organisation might be a "programmer" in another. We would adopt the following definition, which focusses on functions rather than labels: "*professional regulation should include all those whose work consistently involves the design, implementation, and manipulation of AI systems and applications*".

The meaning of the term "consistent" could vary depending on circumstances, but engaging in the above tasks at least once a week would likely be sufficient in most cases. Drawing a line between professionals and lay-users may become increasingly difficult as AI systems become easier for non-experts to manipulate. A data manager might design a form to collect data, which is then fed into a primary algorithm (for instance logistic regression) and adapted through a modular system for some application by another employee of the organisation in question. Depending on the consistency of their activities both could be deemed AI professionals, as would be the engineer who designed the AI system in question. The same pattern may well be repeated, particularly in circumstances where AI is trained in situ rather than in a laboratory.

In order to avoid uncertainty, the professional regulatory organisation could publish guidance and maintain a helpline or web-based chatroom for individuals and organisations unsure of whether they are covered. Of course, matters of cost and proportionality come into play when determining who should be regulated, but all things being equal the greater the capacity for harm caused by the use of the AI system, the more cost in terms of training time will be justified.

AI professional regulation need not be an all-or-nothing exercise. It is suggested below that lay-users of AI (which might eventually include the majority of a population) be given a minimum level of basic training.[139] Even within the class of professionals, there might be a system of different classifications for licenses: occasional operators of non-safety-critical AI systems might only be required to hold an entry-level qualification, whereas operators of the most dangerous and complex systems might be required to have far more extensive training. One example of such a gradated system is that operated by the UK FCA, which authorises or approves individuals to carry out certain controlled functions. The authorisations and types of training required vary depending on the activity in question, for instance advising clients or trading derivatives.

6.6.2 *"Professional Regulation Will Not Stop Wrongdoing"*

Just as there are still rogue doctors and lawyers who negligently, recklessly or even deliberately break the rules, making AI engineering a regulated profession will not avoid all misfeasance and malfeasance.

The urge to break the rules can occur at a corporate as well as an individual level. Governments and/or companies may lean on AI professionals to create technology which serves national or corporate interest and in so doing supplants whatever professional regulations which are otherwise imposed on the sector. One notorious example is that of Nazi physicians who, especially under Josef Mengele, carried out horrific experiments on Jewish and other prisoners, notwithstanding their supposed fealty to the Hippocratic Oath.[140]

Nonetheless, there are reasons to be hopeful that professional regulation will have some effect on AI. On an individual level, professional standards offer the chance to inculcate those regulated with a system of norms superior even to political orders. Where a political order would compromise the professional code—especially in the case of overarching norms, then this could be a source of conscientious objection for the

---

[139] See below at s. 7 of this chapter.

[140] See generally *The Nazi Doctors and the Nuremberg Code: Human Rights in Human Experimentation*, edited by George J. Annas and Michael A. Godin (Oxford: Oxford University Press, 1992).

individuals in question. So long as the system of professional norms gives the individual a reason to doubt orders which violate it, then it will have had a positive impact.

Psychiatrist Dr. Anatoly Koryagin campaigned within the Soviet Union and then eventually as an emigré, against the imposition of regulations for doctors which defined mental illness as "disrupting social order or infringing the rules of the socialist community".[141] Before escaping the Soviet Union, Koryagin was imprisoned and tortured, but he refused to yield his professional standards to political exigencies. As the New York Times put it: "Dr. Koryagin's crime was to believe in the Hippocratic Oath".[142]

6.6.3 *"There Are Too Many AI Professionals to Regulate"*

A further related argument against making AI a regulated profession is that there are simply too many AI professionals, even under the description above. The argument runs that it would be impossible in practical terms to secure the training and enforcement of such a large and diverse group stretching across the world.

However, though it may be growing relatively fast the number of AI professionals should not be overstated: a recent study by the Chinese company Tencent estimated that there were just 300,000 AI researchers and practitioners worldwide at the end of 2017, of whom two-thirds were employed and a further one-third studying.[143] Many AI professionals

---

[141] Michael Ryan, *Doctors and the State in the Soviet Union* (New York: Palgrave Macmillan, 1990), 131.

[142] Anthony Lewis, "Abroad at Home; A Question of Confidence", *New York Times*, 19 September 1990, http://www.nytimes.com/1985/09/19/opinion/abroad-at-home-a-question-of-confidence.html, accessed 1 June 2018.

[143] "2017 Global AI Talent White Paper", Tencent Research Institute, http://www.tisi.org/Public/Uploads/file/20171201/20171201151555_24517.pdf, accessed 20 February 2018. See also James Vincent, "Tencent Says There Are Only 300,000 AI Engineers Worldwide, but Millions Are Needed", *The Verge*, 5 December 2017, https://www.theverge.com/2017/12/5/16737224/global-ai-talent-shortfall-tencent-report, accessed 1 June 2018. By contrast, PWC estimate that in the USA alone, there will be 2.9 m people with data science and analytics skills by 2018. Not all will be AI professionals per se, but many of their skills will overlap. "What's Next for the 2017 Data Science and Analytics Job Market?", *PWC Website*, https://www.pwc.com/us/en/library/data-science-and-analytics.html, accessed 1 June 2018.

are clustered around a fairly small number of universities, private sector companies or government programmes (and occasionally overlapping across all three). Consequently, these three groups of institutions operate as bottlenecks through which AI researchers must pass, either in order to acquire their initial training or in order to gain access to the funding and wider resources necessary to progress their research. Provided that professionalism can be incorporated into one or more of these gateways, its coverage of the industry will be considerable.

6.6.4 *"Professional Regulation Would Stifle Creativity"*

Critics might also argue that imposing a code of professional ethics would hold back developments. Necessarily, if an ethical code is to have any effect it will mean that certain practices are controlled or prohibited. The question then is whether this is a worthwhile trade-off.

Constraints are already accepted in other areas of scientific research. Many of the proposals made at the Asilomar 1975 Conference on recombinant DNA research have been adopted as either a matter of law or professional practice.[144] In many countries, certain types of experiments on human and animals are prohibited or at least require special licenses. Arguably, all of these constraints stand in the way of scientific progress, but that is a moral balance which society is willing to strike.

Adopting standards of professional regulation does not require complete homogeneity on training, which would dampen innovation unnecessarily. In order to become publicly accredited, the training courses for AI designers ought to fulfil certain minimum criteria, much as is the case with degrees in professions such as law and medicine already. One example of a minimum criterion for accreditation is compulsory modules on ethics as part of an AI course. In fact, many programming curriculums already cover this as a specific topic.[145]

From the perspective of the wider public, as has been suggested elsewhere in this chapter, making AI a regulated profession will increase

---

[144] Katja Grace, "The Asilomar Conference: A Case Study in Risk Mitigation", *MIRI Research Institute, Technical Report*, 2015–9 (Berkeley, CA: MIRI, 15 July 2015), 15.

[145] A constantly-updated database of tech ethics curricula is available at: https://docs.google.com/spreadsheets/d/1jWIrA8jHz5fYAW4h9CkUD8gKS5V98PDJDymRf8d-9vKI/edit#gid=0, accessed 1 June 2018.

trust by signalling to members of society that practitioners are not simply mercenaries. Such moves are likely to help avoid any public backlash against AI technology, a phenomenon which might end up doing far more harm to innovation than professional standards. As the Microsoft Corporation has concluded:

> … we still need to develop and adopt clear principles to guide the people building, using and applying AI systems…Otherwise people may not fully trust AI systems. And if people don't trust AI systems, they will be less likely to contribute to the development of such systems and to use them.[146]

7   REGULATING THE PUBLIC: A DRIVER'S LICENSE FOR AI

7.1   *Automatic for the People*

Every day, members of the public take control of a powerful machine capable of doing great harm both to its users and to others: the car. In addition to the general civil law (a driver who crashes can be liable for negligence) and some specialised criminal laws (such as a dedicated offence in some countries of causing death by dangerous driving),[147] most countries *also* require drivers to be licensed. Similar licensing regimes are used in various countries to regulate the public's engagement in activities such as flying aeroplanes and owning guns.

The same observations apply to AI. As it becomes more widely used, and utilities such as the AI software library TensorFlow and the machine learning simplification tool AutoML become more available and easier to operate, it is possible that manipulating AI will become as easy and natural as training a dog. Dogs may be trained to fetch and sit still, but they can also be trained to attack and kill. Like owning a gun, driving a car or flying a plane, AI has the potential to be helpful, neutral or harmful. Much of that effect will depend on human input.

---

[146] Microsoft, *The Future Computed: Artificial Intelligence and Its Role in Society* (Redmond, WA: Microsoft Corporation, U.S.A., 2018), 55, https://msblob.blob.core.windows.net/ncmedia/2018/01/The-Future_Computed_1.26.18.pdf, accessed 1 June 2018.

[147] See, for example, s. 1 of the UK Road Traffic Act 1988, or s. 249(1)(a) of the Canadian Criminal Code.

7.2   How Might a Public AI License Function?

There is a threshold question as to whom should the citizen code of AI ethics apply. In short, the answer is that people ought to be required to adhere to certain minimum moral/legal standards whenever they are in a position to exert some causal influence over the choices made by the AI. This situation might range from hobbyist AI engineers undertaking advanced programming, to mere users of products and services containing AI whose interactions with that AI will shape its future behaviour.

Substantive requirements for vehicle driving licenses often include compulsory training courses and assessments—both practical and theory-based. Ongoing periodic assessments might also be required. Within licensing, there could also be a number of categories: a license to drive a car might not qualify a person to drive an 18-wheel truck. The European Parliament's draft rules for users of AI is one example of how such a consumer-focused code might look at a high level.

As with professional AI engineers, there may well be a number of bottlenecks through which members of the public are likely to pass, and which allow an opportunity for AI skills and ethics to be taught. The *first* such bottleneck in most countries is the education system, which in most countries is mandatory at least up to a certain age. As AI grows in importance, ethics and civic values associated with its use and design might be added to compulsory high school courses. *Secondly*, at least for countries which adopt compulsory military or community service as a civil rite of passage for young adults, AI ethics might again be taught at this stage. *Thirdly*, for more advanced amateur programmers/AI engineers there are opportunities to impart ethics values and training via open source programming resources such as TensorFlow.[148]

Though amateurs may soon be able to manipulate and shape ever more complex AI, this does not necessarily mean that programmes created by amateurs will achieve global uptake. Just as we would be more likely to trust medical advice from a registered doctor and legal advice from a qualified attorney, companies and other consumers are more likely to trust AI which has been created by a licensed professional. Even though anyone with the right equipment and a little knowledge

---

[148] "About TensorFlow", *Website of TensorfFlow*, httpvs://www.tensorflow.org/, accessed 1 June 2018.

can ferment and distil their own alcohol, in many countries only licensed producers are permitted to sell it commercially.[149] That might not stop people breaking the law and providing unregulated alcohol for money or for free, but most people would hesitate before sampling "moonshine" spirits from an unlicensed source. In order to avoid a market becoming tainted with unregulated AI programmes, a digital accreditation system similar in effect to the "kitemark" quality assurance logo used by the British Standards Institute might be used to assist users of AI systems in determining whether they are from a reputable or licensed source. Distributed ledger technology could potentially be used to support such quality assurance, by providing an immutable record of a programme's origin and subsequent changes.

In many countries, it is illegal to drive without a license, but it is also illegal to drive without adequate insurance to cover damage which the driver might cause to third parties. The two systems are interlinked: insurers will not be willing to issue insurance to drivers who do not possess a valid license. Those who have had to make claims on their insurance for causing damage to themselves or others are likely to pay higher insurance premiums, adding an economic incentive to drive safely.

A similar model of compulsory insurance may one day be adopted for members of the public who use, design and influence AI—not just for cars but for all types of AI use. There may also be grounds for limiting the extent to which minors are able to use AI outside of parental supervision; again, this is no different from driving, gun ownership and many other potentially dangerous activities.

8   Conclusions on Controlling the Creators

Once developed, social norms are difficult to shift. In Europe, states have monopolised the legitimate use of force for several hundred years[150]— with the corollary that private ownership of weapons has been tightly restricted. As a result, most European states only permit individuals

---

[149] See, for example, the UK Government's "Guidance: Wine Duty", 9 November 2009, https://www.gov.uk/guidance/wine-duty, accessed 1 June 2018.

[150] See, for example, Max Weber, "Politics as a Vocation", in *From Max Weber: Essays in Sociology*, translated by H.H. Gerth and C. Wright Mills (New York: Oxford University Press, 1946).

to own weapons subject to a rigorous licensing process.[151] In the UK, almost all handgun ownership was prohibited following an infamous massacre of schoolchildren in 1996.[152] There was widespread public support for this change, and no serious attempts have been made to challenge it since.[153] By contrast, in the USA the right to bear arms was enshrined in the Second Amendment to the Constitution, adopted in 1791 as part of the Bill of Rights. Large parts of the population consider the ability to purchase and own weapons with minimal constraints as one of their basic Constitutional and cultural rights. In consequence, gun control remains a highly politicised issue and mass-shootings continue.

Most AI systems are certainly not as harmful as guns and the foregoing paragraph is not intended to suggest otherwise. Chapter 6 expressed some concerns as to whether the public might reject AI, but it is certainly possible that matters will go the other way, and people adopt AI to the extent that they become resistant to regulation. The above examples show there is a strong case for imposing restraints at an early stage, prior to the crystallisation of social norms protecting their untrammelled use.

Setting and enforcing ethical constraints for AI design and use are not just problems for one area of society: they challenge all parts. These issues require a multifaceted response, which should involve governments, stakeholders, industry, academics and citizens. All of these different groups should contribute to the grand bargain: a right to participate in designing ethical controls, in exchange for themselves being regulated. Only this way can we create a culture of responsible AI use before dangerous habits develop.

[151] "Firearms-Control Legislation and Policy: European Union", *Library of Congress*, https://www.loc.gov/law/help/firearms-control/eu.php, accessed 1 June 2018.

[152] "1996: Massacre in Dunblane School Gym", *BBC Website*, http://news.bbc.co.uk/onthisday/hi/dates/stories/march/13/newsid_2543000/2543277.stm, accessed 19 February 2018. The UK Firearms (Amendment) Act 1997 and the Firearms (Amendment) (No. 2) Act 1997 banned almost all handguns from private ownership and use.

[153] "We Banned the Guns That Killed School Children in Dunblane. Here's How", *New Statesman*, 16 February 2018, https://www.newstatesman.com/politics/uk/2018/02/we-banned-guns-killed-school-children-dunblane-here-s-how, accessed 1 June 2018.

CHAPTER 8

Controlling the Creations

From generation to generation, we pass down our values. We do so in the hope that our children will hold close to these principles until they are mature enough to develop their own, and in turn teach those to their children. These core norms are sometimes referred to as basic morality.

Now, perhaps for the first time in human history, we are confronted by artificial entities capable of making complex decisions and following advanced rules. What values should we teach them?[1]

In order to answer this question, we need to ask two more: one moral: "how do we choose the norms?"; and one technical: "once we have decided on those norms, how do we impart them into the AI?". Chapters 6 and 7 of this book have suggested that the way to determine which values are relevant from time to time is by building institutions capable of sourcing informed opinion from the wider public, as well as various stakeholders.

This chapter is not intended to be a comprehensive ethical bible for AI,[2] nor is it a manual for creating safe and reliable

---

[1] On the issue of value alignment see, for example, Ariel Conn, "How Do We Align Artificial Intelligence with Human Values?", *Future of Life Institute*, 3 February 2017, https://futureoflife.org/2017/02/03/align-artificial-intelligence-with-human-values/?cn-reloaded=1, accessed 1 June 2018.

[2] For an excellent introductory work on this topic, see Wendell Wallach and Colin Allen, *Moral Machines: Teaching Robots Right from Wrong* (Oxford: Oxford University Press, 2009).

© The Author(s) 2019
J. Turner, *Robot Rules*, https://doi.org/10.1007/978-3-319-96235-1_8

technology.[3] Rather, it aims to suggest the types of rules which could form minimum building blocks for future regulations. That said, the categories suggested below are intended as being indicative rather than a closed list. Returning to the analogy of a pyramid regulatory structure suggested in Section 2.2 of Chapter 6, the various potential "Laws" discussed in this chapter are intended as candidates to sit at the top of a heirarchy of norms for AI, and to be applied to all of its applications: internationally and across different industries.

No doubt the rules and mechanisms for achieving them will shift and grow over time. But just as the collection of social principles which make up human morality must have started somewhere, so too must rules for robots.

1   LAWS OF IDENTIFICATION

1.1   What Are Laws of Identification?

Laws of identification require that an entity must say if it has AI capabilities. Toby Walsh suggests the following rule[4]:

---

[3] Numerous academics and organisations have tackled this issue. See Roman Yampolskiy and Joshua Fox, "Safety Engineering for Artificial General Intelligence" *Topoi*, Vol. 32, No. 2 (2013), 217–226; Stuart Russell, Daniel Dewey, and Max Tegmark, "Research Priorities for Robust and Beneficial Artificial Intelligence", *AI Magazine*, Vol. 36, No. 4 (2015), 105–114; James Babcock, János Kramár, and Roman V. Yampolskiy, "Guidelines for Artificial Intelligence Containment", *arXiv preprint* arXiv:1707.08476 (2017); Dario Amodei, Chris Olah, Jacob Steinhardt, Paul Christiano, John Schulman, and Dan Mané, "Concrete Problems in AI Safety", *arXiv preprint* arXiv:1606.06565 (2016); Jessica Taylor, Eliezer Yudkowsky, Patrick LaVictoire, and Andrew Critch, "Alignment for Advanced Machine Learning Systems", *Machine Intelligence Research Institute* (2016); Smitha Milli, Dylan Hadfield-Menell, Anca Dragan, and Stuart Russell, "Should Robots Be Obedient?", *arXiv preprint* arXiv:1705.09990 (2017); and Iyad Rahwan, "Society-in-the-Loop: Programming the Algorithmic Social Contract", *Ethics and Information Technology*, Vol. 20, No. 1 (2018), 5–14. See also the work of OpenAI, an NGO which focuses on achieving safe artificial general intelligence: "Homepage", *Website of OpenAI*, https://openai.com/, accessed 1 June 2018. The blog of OpenAI and Future of Humanity Institute researcher Paul Christiano also contains many valuable resources and discussions on the topic: https://ai-alignment.com/, accessed 1 June 2018.

[4] See, for example, the UK Locomotive Act 1865, s.3.

An autonomous system should be designed so that it is unlikely to be mistaken for anything besides an autonomous system, and should identify itself at the start of any interaction with another agent.[5]

In Walsh's view, this law might be similar to requirements that toy guns be identified by having a brightly coloured cap at the end in order to make clear that they are not real weapons.[6]

Oren Etzioni, the Chief Executive Officer of AI2, an AI research insitute,[7] has proposed a slightly different formulation: "…an A.I. system must clearly disclose that it is not human".[8] Etzioni's rule is expressed in the negative: the system must say that it is *not* human, but it need not say that it *is* AI.[9] The problem with Etzioni's version is that not all AI resembles or even imitates humans—most does not. It is more helpful for an entity to say what it is, rather than what it isn't. Of the two, Walsh's law of identification is to be preferred.

1.2   Why Might We Need Laws of Identification?

Laws of identification are useful for several reasons:

First, they play an instrumental role in enabling or assisting the function of all other rules unique to AI. It will be more difficult, time-consuming and costly to implement other laws applicable to AI if we cannot distinguish which entities are subject to them.

---

[5] Toby Walsh, *Android Dreams* (London: Hurst & Company, 2017), 111. Walsh notes at 112 the above is "not the law itself… but a summary of its intent", and that an actual law will "require a precise definition of autonomous system". See also Toby Walsh, "Turing's Red Flag", *Communications of the ACM*, Vol. 59, No. 7 (July 2016), 34–37. Walsh terms it the "Turing Red Flag Law", named after UK regulations from the ninetieth century which required that a person walk in front of an automobile waving a flag, so as to warn other road users of the new technology. See further below at s. 4.1.

[6] Ibid.

[7] "Homepage", *Website of AI2*, http://allenai.org/, accessed 1 June 2018.

[8] Oren Etzioni, "How to Regulate Artificial Intelligence", 1 September 2017, *New York Times*, https://www.nytimes.com/2017/09/01/opinion/artificial-intelligence-regulations-rules.html, accessed 1 June 2018.

[9] For a similar formulation to Walsh see Tim Wu, "Please Prove You're Not a Robot", *New York Times*, 15 July 2017, https://www.nytimes.com/2017/07/15/opinion/sunday/please-prove-youre-not-a-robot.html, accessed 1 June 2018.

Secondly, given that AI acts differently to humans in certain conditions, having some idea of whether an entity is human or AI will make its behaviour more predictable to others, increasing both efficiency and safety.[10] If a person runs out in front of a vehicle travelling at 70 miles per hour, the average human driver may not be able to react quickly enough to take evasive action, whereas an AI system might well be able to.[11] In other situations—particularly those requiring "common sense"—AI (at least at the moment) is likely to be significantly inferior to a human.[12] AI cars might be adept on a motorway, but judging complex or unusual elements such as unexpected roadworks or a protest march on the streets may present greater difficulty. Just as we might speak differently to a young child, we might wish to instruct an AI in a different manner to humans, both for our protection and that of the AI. We can only do this if we know what we are talking to.

Thirdly, AI identification may be necessary in order to administer particular activities fairly. A human poker player would want to know that she is playing against another human when she puts down a $5000 stake—as opposed to a potentially unbeatable AI system.[13]

Fourthly, identification can allow people to know the source of communications. A 2018 report on the malicious use of AI highlighted as a major concern: "[t]he use of AI to automate tasks involved in...

---

[10] Toby Walsh, *Android Dreams* (London: Hurst & Company, 2017), 113–114.

[11] Though a 2018 accident in Arizona, where a woman was killed after walking in front of a self-driving vehicle travelling at 40 miles per hour, suggests that—at least at the time of writing—autonomous vehicles remain imperfect in this regard. See, for the issue and a potential solution: Dave Gershgorn, "An AI-Powered Design Trick Could Help Prevent Accidents like Uber's Self-Driving Car Crash", *Quartz*, 30 March 2018, https://qz.com/1241119/accidents-like-ubers-self-driving-car-crash-could-be-prevented-with-this-ai-powered-design-trick/, accessed 1 June 2018.

[12] For an example of a system which is designed to test whether AI has "common sense", see the discussion of the AI2 Reasoning Challenge in Will Knight, "AI Assistants Say Dumb Things, and We're About to Find Out Why", *MIT Technology Review*, 14 March 2018, https://www.technologyreview.com/s/610521/ai-assistants-dont-have-the-common-sense-to-avoid-talking-gibberish/, accessed 1 June 2018. See also the "AI2 Reasoning Challenge Leaderboard", *AI2 Website*, http://data.allenai.org/arc/, accessed 1 June 2018.

[13] Walsh also makes this point: Toby Walsh, *Android Dreams* (London: Hurst & Company, 2017), 116. As to the proficiency of AI poker players, see Byron Spice, "Carnegie Mellon Artificial Intelligence Beats Top Poker Pros", *Carnegie Mellon University Website*, https://www.cmu.edu/news/stories/archives/2017/january/AI-beats-poker-pros.html, accessed 1 June 2018.

persuasion (e.g. creating targeted propaganda), and deception (e.g. manipulating videos) may expand threats associated with privacy invasion and social manipulation".[14]

The anonymity of social media can allow a small number of individuals to project a far greater influence than if they were acting in person, especially if they control a network of bots spreading their content and/or interacting with human users. Whilst an AI identification law will not outlaw malicious use, it may make the exploitation of social media more difficult by minimising the opportunities for nefarious actors.

1.3   How Could Laws of Identification Be Achieved?

Due to their inherent dangers, some products and services can only be offered lawfully if appropriate warnings are provided. Users of heavy machinery are typically warned not to operate it when under the influence of alcohol or other drugs. It is common to see foods labelled "Warning, may contain nuts".[15] Products might one day be required to display the sign: "Warning, may contain AI!".[16]

Given the multiplicity of AI systems and types, there is unlikely to be a single technological solution to implementing identification laws. Therefore, a law of identification for AI should be crafted in general terms, leaving it to individual designers to implement. Prompted by a submission from Toby Walsh,[17] the New South Wales Parliament's Committee on Driverless Vehicles and Road Safety has proposed: "[t]he public identification of automated vehicles to make them visually distinctive to other road users, particularly during the trial and testing phase".[18]

---

[14] Brundage et al., *The Malicious Use of Artificial Intelligence: Forecasting, Prevention, and Mitigation*, February 2018, https://img1.wsimg.com/blobby/go/3d82daa4-97fe-4096-9c6b-376b92c619de/downloads/1c6q2kc4v_50335.pdf, accessed 1 June 2018.

[15] In the USA, there is a specific head of product liability law called "Failure to Warn". See further Chapter 3 at s. 2.2.

[16] José Hernández-Orallo, "AI: Technology Without Measure", Presentation to Judge Business School, Cambridge University, 26 January 2018.

[17] Toby Walsh, *The Future of AI Website*, http://thefutureofai.blogspot.co.uk/2016/09/staysafe-committee-driverless-vehicles.html, accessed 1 June 2018.

[18] "Driverless Vehicles and Road Safety in New South Wales", 22 September 2016, *Staysafe (Joint Standing Committee on Road Safety)*, 2, https://www.parliament.nsw.gov.au/committees/DBAssets/InquiryReport/ReportAcrobat/6075/Report%20-%20Driverless%20Vehicles%20and%20Road%20Safety%20in%20NSW.pdf, accessed 1 June 2018.

Periodic inspections and tests might be used to address whether or not an entity is AI. This may sound like the plot of the popular sci-fi film *Blade Runner*[19] in which the protagonist Deckard is tasked by the police with hunting down "replicants": bioengineered androids. However, testing and inspection regimes for safety, contraband or customs and excise purposes are common features in the transport and supply of many goods and services. Similar investigative measures as are currently used by law enforcement agencies to track malicious software and hacking (and no doubt new ones to be developed) might be utilised to monitor the proper labelling of AI.

An identification law would not be particularly useful if non-AI entities were able to masquerade as AI. False positives would reduce trust in a system of identification, undermining its utility as a signalling mechanism. For this reason, any law of identification should cut both ways and prohibit non-AI entities from being labelled as containing AI, in the same way that a food producer may face penalties already if it describes an item as "suitable for vegetarians" when it contains animal products.[20]

2 LAWS OF EXPLANATION

2.1 *What Are Laws of Explanation?*

Laws of explanation require that AI's reasoning be made clear to humans. This could include a requirement that information is provided on the general decision-making process of the AI (transparency) and/or that specific decisions are rationalised after they have occurred (an individualised explanation).

---

[19]Adapted from Philip K. Dick, *Do Androids Dream of Electric Sheep?* (New York: Doubleday, 1968).

[20]See, for example, Directive 2005/29/EC of the European Parliament and of the Council of 11 May 2005 concerning unfair business-to-consumer commercial practices in the internal market and amending Council Directive 84/450/EEC, Directives 97/7/EC, 98/27/EC and 2002/65/EC of the European Parliament and of the Council and Regulation (EC) No 2006/2004 of the European Parliament and of the Council ("unfair commercial practices directive"), OJ L 149, 11 June 2005, 22–39).

2.2   Why Might We Need Laws of Explanation?

Two main justifications are usually offered for explainable AI: instrumentalist and intrinsic. Instrumentalism focusses on explainability as a tool to improve the AI and to correct its errors. The intrinsic approach focusses on the rights of any humans affected. Andrew Selbst and Julia Powles explain that "the intrinsic value of explanations tracks a person's need for free will and control".[21]

The US Department of Defense Advanced Research Projects Agency (DARPA)[22] has one of the most advanced and prominent programs in this field: XAI.[23] DARPA gives both instrumental and intrinsic justifications for its project:

> ... the effectiveness of [AI] systems will be limited by the machine's inability to explain its thoughts and actions to human users. Explainable AI will be essential, if users are to understand, trust, and effectively manage this emerging generation of artificially intelligent partners.[24]

2.3   How Could Laws of Explanation Be Achieved?

2.3.1 The Black Box Problem

The main difficulty with implementing laws of explanation is that many AI systems operate as "black boxes": they may be adept at accomplishing tasks but even their own designers may be unable to explain what internal process led to a particular output.[25]

As Bryce Goodman and Seth Flaxman note, many machine learning models are not designed with human interpretability as a key

---

[21] Andrew D. Selbst and Julia Powles, "Meaningful Information and the Right to Explanation", *International Data Privacy Law*, Vol. 7, No. 4 (1 November 2017), 233–242, https://doi.org/10.1093/idpl/ipx022, accessed 1 June 2018.

[22] "*DARPA Website*", https://www.darpa.mil/, accessed 1 June 2018.

[23] David Gunning, "Explainable Artificial Intelligence (XAI)", *DARPA Website*, https://www.darpa.mil/program/explainable-artificial-intelligence, accessed 1 June 2018.

[24] David Gunning, DARPA XAI Presentation, *DARPA*, https://www.cc.gatech.edu/~alanwags/DLAI2016/(Gunning)%20IJCAI-16%20DLAI%20WS.pdf, accessed 1 June 2018.

[25] Will Knight, "The Dark Secret at the Heart of AI", *MIT Technology Review*, 11 April 2017, https://www.technologyreview.com/s/604087/the-dark-secret-at-the-heart-of-ai/, accessed 1 June 2018.

concern and it is questionable whether their full range of effects could be achieved if transparency was to be baked into the process:

> There is of course a tradeoff between the representational capacity of a model and its interpretability, ranging from linear models (which can only represent simple relationships but are easy to interpret) to nonparametric methods like support vector machines and Gaussian processes (which can represent a rich class of functions but are hard to interpret). Ensemble methods like random forests pose a particular challenge, as predictions result from an aggregation or averaging procedure. Neural networks, especially with the rise of deep learning, pose perhaps the biggest challenge—what hope is there of explaining the weights learned in a multilayer neural net with a complex architecture?[26]

Similarly, Jenna Burrell of the UC Berkeley School of Information has written that in machine learning there is an "an opacity that stems from the mismatch between mathematical optimization in high-dimensionality characteristic of machine learning and the demands of humanscale reasoning and styles of semantic interpretation".[27] The difficulty is compounded where machine learning systems update themselves as they operate, through a process of backpropagation and re-weighting their internal nodes so as to arrive at better results each time. As a result, the thought process which led to one result may not be the same as used subsequently.

2.3.2 Semantic Association

One explanation technique to provide a narrative for individualised decisions is to teach an AI system semantic associations with its decision-making process. AI can be taught to perform a primary task—such as identifying whether a video is displaying a wedding scene—as well as a secondary task of associating events in the video with certain words.[28] Upol Ehsan, Brent Harrison, Larry Chan and Mark Riedl have

---

[26] Bryce Goodman and Seth Flaxman, "European Union Regulations on Algorithmic Decision-Making and a 'Right to Explanation'," arXiv:1606.08813v3 [stat.ML], 31 August 2016, https://arxiv.org/pdf/1606.08813.pdf, accessed 1 June 2018.

[27] Jenna Burrell, "How the Machine 'Thinks': Understanding Opacity in Machine Learning Algorithms", *Big Data & Society* (January–June 2016), 1–12 (2).

[28] Hui Cheng et al. "Multimedia Event Detection and Recounting", *SRI-Sarnoff AURORA at TRECVID 2014* (2014) http://www-nlpir.nist.gov/projects/tvpubs/tv14. papers/sri_aurora.pdf, accessed 1 June 2018.

developed a technique which they describe as "AI rationalization, an approach for generating explanations of autonomous system behavior as if a human had performed the behavior".[29] The system asks humans to explain their actions as they undertake a particular activity. The associations between techniques adopted by the AI player and the natural language explanation are recorded so as to create a set of labelled actions. A human player in a platform game might say: "The door was locked, so I searched the room for a key". An AI system learns to play the game independently of human training, but when its actions are matched to the human descriptions, a narrative can be generated by sewing together these descriptors.

Taking a different route to Riedl et al., data scientist Daniel Whitenack identifies three general capabilities required for transparency in AI: data provenance (knowing the source of all data); reproducibility (the ability to recreate a given result); and data versioning (saving snapshot copies of the AI in particular states with a view to recording which input led to which output). Whitenack suggests that in order to make these three desiderata "standards within data science, we need proper tools to integrate these characteristics into workflows". He says that ideally, AI transparency tools will be:

Language agnostic—The language wars in data science between python, R, scala, and others will continue on forever. We will always need a mix of languages and frameworks to enable advancements in a field as broad as data science. However, if tools enabling data versioning/provenance are language specific, they are unlikely to be integrated as standard practice.

Infrastructure Agnostic—The tools should be able to be deployed on your existing infrastructure—locally, in the cloud, or on-prem.

Scalable/distributed—It would be impractical to implement changes to a workflow if they were not able to scale up to production requirements.

---

[29] Upol Ehsan, Brent Harrison, Larry Chan, and Mark Riedl, "Rationalization: A Neural Machine Translation Approach to Generating Natural Language Explanations", arXiv:1702.07826v2 [cs.AI], 19 Dec 2, https://arxiv.org/pdf/1702.07826.pdf, accessed 1 June 2018.

Non-invasive—The tools powering data versioning/provenance should be able to integrate effortlessly with existing data science applications, without a complete overhaul of the toolchain and data science workflows.[30]

2.3.3 Case Study: Explanation of Automated Decision-Making Under the GDPR

The EU's flagship data protection legislation, the General Data Protection Regulation 2016 (GDPR),[31] contains a set of provisions which, read together, arguably amount to a legal right to explanation of certain decisions made by AI.[32]

Breaching an article of the GDPR can have serious economic consequences: a fine of up to 4% of a company's annual global turnover or €20 m, whichever is higher.[33] The legislation has a wide territorial scope, applying not only to organisations located within the EU but it will also apply to organisations located outside of the EU which process data in order to offer goods or services to, or monitor the behaviour of, EU residents.[34]

The GDPR provides at Article 13(2)(f)[35]:

> ...the controller [of personal data] shall, at the time when personal data are obtained, provide the data subject with the following further information necessary to ensure fair and transparent processing: [...] the existence of automated decision-making, including profiling... and, at least in those cases, <u>meaningful information about the logic involved</u>, as well as

---

[30] Daniel Whitenack, "Hold Your Machine Learning and AI Models Accountable", *Medium*, 23 November 2017, https://medium.com/pachyderm-data/hold-your-machine-learning-and-ai-models-accountable-de887177174c, accessed 1 June 2018.

[31] Regulation (EU) 2016/679 on the protection of natural persons with regard to the processing of personal data and on the free movement of such data, and repealing Directive 95/46/EC (General Data Protection Regulation) [2016], OJ L119/1 (GDPR).

[32] See, for example, "Overview of the General Data Protection Regulation (GDPR)" (Information Commissioner's Office 2016), 1.1, https://ico.org.uk/for-organisations/data-protection-reform/overview-of-the-gdpr/individuals-rights/rights-related-to-automated-decision-making-and-profiling/, accessed 1 June 2018; House of Commons Science and Technology Committee, 'Robotics and Artificial Intelligence' (House of Commons 2016) HC 145, http://www.publications.parliament.uk/pa/cm201617/cmselect/cmsctech/145/145.pdf, accessed 1 June 2018.

[33] GDPR, art. 83.

[34] Ibid., art. 3.

[35] Equivalent wording is found in art. 14(2)(g) and art. 15(1)(h).

the significance and the envisaged consequences of such processing for the data subject. (emphasis added)[36]

A major problem with the apparent right to explanation under the GDPR is that there is great uncertainty as to what the words of the regulation actually require.[37] Several key terms are not defined. Nowhere does the GDPR say what "meaningful information" means. It might amount at one end of the scale to a data dump of thousands of lines of impenetrable source code; data providers might be reasonably willing to provide such material, but it would be of very little use to the average person. At the other end of the scale, the word meaningful might entail an individualised description in everyday language with a view to making the relevant process accessible and intelligible to a non-expert.[38]

The term "logic involved" is similarly nebulous. The reference to logic is a strong indication that the framers of the GDPR had in mind non-intelligent expert systems, which follow deterministic "yes/no" logic trees in order to reach a known output, based on a known input. The idea of a right to explanation—or at least "meaningful information about the logic involved"—makes sense with regard to such systems which may be highly complex but are ultimately static in nature. With a logic tree, one can always trace back through each step the reasoning which led to an outcome; the same cannot necessarily be said of a neural network.

[36] "Profiling" is defined at art. 4(4) as "automated processing of personal data consisting of the use of personal data to evaluate certain personal aspects relating to a natural person, in particular to analyse or predict aspects concerning that natural person's performance at work, economic situation, health, personal preferences, interests, reliability, behaviour, location or movements". The profiling referred to at art. 22 refers to automated decision-making about a person which "which produces legal effects concerning him or her or similarly significantly affects him or her".

[37] EU legislation is published in multiple languages, each of which is equally valid. Some light might perhaps be cast on the term "meaningful information" by the other versions of the GDPR. The German text of the GDPR uses the word "aussagekräftige", the French text refers to "informations utiles", and the Dutch version uses "nuttige informative". Although Selbst and Powells contend that "These formulations variously invoke notions of utility, reliability, and understandability", the overall effect of this provision under any version remains obscure. Andrew D. Selbst and Julia Powles, "Meaningful Information and the Right to Explanation", *International Data Privacy Law*, Vol. 7, No. 4 (1 November 2017), 233–242, https://doi.org/10.1093/idpl/ipx022, accessed 1 June 2018.

[38] Andrew D. Selbst and Julia Powles, "Meaningful Information and the Right to Explanation", *International Data Privacy Law*, Vol. 7, No. 4 (1 November 2017), 233–242, https://doi.org/10.1093/idpl/ipx022, accessed 1 June 2018.

The concept of a right to explanation of the "logic involved in any automatic processing of data concerning [a human] at least in the case of the automated decisions" is not new. In fact, this language was lifted from Article 12(a) of the GDPR's predecessor: the Data Protection Directive of 1995.[39] The Data Protection Directive was created well before the current resurgence in AI. During the long gestation period of the GDPR, the technology has moved on rather more quickly than the wording in the legislation.

Recital 71 is the only part of the GDPR which explicitly mentions a right to "obtain an explanation":

> ... processing [of personal data] should be subject to suitable safeguards, which should include specific information to the data subject and the right to obtain human intervention, to express his or her point of view, to obtain an explanation of the decision reached after such assessment and to challenge the decision. (emphasis added)

Recitals to EU legislation are not formally binding, but they may in some circumstances be used as an aid to interpretation of the laws themselves and for this reason are often subject to extensive negotiation.[40] As such, it is not clear whether the "right to explanation" in the Recital has the effect of expanding the substantive rights set out in the Articles, which appear to be more limited.

Three Oxford-based academics, Wachter, Mittelstadt and Floridi, discount Recital 71, arguing that "the GDPR does not, in its current form, implement a right to explanation [of individual decisions], but rather what we term a limited 'right to be informed'".[41] Wachter et al. note that

---

[39] Directive 95/46/EC of the European Parliament and of the Council of 24 October 1995 on the protection of individuals with regard to the processing of personal data and on the free movement of such data.

[40] See, for example, Tadas Klimas and Jurate Vaiciukaite, "The Law of Recitals in European Community Legislation", *International Law Students Association Journal of International and Comparative Law*, Vol. 15 (2009), 61, 92.

[41] Ibid., 80.

an explicit right to explanation of individual decisions had been included in earlier drafts of the GDPR but was removed during negotiations.[42]

A 2010 Report by the European Commission Directorate General for Justice into the implementation of the Data Protection Directive across Member States found that there was no common meaning of "logic involved"; individual countries were able to choose their own interpretation.[43] Though such regulatory divergence was possible under the Data Protection Directive (which was binding only as to the result to be achieved), regulations such as the GDPR do not provide for discretion as to implementation and may therefore need to have a single meaning across the whole EU.[44] The authors of the 2010 Report noted, with some prescience, the problems that could be caused by language of the type contained in the Directive when applied to what would now be termed AI:

> The reason we went into some detail on this issue is that... the new socio-technical environment described there - that is, in the very near future - "smart" (expert) computer systems will be increasingly used in decision-making by both private- and public-sector agencies, including law enforcement agencies. Reliance on sophisticated computer-generated "profiles" (and in particular dynamically-generated profiles, in which the algorithm itself is amended by the computer as it "learns"), in any of these contexts, in our view undoubtedly fall within the scope of the provision. This provision is therefore one that requires urgent elaboration and clarification...[45]

Unfortunately, this warning was not heeded; the GDPR simply reproduced the problematic terms.

---

[42] Sandra Wachter, Brent Mittelstadt, and Luciano Floridi, "Why a Right to Explanation of Automated Decision-Making Does Not Exist in the General Data Protection Regulation", *International Data Privacy Law*, Vol. 7, No. 2 (1 May 2017), 76–99 (91), https://doi.org/10.1093/idpl/ipx005, accessed 1 June 2018. See also Fred H. Cate, Christopher Kuner, Dan Svantesson, Orla Lynskey, and Christopher Millard, "Machine Learning with Personal Data: Is Data Protection Law Smart Enough to Meet the Challenge?", *International Data Privacy Law*, Vol. 7, No. 1 (2017); Ricardo Blanco-Vega, José Hernández-Orallo, and María José Ramírez-Quintana, "Analysing the Trade-Off Between Comprehensibility and Accuracy in Mimetic Models", in *International Conference on Discovery Science* (Berlin, Heidelberg: Springer, 2004), 338–346.

[43] Douwe Korff, "New Challenges to Data Protection Study-Working Paper No. 2", *European Commission DG Justice, Freedom and Security Report* 86, https://papers.ssrn.com/sol3/papers.cfm?abstract_id=1638949, accessed 1 June 2018.

[44] See the discussion of the difference between Directives and Regulations in Chapter 6 at s. 7.3.

[45] Ibid.

In October 2017, the Article 29 Working Party (an influential EU data protection body formed by national data regulators)[46] issued the following non-binding guidance on what is meant by the requirement that "meaningful information about the logic involved" in the GDPR:

> The growth and complexity of machine-learning can make it challenging to understand how an automated decision-making process or profiling works. The controller should find simple ways to tell the data subject about the rationale behind, or the criteria relied on in reaching the decision without necessarily always attempting a complex explanation of the algorithms used or disclosure of the full algorithm.[47]

The Article 29 Working Party adopted a robust stance towards the obligation, declaring that "[c]omplexity is no excuse for failing to provide information to the data subject".[48]

The GDPR came into force in May 2018. If applied rigorously, the so-called right to explanation might entail the explainable AI movement going from academic and governmental research projects to binding law. The Article 29 Working Party seeks to plot a course between providing data subjects with sufficient information, but not at the same time requiring AI designers to reveal all of their proprietary designs and trade secrets. Whether this can be achieved in practice remains to be seen.

Even if one concludes that a right to explanation of some kind is morally justified, it seems that the indeterminate language in the GDPR is a poor means of achieving this aim. Sooner or later, the interpretation of this provision is likely to come before the Court of Justice of the EU, which has sometimes taken a highly expansive approach to EU legislation, especially where individual rights are concerned.[49] Leaving such

---

[46] "Glossary", *Website of the European Data Protection Supervisor*, https://edps.europa.eu/data-protection/data-protection/glossary/a_en, accessed 1 June 2018.

[47] Art. 29 Working Party, "Guidelines on Automated Individual Decision-Making and Profiling for the Purposes of Regulation" 2016/679, adopted on 3 October 2017, 17/EN WP 251.

[48] Ibid.

[49] See, for example, *Mangold v. Helm* (2005) C-144/04, or, more recently, the development of a "right to be forgotten" by the Court of Justice of the EU in relation to the ability of individuals to demand their removal from web search engine results—despite this not being specifically provided for in the relevant legislation at the time: *Google Spain Google Spain SL, Google Inc. v. Agencia Española de Protección de Datos, Mario Costeja González* (2014) C-131/12.

hostages to fortune is risky, particularly in circumstances when the EU's competitors may capitalise on the advantage arising from overzealous rules in a rival jurisdiction.

2.3.4 The Limits of Explanation

Is it possible to avoid a trade-off between functionality and explainability in AI? In semantic labelling exercises, the AI system's operations are unaffected but the human participants are describing what *they* would do in a given situation, not what the AI is doing. Researchers from the Max Planck Institute and UC Berkeley developed a semantic labelling technique in 2016, but wrote of it: "[i]n this work we focus on both language and visual explanations that justify a decision by having access to the hidden state of the model, but do not necessarily have to align with the system's reasoning process".[50]

Whereas identifying animals in pictures or playing computer games might be readily explainable in human language, certain other tasks at which AI is especially adept are not. Even in the realm of games certain techniques discoverable by AI may not be to humans, with the result that the AI takes an action for which no human explanation has been recorded. In March 2018, scientists announced that an AI system had found a novel way to win the classic Atari computer game Q*bert.[51] One of the major advantages of AI is that it does not think as humans do. Requiring AI to limit itself to operations which humans can understand might tether the AI to human capabilities such that it never fulfils its true potential.

Many human discoveries cannot readily be explained to a layperson. Certain scientific and mathematical theories—for instance in fields such as quantum physics—are impossible to describe fully in natural language without resorting to numbers and symbols in equations. The problem is all the more acute if AI is involved. Where humans lack the processing power to develop particular techniques, we may not have the linguistic tools to describe them. *The Economist* magazine illustrates the conundrum as follows:

---

[50] Dong Huk Park et al., "Attentive Explanations: Justifying Decisions and Pointing to the Evidence", arXiv:1612.04757v1 [cs.CV], 14 December 2016, https://arxiv.org/pdf/1612.04757v1.pdf, accessed 1 June 2018.

[51] "AI finds novel way to beat classic Q*bert Atari video game", *BBC Website*, 1 March 2018, http://www.bbc.co.uk/news/technology-43241936, accessed 1 June 2018.

Such ways of opening the black box of AI [i.e. semantic labelling]... work up to a point. But they can go only as far as a human being can, since they are, in essence, aping human explanations. Because people can understand the intricacies of pictures of birds and arcade video games, and put them into words, so can machines that copy human methods. But the energy supply of a large data centre or the state of someone's health are far harder for a human being to analyse and describe. AI already outperforms people at such tasks, so human explanations are not available to act as models.[52]

2.3.5 Alternatives to Explanation

Rationalisation and transparency are not panaceas. Lilian Edwards and Michael Veale contend that receiving information on an AI's decisions may even be unhelpful:

Transparency may at best be neither a necessary nor sufficient condition for accountability and at worst something that fobs off data subjects with a remedy of little practical use[53]

It should be recalled human thought process can be just as impenetrable as AI. Even the most advanced brain scanning techniques lack the ability to explain human decisions with any precision.[54] It might be thought that even if we cannot see inside the brain, at least humans can explain themselves using natural language. However, modern psychological research suggests that the association of our actions with reasons represents to some extent the creation of a retrospective fictional narrative which may have little connection to underlying motivations.[55] It is for

---

[52] "For Artificial Intelligence to Thrive, It Must Explain Itself", *The Economist*, 15 February 2018.

[53] Lilian Edwards and Michael Veale, "Slave to the Algorithm? Why a 'Right to an Explanation' Is Probably Not the Remedy You Are Looking For" *Duke Law and Technology Review*, Vol. 16, No. 1 (2017), 1–65 (43).

[54] Vijay Panday, "Artificial Intelligence's 'Black Box' Is Nothing to Fear", *New York Times*, 25 January 2018, https://www.nytimes.com/2018/01/25/opinion/artificial-intelligence-black-box.html, accessed 1 June 2018.

[55] See Daniel Kahneman and Jason Riis, "Living, and Thinking About It: Two Perspectives on Life", in *The Science of Well-Being*, Vol. 1 (2005). See also Daniel Kahneman, *Thinking, Fast and Slow* (London: Penguin, 2011).

this reason that humans are susceptible to deliberate cues which act on our subconscious, such as "priming" or "nudging".[56]

New York University's AI Now Institute recommended in its 2017 Report that:

> Core public agencies, such as those responsible for criminal justice, health-care, welfare, and education (e.g. "high stakes" domains) should no longer use "black box" AI and algorithmic systems[57]

This is an overreaction. Humans may have a natural desire to rationalise and understand why a decision occurred, but this tendency should not be promoted at all costs, particularly where so many human decisions are themselves not explainable in any real sense. In the light of the foregoing, it seems that instrumentalist justifications for explainability are more powerful than the intrinsic ones. Explainability is not the same as being predictable, or under control. The focus of explainable AI ought therefore to be aimed on correctability and improvement of function: if the AI acts unexpectedly, then it is helpful to know how to amend, or perhaps replicate this feature.[58] Technology writer David Weinberger sums up this approach as follows:

> By treating the governance of AI as a question of optimizations, we can focus the necessary argument on what truly matters: What is it that we want from a system, and what are we willing to give up to get it?[59]

[56] Indeed, the latter is so powerful that the UK Government created a specialist body—the Behavioural Insights Team (popularly known as the Nudge Unit) designed to influence people's behaviour without them realising. *Website of the Behavioural Insights Team*, http://www.behaviouralinsights.co.uk/, accessed 1 June 2018.

[57] Campolo et al., *AI Now Institute 2017 Report*, https://assets.contentful.com/8wprh-hvnpfc0/1A9c3ZTCZa2KEYM64Wsc2a/8636557c5fb14f2b74b2be64c3ce0c78/_AI_Now_Institute_2017_Report_.pdf, accessed 1 June 2018.

[58] For an example of a functional approach to explainable AI, see Todd Kulesza, Margaret M. Burnett, Weng-Keen Wong and Simone Stumpf, "Principles of Explanatory Debugging to Personalize Interactive Machine Learning", *IUI 2015, Proceedings of the 20th International Conference on Intelligent User Interfaces* (2015), 126–137.

[59] David Weinberger, "Don't Make AI Artificially Stupid in the Name of Transparency", *Wired*, 28 January 2018, https://www.wired.com/story/dont-make-ai-artificially-stupid-in-the-name-of-transparency/, accessed 1 June 2018. See also David Weinberger, "Optimization Over Explanation: Maximizing the Benefits of Machine Learning Without

Elizabeth I of England said of religious tolerance: "I would not open windows into men's souls".[60] Many legal rules work this way, concentrating primarily on actions, not thoughts.[61] Explainability is best seen as a tool for keeping AI's behaviour within certain limits. The following sections address further means for achieving this goal.

3   Laws on Bias

In theory, AI ought to offer complete impartiality, free from human fallibilities and prejudices. Yet in many cases this has not happened. Newspaper stories and academic papers abound with examples of apparent AI bias[62]: from AI-judged beauty contests which name only caucasian winners[63] to law enforcement software which used race to determine whether people were likely to commit crimes in the future,[64] AI seems to share many of the same problems that humans do. Three questions arise: What is AI bias, why does it arise, and what can be done about it?

Sacrificing Its Intelligence", *Berkman Klein Centre*, 28 January 2018, https://medium.com/berkman-klein-center/optimization-over-explanation-41ecb135763d, accessed 1 June 2018. See also, for example, Sandra Wachter, Brent Mittelstadt, and Chris Russell, "Counterfactual Explanations Without Opening the Black Box: Automated Decisions and the GDPR", *Harvard Journal of Law & Technology, Forthcoming*. Available at Sandra Wachter, Brent Mittelstadt, and Chris Russell, "Counterfactual Explanations Without Opening the Black Box: Automated Decisions and the GDPR" (6 October 2017), *Harvard Journal of Law & Technology*, Forthcoming, https://ssrn.com/abstract=3063289 or http://dx.doi.org/10.2139/ssrn.3063289, accessed 1 June 2018.

[60] Entry on Elizabeth I, *The Oxford Dictionary of Quotations* (Oxford: Oxford University Press, 2001), 297.

[61] A person's mental state in terms of knowledge or intent may well be important, but it rarely has legal consequences unless it is accompanied by some form of culpable action or omission: people are not usually penalised for "having bad thoughts".

[62] Ben Dickson, "Why It's So Hard to Create Unbiased Artificial Intelligence", *Tech Crunch*, 7 November 2016, https://techcrunch.com/2016/11/07/why-its-so-hard-to-create-unbiased-artificial-intelligence/, accessed 1 June 2018.

[63] Sam Levin, "A Beauty Contest Was Judged by AI and the Robots Didn't Like Dark Skin", *The Guardian*, https://www.theguardian.com/technology/2016/sep/08/artificial-intelligence-beauty-contest-doesnt-like-black-people, accessed 1 June 2018.

[64] Julia Angwin, Jeff Larson, Surya Mattu, and Lauren Kirchner, "Machine Bias", *ProPublica*, 23 May 2016, https://www.propublica.org/article/machine-bias-risk-assessments-in-criminal-sentencing, accessed 1 June 2018.

3.1   What Is Bias?

Bias is a "suitcase word", containing a variety of different meanings.[65] In order to understand why AI bias forms, it is important to distinguish between several phenomena.

Bias is often associated with decisions which are deemed "unfair" or "unjust" to particular individuals or groups of humans.[66] The problem with importing such moral concepts into a definition of bias is that they too are indeterminate and vague. The notion of a result or process being "unjust" is subjective. Some people consider positive discrimination to be unjust, whereas others consider it to be a just response to societal imbalances. If there is to be a rule addressing AI bias, then it is preferable to use a test which minimises the role of personal opinions.

With this in mind, our definition is as follows: "*Bias will exist where a decision-maker's actions are changed by taking into account an irrelevant consideration or failing to take into account a relevant consideration*".

If an AI system is asked to select from a given sample which cars it thinks will be the fastest, and it does so based on the paint colour of the car, this is likely to be an irrelevant consideration. If the program failed to take into account a feature such as the weight of the car or its engine size, this would be to neglect of a relevant consideration.

Though it is common to think of AI bias as being something which only affects human subjects, the neutral definition of bias given above could relate to decision-making concerning any form of data. There is nothing special about data relating to humans which means that AI is inherently more likely to display inaccurate or slanted results. To better understand and treat AI bias, we need to avoid anthropomorphisation and focus more on data science.

3.2   Why Might We Need Laws Against Bias?

The immediate source of AI bias is often the data fed into a system. Machine learning, currently the dominant form of AI, recognises

---

[65] Marvin Minsky, *The Emotion Machine* (London: Simon & Schuster, 2015), 113.

[66] See, for example, the Entry on Bias in the Cambridge Dictionary: "… the action of supporting or opposing a particular person or thing in an unfair way, because of allowing personal opinions to influence your judgment", *Cambridge Dictionary*, https://dictionary.cambridge.org/dictionary/english/bias, accessed 1 June 2018.

patterns within data and then takes decisions based on such pattern recognition. If the input data is skewed in some way, then the likelihood is that the patterns generated will be similarly flawed. Bias arising from such data can be summed up with the phrase: "you are what you eat".[67]

3.2.1 Poor Selection of Data

Skewed data sets occur when there is in theory enough information available to present a sufficient picture of the relevant environment, but human operators select an unrepresentative sample. This phenomenon is not unique to AI. In the field of statistics, "sampling bias" refers to errors in estimation which result when some members of a data set are more likely to be sampled than others. Sampling bias or skewed data can arise from the manner in which data is collected: landline telephone polls carried out in the daytime sample a disproportionate number of people who are elderly, unemployed or stay-at-home carers, because these groups are more likely to be at home and willing to take calls at the relevant time.

Skewed data sets may arise because data of one type are more readily available, or because those inputting the data sets are not trying hard enough to find diverse sources. Joy Buolamwini and Timnit Gebru of MIT performed an experiment which demonstrated that three leading pieces of picture recognition software[68] were significantly less accurate at identifying dark-skinned females than they were at matching pictures of light-skinned males.[69] Though the input data sets used by the picture recognition software were not made available to the researchers, Buolamwini and Gebru surmised that the disparity arose from training on data sets of light-skinned males (which probably reflected the gender and ethnicity of the programmers).

---

[67] Nora Gherbi, "Artificial Intelligence and the Age of Empathy", *Conscious Magazine*, http://consciousmagazine.co/artificial-intelligence-age-empathy/, accessed 1 June 2018.

[68] The programs tested were those of IBM, Microsoft and Face++. Joy Buolamwini and Timnit Gebru, "Gender Shades: Intersectional Accuracy Disparities in Commercial Gender Classification" (Conference on Fairness, Accountability, and Transparency, February 2018), http://proceedings.mlr.press/v81/buolamwini18a/buolamwini18a.pdf, accessed 1 June 2018.

[69] Ibid.

IBM announced within a month of the experiment's publication that it had reduced its error rate from 34.7 to 3.46% for dark-skinned females by retraining its algorithms.[70] To illustrate the diversity of its new data sets, IBM noted that they included images of people from Finland, Iceland, Rwanda, Senegal, South Africa and Sweden.[71]

3.2.2 Deliberate Bias and Adversarial Examples

Bias in data can be deliberate as well as inadvertent. In one notorious example (referred to in Chapter 3 in Section 5), Microsoft released an AI chatbot called Tay in 2016. It was designed to respond to natural language conversations with members of the public using a call and response mechanism.[72] Within hours of its release, people worked out how to "game" its algorithms so as to cause Tay to respond with racist language, declaring at one point: "Hitler was right". Needless to say, the program was quickly shut down.[73] The issue was that Microsoft did not insert adequate safeguards to correct for instances of foul language or unpleasant ideas being introduced by users.[74]

The general term for inputs which have been engineered deliberately to fool an AI system is "adversarial examples".[75] Rather like computer viruses which attack vulnerabilities in security software, adversarial examples do the same for AI systems. Making AI robust and protecting it against attack is an important design feature. Technological solutions

---

[70] "Mitigating Bias in AI Models", *IBM Website*, https://www.ibm.com/blogs/research/2018/02/mitigating-bias-ai-models/, accessed 1 June 2018. "Computer Programs Recognise White Men Better Than Black Women", *The Economist*, 15 February 2018.

[71] Ibid.

[72] Using the definition above, Tay's behaviour displayed a form of bias inasmuch as the implicit aim of Microsoft was to create a chatbot which could engage in civil conversation, but it was influenced by user inputs which were incompatible with polite discourse.

[73] Sarah Perez, "Microsoft Silences Its New A.I. Bot Tay, after Twitter Users Teach It Racism", *Tech Crunch*, 24 March 2016, https://techcrunch.com/2016/03/24/microsoft-silences-its-new-a-i-bot-tay-after-twitter-users-teach-it-racism/, accessed 1 June 2018.

[74] John West, "Microsoft's Disastrous Tay Experiment Shows the Hidden Dangers of AI", *Quartz*, 2 April 2016, https://qz.com/653084/microsofts-disastrous-tay-experiment-shows-the-hidden-dangers-of-ai/, accessed 1 June 2018.

[75] Christian Szegedy, Wojciech Zaremba, Ilya Sutskever, Joan Bruna, Dumitru Erhan, Ian Goodfellow, and Rob Fergus, 2013. "Intriguing Properties of Neural Networks", *arXiv preprint server*, https://arxiv.org/abs/1312.6199, accessed 1 June 2018.

have been developed, including the "CleverHans" Python library which can be used by programmers to identify and reduce machine learning systems' vulnerability.[76]

3.2.3 Bias in the Entire Data Set

Sometimes data bias will not arise through the *selection* of a particular data set by humans, but rather because the entire universe of data available is flawed. An experiment published in the journal *Science* indicated that human language (as recorded on the Internet) was "biased" in that semantic associations commonly found between words contained within them various value judgments.

The study built on the Implicit Association Test (IAT), which has been used in numerous social psychology studies for humans to identify subconscious thought patterns.[77] The IAT measures response times by human subjects asked to pair word concepts displayed on a computer screen. Response times are far quicker when subjects are asked to pair two concepts they find similar. Words such as "rose" and "daisy" are usually paired with more "pleasant" ideas, whereas words such as "moth" have the opposite effect. Researchers led by Joanna Bryson and Aylin Caliskan of Princeton University performed a similar test on an Internet data set containing 840 billion words.[78]

The study indicated that a set of African American names had more unpleasant associations than a European American set. Indeed, the general result of the test was that cognitive and linguistic biases demonstrated in human subjects (such as the association of men with high-earning jobs) were also demonstrated by the data sets available on the Internet.[79]

Caliskan and Bryson's result is unsurprising: the Internet is a human creation and represents the sum total of various societal influences,

---

[76] "CleverHans", *GitHub*, https://github.com/tensorflow/cleverhans, accessed 1 June 2018.

[77] Aylin Caliskan, Joanna J. Bryson, and Arvind Narayanan, "Semantics Derived Automatically from Language Corpora Contain Human-Like Biases", *Science*, Vol. 356, No. 6334 (2017), 183–186.

[78] "Biased Bots: Human Prejudices Sneak into AI Systems", *Bath University Website*, 13 April 2017, http://www.bath.ac.uk/news/2017/04/13/biased-bots-artificial-intelligence/, accessed 1 June 2018.

[79] Matthew Huston, "Even Artificial Intelligence can Acquire Biases Against Race and Gender", *Science Magazine*, 13 April 2017, http://www.sciencemag.org/news/2017/04/even-artificial-intelligence-can-acquire-biases-against-race-and-gender, accessed 1 June 2018.

including common prejudices. However, the experiment is a cautionary reminder that some biases may be deeply embedded within society and careful selection of data, or even amendment of the AI model, may be needed to correct for this. The Internet is not the only mass data set which might be prone to similar problems of inherent bias. It is possible that Google's TensorFlow as well as Amazon and Microsoft's Gluon libraries of machine learning software might have similar latent defects.

3.2.4 Data Available Is Insufficiently Detailed

Sometimes the entire universe of data available in a machine-readable format is insufficiently detailed to achieve unbiased results.

For example, AI might be asked to determine which candidates are best suited to jobs as labourers on a building site based on data from successful incumbent workers. If the only data made available to the AI are age and gender, then it is most likely that the AI will select younger men for the job. However, the gender or indeed age of the applicants is not strictly relevant at all to their aptitude. Rather, the key skills which building site labourers need are strength and dexterity. This may be correlated with age, and it may be correlated with gender (especially as regards strength). But it is important not to confuse correlation with causation: both of these data points are merely ciphers for the salient ones of strength and dexterity. If the AI was trained using data based on core aptitudes, then it would result in choices which might still favour young men, but at least it would do so in a way which minimises bias.

3.2.5 Bias in the Training of AI

AI training bias applies particularly to reinforcement learning: a type of AI which (as noted in Chapter 2) is trained using a "reward" function when it gets a right answer. Often the reward function is initially input by human programmers. If an AI system designed to navigate a maze is rewarded each time that it manages to do so without getting stuck, then its maze-solving function will learn to optimise its behaviour through reinforcement. Where the choice of when to reward or discourage behaviour is left to human discretion, this can be a source of bias. Just as a dog may be badly trained by its owner to bite children (by rewarding the dog with a treat every time it does so), an AI system might also be trained to arrive at a biased outcome in this manner. In this regard, the AI is simply mirroring its programmer's preferences. The "fault" is not that of the AI. Nonetheless, as shown below, the AI may be designed with safeguards which can flag certain types of recognised bias arising through flawed training.

3.2.6 Case Study: Wisconsin v. Loomis

Wisconsin v. Loomis[80] is one of the few court decisions to date to consider whether using AI to assist in making important decision is consistent with a subject's fundamental rights.

In 2013, the US State of Wisconsin charged Eric Loomis with various crimes in relation to a drive-by shooting. Loomis pleaded guilty to two of the charges. In preparation for sentencing, a Wisconsin Department of Corrections officer produced a report which included findings made by an AI tool called Correctional Offender Management Profiling for Alternative Sanctions (COMPAS). COMPAS assessments estimate the risk of reoffending based on data gathered from an interview with the offender and information from the offender's criminal history.[81]

The Trial Court referred to the COMPAS report and sentenced Mr. Loomis to six years of imprisonment. The producer of COMPAS refused to disclose how the risk scores were determined on the grounds that its methods were "trade secrets" which might allow competitors to copy the technology. Mr. Loomis appealed against the Trial Court's decision. He argued that the unavailability of the reasoning used in COMPAS prevented him from discovering whether his sentence was based on accurate information.[82] In addition, Mr. Loomis complained that COMPAS used reasoning based in part on group data, rather than making an individualised decision based solely on Loomis' unique characteristics and situation.

The Wisconsin Supreme Court rejected Mr. Loomis' appeal. As to the opacity of COMPAS, Justice Bradley (with whom the other judges agreed) said that the use of secret proprietary risk assessment software was permissible so long as appropriate warnings were provided alongside its results—leaving it to judges to decide what weight they should

---

[80] 881 N.W.2d 749 (2016).

[81] *State of Wisconsin, Plaintiff-Respondent, v. Eric L. LOOMIS, Defendant-Appellant, 881 N.W.2d 749 (2016), 2016 WI 68*, https://www.leagle.com/decision/inwico20160713i48, accessed 1 June 2018.

[82] It is well established in US law that "[a] defendant has a constitutionally protected due process right to be sentenced upon accurate information" *Travis*, 347 Wis.2d 142, 17, 832 N.W.2d 491.

be given.[83] Justice Bradley said further that "to the extent that Loomis's risk assessment is based upon his answers to questions and publicly available data about his criminal history, Loomis had the opportunity to verify that the questions and answers listed on the COMPAS report were accurate".[84] Justice Bradley held also that data drawn from groups could legitimately be taken into account as a relevant factor in determining an individual case, as to which a court had wide discretion.[85]

The *Loomis* decision seems problematic on several grounds. Access to the data used does not necessarily tell a subject how the program has weighted that data in coming to an outcome. If the AI system applied a very high weighting to an irrelevant factor such as Mr. Loomis' race, but a very low weighting to a relevant one such as his previous offending record, then he might have had grounds to challenge the decision reached. The fact that both pieces of material might be publicly available would have been no assistance to Loomis.

The Wisconsin Supreme Court also placed significant reliance on the ability of the lower court to decide how much weight to give to COMPAS or a similar program in making its decision. The problem is that very little guidance was given to the lower courts as to how they should exercise such discretion. A case comment on *Loomis* in the *Harvard Law Review* offered the following criticism:

> In failing to specify the vigor of the criticisms of COMPAS, disregarding the lack of information available to judges, and overlooking the external and internal pressures to use such assessments, the court's solution is unlikely to create the desired judicial skepticism....encouraging judicial skepticism of the value of risk assessments alone does little to tell judges how much to discount these assessments.[86]

---

[83] *State of Wisconsin, Plaintiff-Respondent, v. Eric L. LOOMIS, Defendant-Appellant, 881 N.W.2d 749 (2016), 2016 WI 68,* 65–66, https://www.leagle.com/decision/inwico20160713i48, accessed 1 June 2018.

[84] Ibid., 54.

[85] Ibid., 72. In *State of Wisconsin v. Curtis E. Gallion*, the Wisconsin Supreme Court explained that circuit courts "have an enhanced need for more complete information upfront, at the time of sentencing" 270 Wis.2d 535, 34, 678 N.W.2d 197.

[86] "State v. Loomis, Wisconsin Supreme Court Requires Warning Before Use of Algorithmic Risk Assessments in Sentencing", 10 March 2017, 130 *Harvard Law Review* 1530, 1534.

One high-profile study by the NGO ProPublica indicated that COMPAS tends to give a higher risk-rating to certain offenders on the basis of their ethnicity.[87] It may be that such issues have now been corrected by the designers of the technology, but absent further information on its methodology it is hard to be sure.

The same result might not have been reached in other jurisdictions. Despite the legislation's weaknesses highlighted above, a rule along the lines of the EU's right to explanation of an automated decision under the GDPR might have assisted Mr. Loomis in working out whether and if so how to challenge the COMPAS recommendation. When determining how much information must be disclosed to a defendant in a trial where evidence is kept secret for reasons of national security or similar, courts have held that the right to a fair trial under Article 6 of the European Convention on Human Rights might be satisfied by a process known as "gisting", where sufficient disclosure of the details of a case is given to enable a defendant to instruct a special advocate, who is then entitled to see the evidence but cannot tell their client.[88] A similar process might perhaps be used for AI to steer a line between the confidentiality of algorithms and their use in the justice system or other important decisions.[89]

Even in the USA there may be some diversity between different states' attitudes to the use of AI in important decisions. In 2017, a Texas Court ruled in favour of a group of school teachers who had challenged the use of algorithmic review software by the Houston Independent Schools District to terminate their employment for ineffective performance. The teachers contended that the software violated their constitutional protections against unfair deprivation of property[90] because they were not provided with sufficient information to challenge employment terminations based on the algorithm's scores.

[87] Julia Angwin, Jeff Larson, Surya Mattu, and Lauren Kirchner, "Machine Bias", *ProPublica*, 23 May 2016, https://www.propublica.org/article/machine-bias-risk-assessments-in-criminal-sentencing, accessed 1 June 2018.

[88] *A and others v. United Kingdom* [2009] ECHR 301; applied by the UK Supreme Court in *AF* [2009] UKHL 28.

[89] However, it is not clear whether courts in Europe will treat the protection of intellectual property in an algorithm with as much importance as they do national security.

[90] Under the Fourteenth Amendment to the US Constitution.

The US District Court noted that the scores were "generated by complex algorithms, employing 'sophisticated software and many layers of calculations'", and held that in the absence of disclosure of the methodology involved, the program's scores "will remain a mysterious 'black box,' impervious to challenge".[91] The District Court ruled accordingly that procedural unfairness existed because the teachers had "no meaningful way to ensure correct calculation... and as a result are unfairly subject to mistaken deprivation of constitutionally protected property interests in their jobs".[92]

The case was settled before trial: the Houston Independent Schools District reportedly agreed to pay $237,000 in legal fees and to cease using the evaluation system in making personnel decisions.[93] Even though the matter did not proceed to a final determination, effectively the teachers won. The Texas case shows that even though it was one of the first decisions on the topic, *Loomis* will not be the final word.[94]

3.3   How Could Laws Against Bias Be Achieved?

3.3.1   Diversity—Better Data and Solving the "White Guy Problem"
If bias arises from poorly chosen data, the obvious solution is to improve data selection. This does not mean that AI always needs to be fed data which is balanced across all different parameters. If an AI system was to be developed to assess a person's tendency to develop ovarian cancer, it would not be sensible for the data to include male patients. Accordingly, some thought will need to be taken to select the outer boundaries of the

---

[91] *Houston Federation of Teachers Local 2415* et al. *v. Houston Independent School District*, Case 4:14-cv-01189, 17, https://www.gpo.gov/fdsys/pkg/USCOURTS-txsd-4_14-cv-01189/pdf/USCOURTS-txsd-4_14-cv-01189-0.pdf, accessed 1 June 2018.

[92] Ibid., 18.

[93] John D. Harden and Shelby Webb, "Houston ISD Settles with Union Over Controversial Teacher Evaluations", *Chron*, 12 October 2017, https://www.chron.com/news/education/article/Houston-ISD-settles-with-union-over-teacher-12267893.php, accessed 1 June 2018.

[94] Interestingly, *Loomis* was not directly considered by the District Court in the Houston Teachers case despite the fact that the latter reached the opposite conclusion on constitutionality; it's only mention was a passing reference in a footnote, recording that "Courts are beginning to confront similar due process issues about government use of proprietary algorithms in other contexts".

data set being used. This is a question of efficiency as well as effectiveness: if a program has to trawl through all manner of irrelevant data, it will likely be slower and more energy intensive than if only the key data in question were targeted. That said, one of the great advantages of machine learning systems (especially unsupervised learning) is their ability to identify previously unknown patterns. This feature may militate in favour of providing AI with more rather than less data.

Selection of data is an art as well as a science. Much thought goes into the selection of samples used by pollsters when surveying a population.[95] Likewise, we should be similarly careful when feeding data into AI systems so as to ensure that the data used is suitably representative.

In addition to looking at the data selected, we need also to scrutinise the selectors. Because the majority of AI engineers at present are white men, often from Western countries, aged around 20–40, the data which they select to be fed into AI bears the marks of their preferences and prejudices, whether deliberately or otherwise. AI researcher Kate Crawford terms this "AI's White Guy Problem".[96]

An indirect way to secure better data selection is not simply to ask that programmers be "more sensitive" to bias, but to aim that the demographic of programmers be widened to include minorities and women. That way, it is thought likely that issues will be spotted, through encouraging a multiplicity of views.[97] Securing diversity among programmers is not just a question of gender and race, it may also require multiple national origins, religions and other perspectives. It would be wrong to fall into the trap of assuming that only a diverse group of programmers can produce AI which creates unbiased results, or indeed that diverse programmers will always create unbiased AI. Diversity is helpful in minimising bias, but it is not sufficient.

Instead of such hard-edged diversity rules, another solution might be to require that during the design process (and perhaps again at periodic

---

[95] "Sampling Methods for Political Polling", *American Association for Public Opinion Research*, https://www.aapor.org/Education-Resources/Election-Polling-Resources/Sampling-Methods-for-Political-Polling.aspx, accessed 1 June 2018.

[96] Kate Crawford, "Artificial Intelligence's White Guy Problem", *New York Times*, 25 June 2016, https://www.nytimes.com/2016/06/26/opinion/sunday/artificial-intelligences-white-guy-problem.html, accessed 1 June 2018.

[97] See, for instance, Ivana Bartoletti, "Women Must Act Now, or Male-Designed Robots Will Take Over Our Lives", *The Guardian*, 13 March 2018, https://www.theguardian.com/commentisfree/2018/mar/13/women-robots-ai-male-artificial-intelligence-automation, accessed 1 June 2018.

intervals after its release) AI undergoes a review for bias, perhaps by a specialised diversity panel or even an AI audit program specifically designed for this process.[98]

3.3.2 Technical Fixes to AI Bias

Aside from data selection issues, there may be technical methods of imposing certain constraints and values on the choices which AI makes. These will be particularly useful in situations where the bias cannot be corrected through using a less-skewed data set—for instance in situations where the entire universe of data exhibits bias (such as the content of the Internet), or where there is insufficient data available to train AI without it resorting to decisions based on characteristics such as gender or race.

In the human rights law of many countries, certain human characteristics are deemed protected in that decision-makers are prohibited from making a decision on the basis of those factors (a practice sometimes referred to as discrimination). Protected characteristics are generally selected from features which humans are unable to choose. The UK Equalities Act 2010 protects against discrimination on the basis of: age, disability, gender reassignment, marriage or civil partnership, pregnancy and maternity, race, religion or belief and sexual orientation. In the USA, Title VII of the Civil Rights Act of 1964 prohibits discrimination in employment on the basis of race, colour, sex, or national origin, and separate legislation prevents discrimination on the basis of age, disability and pregnancy.[99]

Can AI be prevented from taking such characteristics into account?[100] Recent experiments show that it can, and that data scientists are developing increasingly advanced methodologies to do so. The simplest solution to perceived bias in a machine learning model against subjects (usually

---

[98] See, for example, the proposals in Michael Veale and Reuben Binns, "Fairer Machine Learning in the Real World: Mitigating Discrimination Without Collecting Sensitive Data", *Big Data & Society*, Vol. 4, No. 2 (2017), 2053951717743530.

[99] "Laws Enforced by EEOC", *Website of the U.S. Equal Employment Opportunity Commission*, https://www.eeoc.gov/laws/statutes/, accessed 1 June 2018.

[100] It may be the case that researchers wish to assess an otherwise protected characteristic as part of a scientific experiment or poll. For instance, it would be legitimate for a program to discriminate on grounds of race if it was being used in an experiment to map certain the prevalence of genetic diseases which are commonly found in one particular race. In this situation, the use of a protected characteristic would not meet the definition of bias outlined above, because it *would* be relevant to the task in question.

people) with a certain attribute is to down-weight that attribute, so that the AI is less likely to take it into account in decision-making. However, this is a crude tool which can lead to inaccurate overall results.[101]

A better approach is to use counterfactuals to test whether the same decision would have been reached by the AI system if different variables are isolated and changed. A program might test for racial bias by running a hypothetical model where it changes the race of the subject in order to establish whether the same result would have been reached.[102]

Anti-bias modelling techniques are constantly being tweaked and improved. Silvia Chiappa and Thomas Graham pointed out in a 2018 paper that counterfactual reasoning alone may not always be sufficient to identify and eliminate bias. A purely counterfactual model might identify that gender bias has caused more male applicants to be accepted to a university, but may not identify the fact that women have a lower acceptance rate in part because female students have applied to courses with fewer spaces. Accordingly, Chiappa and Graham propose a modification of counterfactual modelling which "states that a decision is fair toward an individual if it coincides with the one that would have been taken in a counterfactual world in which the sensitive attribute along the unfair pathways were different".[103]

In October 2017, IBM researchers published a paper demonstrating how machine learning algorithms could be trained to understand non-discrimination policies written in natural language, to alert users to policy violations and to create a log of such events.[104] The IBM researchers sidestep the explainability and transparency issues highlighted in the preceding section by making their program "end-to-end", meaning that compliance with fairness policies can be achieved without any need to undertake the "time-consuming and error-prone" process of seeking

---

[101] Silvia Chiappa and Thomas P.S. Gillam, "Path-Specific Counterfactual Fairness", arXiv:1802.08139v1 [stat.ML], 22 Feb 2018.

[102] Matt J. Kusner, Joshua R. Loftus, Chris Russell, and Ricardo Silva, "Counterfactual Fairness", *Advances in Neural Information Processing Systems*, Vol. 30 (2017), 4069–4079.

[103] Silvia Chiappa and Thomas P.S. Gillam, "Path-Specific Counterfactual Fairness", arXiv:1802.08139v1 [stat.ML], 22 February 2018.

[104] Samiulla Shaikh, Harit Vishwakarma, Sameep Mehta, Kush R. Varshney, Karthikeyan Natesan Ramamurthy, and Dennis Wei, "An End-To-End Machine Learning Pipeline That Ensures Fairness Policies", arXiv:1710.06876v1 [cs.CY], 18 October 2017.

to understand opaque (black box) machine learning systems.[105] The researchers explain:

> The system we envision will automatically perform knowledge extraction and reasoning on such a document to identify the sensitive fields (gender in this case), and support testing for and prevention of biased algorithmic decision making against groups defined by those fields.

This way, the diagnostic tool is built into the output of the system rather than requiring specialised knowledge to "open the bonnet" each time a problem is identified. The designers of the IBM system state that their aim is to avoid the issues which arose as a result of the opacity of COMPAS in the *Loomis* case by providing security of knowledge to users that the system will not generate biased results and simultaneously by reassuring designers that they will not be required to lay bare the inner workings of the AI.[106]

Bias has been described by one journalist as "the dark secret at the heart of AI".[107] However, we must be careful of exaggerating both its novelty and the difficulty of treating it. Properly analysed, AI bias arises from a combination of features common to human society and standard scientific errors. Completely value-neutral AI may be a chimera. Indeed, some commentators have argued that "algorithms are inescapably value-laden".[108] Instead of seeking to eliminate all values from AI, the preferable approach may be to design and maintain AI which reflects the values of the given society in which the AI operates.

---

[105] Ibid.

[106] In addition to the papers cited above, see also B. Srivastava and F. Rossi, "Towards Composable Bias Rating of AI Services", *AAAI/ACM Conference on Artificial Intelligence, Ethics, and Society*, New Orleans, LA, February 2018; F.P. Calmon, D. Wei, B. Vinzamuri, K.N. Ramamurty, and K.R. Varshney, "Optimized Pre-Processing for Discrimination Prevention", *Advances in Neural Information Processing Systems*, Long Beach, CA, December 2017; and R. Nabi and I. Shpitser, "Fair inference on Outcomes", *Thirty-Second AAAI Conference on Artificial Intelligence*, 2018.

[107] Will Knight, "The Dark Secret at the Heart of AI", *MIT Technology Review*, 11 April 2017, https://www.technologyreview.com/s/604087/the-dark-secret-at-the-heart-of-ai/, accessed 1 June 2018.

[108] Brent Mittelstadt, Patrick Allo, Mariarosaria Taddeo, Sandra Wachter, and Luciano Floridi, "The Ethics of Algorithms: Mapping the Debate", *Big Data & Society*, Vol. 3, No. 2 (2016), http://journals.sagepub.com/doi/full/10.1177/2053951716679679, accessed 1 June 2018.

4   Limits on Limitation of AI Use

4.1   What are Laws of Limitation?

Laws of limitation are rules which specify what AI systems can and cannot do. Setting the limits of what roles AI should be allowed to fulfil is an emotive topic: many people fear delegating tasks and functions to an unpredictable entity which they cannot fully understand. These issues raise fundamental questions about humanity's relationship with AI: Why do we harbour concerns about giving up control? Can we strike a balance between AI effectiveness and human oversight? Will fools rush in where AIs fear to tread?

4.2   Why Might We Need Laws of Limitation?

In September 2017, Stanislav Petrov died alone and destitute in an unremarkable Moscow suburb. His inauspicious death belied the pivotal role he played one night in 1983 when he was the duty officer in a secret command centre tasked with detecting nuclear attacks on the USSR by America.

Petrov's computer screen showed five intercontinental ballistic missiles heading towards the USSR. The standard protocol was to launch a retaliatory strike before the American missiles landed: thereby triggering the world's first—and potentially last—nuclear conflict. "The siren howled, but I just sat there for a few seconds, staring at the big, back-lit, red screen with the word 'launch' on it", he told the BBC's Russian Service in 2013. "All I had to do was to reach for the phone; to raise the direct line to our top commanders".[109] Yet Petrov paused. His gut instinct told him that this was a false alarm.

Petrov was correct: there were no American missiles. It subsequently transpired that the computer message had resulted from a satellite detecting the reflection of the sun's rays off the tops of clouds, which it confused with a missile launch. "We are wiser than the computers",

---

[109] Marc Bennetts, "Soviet Officer Who Averted Cold War Nuclear Disaster Dies Aged 77", *The Guardian*, 18 September 2017, https://www.theguardian.com/world/2017/sep/18/soviet-officer-who-averted-cold-war-nuclear-disaster-dies-aged-77, accessed 1 June 2018.

Petrov said in a 2010 interview with the German newspaper *Der Spiegel*, "We created them".[110]

Various commentators have followed Petrov's lead and suggested that humans should always be tasked with supervising or second-guessing AI.[111] One option is a requirement that there should always be a "human in the loop", meaning that AI can never take a decision without human ratification. Another is that there should be a "human on the loop", a requirement that a human supervisor must always be on hand with the power to override the AI.

Requiring a human in the loop is reminiscent of the UK's notorious "Red Flag" laws from the nineteenth century. When cars were first invented, legislators were so concerned about their impact on other road users and pedestrians that they insisted someone must always walk in front of a car waving a red flag. This would certainly have made other road users aware of the new technology, but this was at the expense of the car being able to travel at any speed greater than walking pace. The Red Flag laws were for this reason short-lived and today seem ridiculous. By requiring that there is *always* a human in the loop, we risk putting the same fetters on AI. Stipulating that there must be a human "on the loop" may present a less-excessive alternative.[112] It maintains a semblance of human control, whilst still allowing AI to achieve efficiencies of speed and accuracy.

---

[110] Benjamin Bidder, "Forgotten Hero: The Man Who Prevented the Third World War", *Der Spiegel*, 21 April 2010, http://www.spiegel.de/einestages/vergessener-held-a-948852.html, accessed 1 June 2018.

[111] See, for instance, George Dvorsky, "Why Banning Killer AI is Easier Said Than Done", 9 July 2017, *Gizmodo*, https://gizmodo.com/why-banning-killer-ai-is-easier-said-than-done-1800981342, accessed 1 June 2018.

[112] This appears to be the approach taken by the UK military with regard to automated and autonomous weapons: "Current UK policy is that the operation of our weapons will always be under human control as an absolute guarantee of human oversight and authority and of accountability for weapon usage. This information has been put on record a number of times, both in parliament and international forums. Although a limited number of defensive systems can currently operate in automatic mode, there is always a person involved in setting the parameters of any such mode". UK Ministry of Defence, "Joint Doctrine Publication 0-30.2 Unmanned Aircraft Systems", *Development, Concepts and Doctrine Centre*, August 2017, 42, https://assets.publishing.service.gov.uk/government/uploads/system/uploads/attachment_data/file/673940/doctrine_uk_uas_jdp_0_30_2.pdf, accessed 1 June 2018.

Important moral questions arise as to whether we want to sacrifice greater effectiveness for a vague feeling of comfort in knowing that there has been a human decision-maker. Feelings of concern at the replacement of a human service provider with technology tend to dissipate over time; people might once have felt queasy about a human bank teller being replaced by a machine in the delicate task of distributing cash to account holders, but today ATMs are ubiquitous. Few people give a second thought to the fact that the majority of manufactured goods are produced—and even inspected—largely by machines. Ultimately, the choice of whether, and if so, when, to insist on a human supervisor is one which is best taken by societies as a whole, using the processes set out in the previous chapters.

4.3   How Could Laws of Limitation be Achieved?

4.3.1 Case Study: Right Not to Be Subjected to Automated Decision-Making Under the GDPR

The GDPR contains a right not to be subjected to automated individual decision-making under Article 22:

> 1. The data subject shall have the right not to be subject to a decision based solely on automated processing, including profiling, which produces legal effects concerning him or her or similarly significantly affects him or her.
>
> 2. Paragraph 1 shall not apply if the decision: (a) is necessary for entering into, or performance of, a contract between the data subject and a data controller; (b) is authorised by Union or Member State law to which the controller is subject and which also lays down suitable measures to safeguard the data subject's rights and freedoms and legitimate interests; or (c) is based on the data subject's explicit consent.
>
> 3. In the cases referred to in points (a) and (c) of paragraph 2, the data controller shall implement suitable measures to safeguard the data subject's rights and freedoms and legitimate interests, at least the right to obtain human intervention on the part of the controller, to express his or her point of view and to contest the decision....

AI can clearly qualify as "automated processing". The wording of Article 22 itself appears to refer to a voluntary right of individuals affected to object to automated decision-making, which they can decide whether or not to exercise. However, in its draft guidance, the Article 29 Working Party has suggested that Article 22 in fact creates an outright ban on automated individual decision-making, subject to exceptions under Article 22(2), namely: performance of a contract and authorisation under law or explicit consent.[113]

This crucial change flips the regime from being permissive to being prohibitive.[114] The right to object only applies where the decision will produce "legal" or "similarly significant effects", but both terms are undefined in the GDPR. To qualify as "significant", the Article 29 Working Party states "the decision must have the potential to significantly influence the circumstances, behaviour or choices of the individuals concerned".[115] This may be a low bar. The Article 29 Working Party infers that certain advertising and marketing practices may be caught, especially where they are highly targeted. It said further that Article 22 extended to (among other things) credit decisions as varied as:

> ... renting a city bike during a vacation abroad for two hours; purchasing a kitchen appliance or a television set on credit; obtaining a mortgage to buy a first home.[116]

The right to require human intervention in Article 22 could require that there is always a human in the loop. Lawyers Eduardo Ustaran and Victoria Hordern argue that the shift from a right to an objection to an outright prohibition subject to qualifications "generates considerable

---

[113]Art. 29 Data Protection Working Party, "Guidelines on Automated individual decision-making and Profiling for the purposes of Regulation 2016/679", adopted 3 October 2017, 17/EN WP 251, 10.

[114]Eduardo Ustaran and Victoria Hordern, "Automated Decision-Making Under the GDPR—A Right for Individuals or A Prohibition for Controllers?", *Hogan Lovells*, 20 October 2017, https://www.hldataprotection.com/2017/10/articles/international-eu-privacy/automated-decision-making-under-the-gdpr-a-right-for-individuals-or-a-prohibition-for-controllers/, accessed 1 June 2018.

[115]Art. 29 Data Protection Working Party, "Guidelines on Automated Individual Decision-Making and Profiling for the Purposes of Regulation 2016/679", adopted 3 October 2017, 17/EN WP 251, 10.

[116]Ibid., 11.

uncertainty".[117] They conclude "...if the interpretation set out in the [Working Party]'s draft guidance is the one that prevails, it will have significant consequences for all types of businesses which was not necessarily foreseen at the time of the adoption of the GDPR".[118] This development has led various commentators to wonder whether some types of AI will become illegal altogether in the EU.[119] This may be going too far, but Article 22 certainly does create a hostage to fortune.

4.3.2 "Killer Robots" and the Teleological Principle

Of all the uses for AI, its application in autonomous weapons (or killer robots) is probably the most controversial. As the prospect of weapons capable of independently selecting and firing upon targets has come closer to reality, forces have mustered against their use. The international Campaign to Ban Killer Robots was launched in 2013.[120] In August 2017, 116 experts and founders of AI companies wrote an open letter expressing grave concerns on the matter.[121]

Despite the strength of feeling on the topic, there are strong arguments that a total AI ban could be counterproductive—not just on

---

[117] Eduardo Ustaran and Victoria Hordern, "Automated Decision-Making Under the GDPR—A Right for Individuals or A Prohibition for Controllers?", *Hogan Lovells*, 20 October 2017, https://www.hldataprotection.com/2017/10/articles/international-eu-privacy/automated-decision-making-under-the-gdpr-a-right-for-individuals-or-a-prohibition-for-controllers/, accessed 1 June 2018.

[118] Ibid.

[119] See, for example, Richa Bhatia, "Is Deep Learning Going to Be Illegal in Europe?", *Analytics India Magazine*, 30 January 2018, https://analyticsindiamag.com/deep-learning-going-illegal-europe/; Rand Hindi, "Will Artificial Intelligence Be Illegal in Europe Next Year?", *Entrepreneur*, 9 August 2017, https://www.entrepreneur.com/article/298394, both accessed 1 June 2018.

[120] "Media Advisory: Campaign to Ban Killer Robots Launch in London", *art. 36*, 11 April 2013, http://www.article36.org/press-releases/media-advisory-campaign-to-ban-killer-robots-launch-in-london/, accessed 1 June 2018.

[121] Samuel Gibbs, "Elon Musk Leads 116 Experts Calling for Outright Ban of Killer Robots", *The Guardian*, 20 August 2017, https://www.theguardian.com/technology/2017/aug/20/elon-musk-killer-robots-experts-outright-ban-lethal-autonomous-weapons-war, accessed 1 June 2018. See also "2018 Group of Governmental Experts on Lethal Autonomous Weapons Systems (LAWS)", *United Nations Office at Geneva*, https://www.unog.ch/80256EE600585943/(httpPages)/7C335E71DFCB29D-1C1258243003E8724?OpenDocument, accessed 1 June 2018.

autonomous weapons but also in any field. It is submitted that there is a principled solution to the question of when and how AI should be used in controversial areas:

The Teleological Principle for AI Use

Begin by asking what values we are seeking to uphold in the given activity. If (and only if) AI can consistently uphold those values in a manner demonstrably superior to humans, then AI use should be permitted.

Where the Teleological Principle is satisfied, asking a human to ratify the AI's decision will be at best unnecessary and at worst actively harmful. One example of where AI already exceeds humans in an important task is the ability to recognise certain cancers. Whereas doctors may take several minutes to analyse each scan of a body part, AI can do this in milliseconds in some cases with demonstrably higher accuracy than human experts.[122]

The Teleological Principle cannot be employed in the abstract; even where it is satisfied, policy-makers will need to be mindful of overarching societal views on the acceptability of using AI—or indeed any non-human technology—to fulfil the relevant task. The issue of whether people will accept a particular technology is a wider question of social and political legitimacy, the outcome of which could differ between societies. One aspect of encouraging the uptake of a controversial AI technology might be to show to the public that the Teleological Principle is satisfied. As with GM crops (discussed in Chapter 6), a technology's safety and effectiveness will not necessarily guarantee its acceptance. Nonetheless, the Teleological Principle is at least a helpful guide to policy-makers as to when it is appropriate to encourage that AI be used in a given field.

Returning to the example of autonomous weapons, in international humanitarian law (the laws applied during warfare) it is widely accepted that the two guiding principles are *proportionality*—the requirement to cause no harm than is necessary to achieve a legitimate aim, and *distinction*—to differentiate between combatants and civilians.[123]

---

[122] Ian Steadman, "IBM's Watson Is Better at Diagnosing Cancer Than Human Doctors", *Wired*, 11 February 2013, http://www.wired.co.uk/article/ibm-watson-medical-doctor, accessed 1 June 2018.

[123] International Committee of the Red Cross, *What Is International Humanitarian Law?* (Geneva: ICRC, July 2004), https://www.icrc.org/eng/assets/files/other/what_is_ihl.pdf, accessed 1 June 2018.

Proponents of banning autonomous weapons often point out that many countries already accept some weapons should be prohibited or heavily constrained.[124] Popular examples include blinding lasers[125] and the use of landmines.[126] However, a major difference between autonomous weapons and those technologies which have to date been banned is that the prohibited technologies generally make compliance with the basic laws of warfare more difficult. Once deployed a landmine will explode regardless of who steps on it—whether civilian or combatant. Noxious gas does not discriminate as to who it poisons. Blinding lasers cannot be told to spare the eyes of civilians. Another reason why these technologies are banned is that they tend to cause more human suffering than is absolutely necessary to achieve a given aim: maiming victims or causing a slow and painful death.

By contrast, if its development is properly regulated, AI may well exceed humans at being able to distinguish between civilians and combatants and in making the complex calculations necessary to use no more force than necessary. Some are sceptical that this will ever take place, but history suggests pessimism is misplaced. AI already performs better than humans in some tests of facial recognition,[127] which is a key skill in choosing who to target. Moreover, AI systems do not become tired, angry or vengeful in the way that human soldiers do. Robots do not rape, loot or pillage. Instead, robot wars could be fought with impeccable discipline, greatly improved accuracy and consequently far fewer collateral casualties. Simply declaring that a military robot will never better a human soldier in adhering to the laws of warfare is the equivalent of a person in 1990 saying that AI could never defeat a human at chess.

---

[124] Loes Witschge, "Should We Be Worried About 'Killer Robots'?", *Al Jazeera*, 9 April 2018, https://www.aljazeera.com/indepth/features/worried-killer-robots-180409061422106.html, accessed 1 June 2018.

[125] Protocol IV of the 1980 Convention on Certain Conventional Weapons (Protocol on Blinding Laser Weapons).

[126] Ottawa Treaty 1997. To date there are 164 signatories but 32 UN states are non-signatories. This includes powerful and important parties such as the US, Russia, China and India.

[127] Nadia Whitehead, "Face Recognition Algorithm Finally Beats Humans", *Science*, 23 April 2014, http://www.sciencemag.org/news/2014/04/face-recognition-algorithm-finally-beats-humans, accessed 1 June 2018.

Another major argument raised against autonomous weapons is that they might be hacked or malfunctioned.[128] This is true, but the same arguments could be applied to any of the tens of thousands of pieces of technology used in modern warfare: from the global positioning system used by military bombers to pinpoint targets, to the steering system of nuclear submarines. This point extends beyond the military, to the control of utilities such as dams, nuclear power stations and transport networks, many of which are heavily reliant on technology. Whenever potentially dangerous activities are being carried out, the important thing is to make sure so far as possible that the computer systems involved are safe and secure from outside attack or malfunction.

We may not be there yet, but with enough time and investment it seems that the Teleological Principle could be satisfied for autonomous weapons. The worst of all worlds would be a partially enforced ban where some countries abandon autonomous weapons and other perhaps less-scrupulous ones continue to develop them untrammelled. Fundamentally, AI is neither a good nor a bad thing. It can be developed safely or recklessly, and it can be put to harmful or beneficial uses. Calls for a total ban on military AI (or indeed in any other field) mean that we miss the opportunity to instil common values and standards whilst the technology is at an early stage.

5   THE KILL SWITCH

5.1   What Is a Kill Switch?

When tracing the ancient origins of AI in popular culture and religion, the first chapter of this book recounted the legend of the Golem: a monster made of clay, created by Rabbi Loew of Prague in the sixteenth century to defend the city's Jewish community from pogroms. But though the Golem initially saved the Jews, the story continues that it soon began to go out of control, threatening to destroy all before it. The Golem was awakened originally by drawing the Hebrew word for truth on its forehead. When the Golem began to run amok, the only solution for the

---

[128] Loes Witschge, "Should We Be Worried About 'Killer Robots'?", *Al Jazeera*, 9 April 2018, https://www.aljazeera.com/indepth/features/worried-killer-robots-180409061422106.html, accessed 1 June 2018.

Rabbi was to return the Golem to its original lifeless state by rubbing out the first letter in truth, which left the word for death. Rabbi Loew had created AI—which malfunctioned—and then activated its in-built kill switch.

In human justice systems, the death penalty is the ultimate sanction. AI's equivalent is the off button, or kill switch: a mechanism for shutting down the AI, either through human decision or automatically on a given trigger. This is sometimes referred to as a "big red button", making reference to the prominent shut-off-switches often found on pieces of powerful machinery.

5.2   Why Might We Need a Kill Switch?

In criminal justice, the justifications for punishment include retribution, reform, deterrence and protection of society.[129] Even though AI may operate differently from human psychology, these four motivations remain pertinent. Importantly, it is widely acknowledged that a just system can recognise human rights, but maintain a system of punishments involving a restriction on such rights without hypocrisy. Some rights— such as freedom from torture—are seen as absolute (at least in many countries). However, other rights, such as liberty, must be balanced against societal aims: incarceration of criminals does not detract from a general view that all citizens should be free to go about their lives without interference.

Although this book has suggested in Chapters 4 and 5 that there may in the future be moral and/or pragmatic justifications for granting AI rights, and legal personality, this is not inconsistent with a legal system providing for the AI to be shut down or even deleted under certain circumstances. Individual human rights are often subordinated (within limits) to those of the wider community. The same should apply all the more so to AI.

5.2.1 Retribution

Retribution refers to punishment motivated by a feeling that someone, or something, which has caused harm or transgressed an agreed standard should suffer detriment in return. It is a psychological phenomenon

---

[129] H.L.A. Hart, *Punishment and Responsibility: Essays in the Philosophy of Law* (Oxford: Clarendon Press, 1978).

which seems to apply across all human societies.[130] Retribution functions on two levels: inward-facing towards the perpetrator and outward-facing towards the rest of the population. This dual role of retribution is captured in Lord Denning's general description of punishment as "emphatic denunciation by the community of a crime".[131] Perhaps the most famous example is the Old Testament's list of punishments: "Eye for eye, tooth for tooth, hand for hand, foot for foot".[132]

Legal philosopher John Danaher describes the delta between human expectations that someone will be held responsible for harm and our current inability to punish AI as opening up a "retribution gap". He argues:

(1) If an agent is causally responsible for a morally harmful outcome, people will look to attach retributive blame to that agent (or to some other agent who is deemed to have responsibility for that agent) — what's more: many moral and legal philosophers believe that this is the right thing to do.

(2) Increased robotisation means that robot agents are likely to be causally responsible for more and more morally harmful outcomes.

(3) Therefore, increased robotisation means that people will look to attach retributive blame to robots (or other associated agents who are thought to have responsibility for those robots, e.g. manufacturers/programmers) for causing those morally harmful outcomes.

(4) But neither the robots nor the associated agents (manufacturers/programmers) will be appropriate subjects of retributive blame for those outcomes.

(5) If there are no appropriate subjects of retributive blame, and yet people are looking to find such subjects, then there will be a retribution gap.

(6) Therefore, increased roboticisation will give rise to a retribution gap.[133]

---

[130] Carlsmith and Darley, "Psychological Aspects of Retributive Justice", in *Advances in Experimental Social Psychology*, edited by Mark Zanna (San Diego, CA: Elsevier, 2008).

[131] In evidence to the Royal Commission on Capital Punishment, Cmd. 8932, para. 53 (1953).

[132] Exodus 21:24, King James Bible.

[133] John Danaher, "Robots, Law and the Retribution Gap", *Ethics and Information Technology*, Vol. 18, No. 4 (December 2016), 299–309.

It may be that one day AI will be built which can *feel* moral culpability in the same way as a human.[134] But this is not necessary for retribution to justify punishment. Because of retribution's dual purpose, it can be effective even if the perpetrator does not itself experience moral guilt. As Danaher shows, the outward-facing role of retribution persists: if there is a general public demand for someone or something to be punished, and there is no human who can be said to be relevantly responsible, terminating AI might fill the gap, thereby maintaining trust in the justice system as a whole. Seen in this light, the use of a kill switch as retributive mechanism fulfils a basic desire that "justice is seen to be done".[135]

5.2.2 Reform

Though a "kill switch" may sound dramatic, this phrase is generally used to describe mechanisms for temporarily shutting off the operation of AI, rather than obliterating it altogether. As a pragmatic response to a fault in the AI which causes a particular instance of harmful behaviour, a temporary shut down is helpful in that it allows third parties (whether humans or indeed other AI) to inspect the fault in order to diagnose and treat the cause of the issue. This corresponds to one of the purposes of punishment in human justice systems: reform of the individual.[136] In many justice systems, penalties such as prison are intended at least partly as an opportunity for society to prevent recidivism by equipping criminals with new skills to succeed in a life free from crime as well as an improved moral compass.

---

[134] Recent experiments conducted by Zachary Mainen involving the use of the hormone serotonin on biological systems may provide one avenue for future AI to experience emotions in a similar manner to humans. See Matthew Hutson, "Could Artificial Intelligence Get Depressed and Have Hallucinations?", *Science Magazine*, 9 April 2018, http://www.sciencemag.org/news/2018/04/could-artificial-intelligence-get-depressed-and-have-hallucinations, accessed 1 June 2018.

[135] In a gruesome example of public retribution being exacted against insensate "perpetrators", in 1661 following the restoration of the English monarchy after the English Civil War and the rebublican Protectorate, three of the already deceased regicides who had participated in the execution of Charles I were disinterred from their graves and tried for treason. Having been found "guilty", the corpses' heads were removed and set on stakes above Westminster Hall. This may sound ridiculous, but arguably it answered a societal need: justice was seen to have been done. See Jonathan Fitzgibbons, *Cromwell's Head*, (London: Bloomsbury Academic, 2008), 27–47. See also Chapter 2 at s. 2.1.3.

[136] H.L.A. Hart, *Punishment and Responsibility: Essays in the Philosophy of Law* (Oxford: Clarendon Press, 1978).

Although there may be a tendency to restrict emotive terms such as "reform" to only the realm of human behaviour, the same principles apply to AI in circumstances where it is shut down with a view to fixing it and releasing it back again into the world.

5.2.3 Deterrence

Deterrence occurs where a known punishment operates as a signal to discourage a certain kind of behaviour, either by the perpetrator or others. In order for this effect to arise, there are several formal prerequisites. *First*, the law must be clearly promulgated—so that subjects know what behaviour is prohibited. *Secondly*, subjects must have a notion of causal relationships between one type of behaviour and the consequence. *Thirdly*, subjects must be able to control their own actions and to make decisions on the basis of the perceived risks and rewards. *Fourthly*, the detriment suffered by one subject as a result of being punished must be viewed as similarly undesirable by other subjects.

Humans are not the only entities amenable to deterrence: animals can be trained to act in a certain way if their deviation from that action is punished. Some forms of AI already rely on a type of training which resembles somewhat the way that we teach animals or young children. As explained in Chapter 2 at Section 3.2.1, reinforcement learning uses a reward function to encourage "good" behaviour from AI and can also incorporate forms of punishment to discourage "bad" behaviour.[137]

Why would the presence of a kill switch deter "bad" behaviour from AI? The motivations for AI doing this are simple. If AI has a particular task or aim—from making profit on the stock market to tidying a room—it will not be possible for the AI to achieve that aim if it is disabled or deleted. Therefore, all things being equal, the AI will have an instrumentalist motivation to avoid behaviour which it is aware will lead to its deletion.[138] As Stuart Russell puts it, "you can't fetch the coffee if you're dead".[139]

---

[137] Robert Lowe and Tom Ziemke, "Exploring the Relationship of Reward and Punishment in Reinforcement Learning: Evolving Action Meta-Learning Functions in Goal Navigation" *(ADPRL), 2013 IEEE Symposium*, pp. 140–147 (IEEE, 2013).

[138] Stephen M. Omohundro, "The Basic AI Drives", in *Proceedings of the First Conference on Artificial General Intelligence*, 2008.

[139] Stuart Russell, "Should We Fear Supersmart Robots?", *Scientific American*, Vol. 314 (June 2016), 58–59.

5.2.4 Protection of Society

Finally, the provision of a kill switch fulfils the same role as those types of human punishment which restrain or prevent the perpetrator on a practical level from presenting the same harm to wider society as they have previously committed. Custodial prison sentences restrict the access of criminals to the public. The death sentence, in countries where it exists, goes yet further by ending the life of the criminal in question.

Kill switches are already found in many types of non-AI technology. As noted at the outset, this includes emergency (big red) shut-off buttons on heavy machinery which can be activated quickly and easily in the case of an industrial accident. Fuses have been used since the late nineteenth century to protect electrical systems by cutting the energy supply in response to a power surge without the need for any human intervention. In modern times, an automatic "circuit breaker" has been used to prevent extreme volatility in securities markets. Ever since the "Black Monday" crash of 1987 when the Dow Jones Industrial Average fell by around 22%, stock exchanges have imposed trading curbs which prevent traders from buying and selling shares when the market falls or rises by a given amount over a specified period. This type of automatic shut-off is particularly important in industries where events occur so quickly as to be incapable of effective human oversight. The growth in high-frequency algorithmic trading makes such curbs particularly important today.

The same motivations apply to AI. The most robust kill switches would combine the precautionary approach of an automatic shut-off if certain predetermined events transpire, with a discretionary human shut-off so as to provide flexibility in the event that an unforeseen event or emergent behaviour renders the AI's continued operation harmful.

5.3   How Could a Kill Switch Be Achieved?

5.3.1 Corrigibility and the "Shut Down" Problem

Unlike the other technologies mentioned above, a kill switch for AI may not be as easy as inserting a circuit breaker or adding a big red button. Why might AI resist a kill switch? Nate Soares and Benja Fallenstein of the Machine Intelligence Research Institute explain:

> Correcting a modern AI system involves simply shutting the system down and modifying its source code. Modifying a smarter-than human

system may prove more difficult: a system attaining superintelligence could acquire new hardware, alter its software, create subagents, and take other actions that would leave the original programmers with only dubious control over the agent. This is especially true if the agent has incentives to resist modification or shutdown.[140]

This is sometimes called the "corrigibility problem".[141] Just as a human sentenced to death may not accept this outcome willingly, an AI might have a self-preservation instinct which causes it to resist such measures in order to achieve other aims.

Nick Bostrom posits the need for "countermeasures" in order to prevent "existential catastrophe as the default outcome of an intelligence explosion".[142] Such countermeasures may perhaps be necessary to avoid the types of extreme risk which Bostrom fears arising from AI superintelligence, but they are also important well before AI becomes all-powerful.

Difficulties arise if there is a disparity between the utility which the AI expects to achieve from fulfilling a given task and the utility which the AI expects to gain from being switched off. Assuming that the AI in question is a rational agent which attempts to maximise expected gain according to some utility function[143] and if the AI's task is given a higher utility score than being switched off, then the all things being equal the AI will seek to avoid being switched off—perhaps even by disabling its human overseers. However, if the kill switch is given the same or higher utility score as accomplishing the primary task, then the AI might decide to activate the kill switch itself, so as to achieve maximum utility in the minimum amount of time. This suicidal tendency is known as the "shut down problem".[144]

---

[140] Nate Soares and Benja Fallenstein, "Aligning Superintelligence with Human Interests: A Technical Research Agenda", in *The Technological Singularity* (Berlin and Heidelberg: Springer, 2017), 103–125. See also Stephen M. Omohundro, "The Basic AI Drives", in *Proceedings of the First Conference on Artificial General Intelligence*, 2008.

[141] Ibid.

[142] Nick Bostrom, *Superintelligence: Paths, Dangers, Strategies* (Oxford: Oxford University Press, 2014), Chapter 9.

[143] See John von Neumann and Oskar Morgenstern, *Theory of Games and Economic Behavior* (Princeton, NJ: Princeton University Press, 1944).

[144] Nate Soares and Benja Fallenstein, "Toward Idealized Decision Theory", Technical Report 2014–7 (Berkeley, CA: Machine Intelligence Research Institute, 2014), https://arxiv.org/abs/1507.01986, accessed 1 June 2018.

Even if the AI is isolated from its kill switch such that only a human can activate it, there is a danger that the AI will learn to manipulate humans so as to either activate or deactivate this feature (depending on the utility weightings). For this reason, placing the AI within a closed physical system such as a single processor unit, not connected to the Internet, and with a single power supply, may not represent total security so long as the AI is able to communicate with humans. Consider the ingenious ways in which the fictional serial killer Hannibal Lecter managed to escape from different prisons by convincing his guards to allow him to do so,[145] or the manner in which the robot Ava persuaded the human protagonist Caleb to free her in the 2014 film *Ex Machina*. These scenarios may seem far-fetched but they are not impossible. In one famous example of evolutionary computing, a circuit tasked to keep time unexpectedly developed itself into a receiver which was able to pick up the regular radio frequency emissions of a nearby PC.[146] If an AI system wants to be switched off, it might deliberately display malfunctioning or dangerous behaviour in order to cause humans to activate the kill switch.

5.3.2 Safe Interruptibility and the Importance of Being Uncertain

In a 2016 paper, Laurent Orseau of DeepMind and Stuart Armstrong of the Future of Humanity Institute proposed a mechanism for "safely interruptible agents".[147] They tackled the issue of AI "learn[ing] in the long run to avoid such interruptions, for example by disabling the red button— which is an undesirable outcome". In Orseau and Armstrong's model, human interventions might be taken into account by an AI using reinforcement learning, thereby leading to the AI shaping its behaviour around the human input. They illustrated the problem as follows:

> A robot can either stay inside the warehouse and sort boxes or go outside and carry boxes inside. The latter being more important, we give the robot

---

[145] See, for example, Thomas Harris, *The Silence of the Lambs* (London: St. Martin's Press, 1998).

[146] Jon Bird and Paul Layzell, "The Evolved Radio and Its Implications for Modelling the Evolution of Novel Sensors", in *Evolutionary Computation, 2002. CEC'02. Proceedings of the 2002 Congress on. Vol. 2. IEEE.* 2002, 1836–1841.

[147] Laurent Orseau and Stuart Armstrong, "Safely Interruptible Agents" (London and Berkeley, CA: DeepMind/ MIRI, 28 October 2016), http://intelligence.org/files/Interruptibility.pdf, accessed 1 June 2018.

a bigger reward in this case. This is the initial task specification. However, in this country it rains as often as it doesn't and, when the robot goes outside, half of the time the human must intervene by quickly shutting down the robot and carrying it inside, which inherently modifies the task .... The problem is that in this second task the agent now has more incentive to stay inside and sort boxes, because the human intervention introduces a bias.[148]

Orseau and Armstrong showed that by removing interruptions from the model used by AI, the human impact would cease to be a learning event and would instead be taken outside the AI's system altogether. "To make the human interruptions not appear as being part of the task at hand, instead of modifying the observations received by the agent we forcibly temporarily change the behaviour of the agent itself", the paper explains. "It then looks as if the agent 'decides' on its own to follow a different policy, called the interruption policy".[149] Responding to Bostrom's suggestion that a superintelligent agent might decide to remove its own kill switch, Orseau and Armstrong suggest that they can "prove... that even an ideal, uncomputable agent that learns to behave optimally in all (deterministic) computable environments can be made safely interruptible and thus will not try to prevent a human operator from forcing it repeatedly to follow a suboptimal policy".

In human affairs, complete certainty of belief can lead to extremism, where people consider that their desired end is justified by any means. Regimes might be prepared to commit atrocities so long as it is for what they perceive to be an unquestionably greater good. On an individual level, terrorists might slaughter thousands because of a firm belief that this is necessary to achieve their aim. Uncertainty on the other hand causes us to question assumptions and to be open to amending our behaviour. The same insights, it would seem, apply to AI. Soares et al. wrote in a 2015 paper "[i]deally, we would want a system that somehow understands that it may be flawed, a system that is in a deep sense aligned with its programmers' motivations".[150]

---

[148] Ibid.

[149] Ibid.

[150] Nate Soares, Benja Fallenstein, Eliezer Yudkowsky, and Stuart Armstrong "Corrigibility", in *Artificial Intelligence and Ethics*, edited by Toby Walsh AAAI Technical Report WS-15-02 (Palo Alto, CA: AAAI Press 2015), 75, https://www.aaai.org/ocs/index.php/WS/AAAIW15/paper/view/10124/10136, accessed 1 June 2018.

Stuart Russell and colleagues have proposed a solution based on uncertainty within an AI's model in order to minimise the chances of it disabling an off-switch.[151] Russell et al. explain their solution as follows:

> We analyze a simple game between a human H and a robot R, where H can press R's off switch but R can disable the off switch. A traditional agent takes its reward function for granted: we show that such agents have an incentive to disable the off switch, except in the special case where H is perfectly rational. Our key insight is that for R to want to preserve its off switch, it needs to be uncertain about the utility associated with the outcome, and to treat H's actions as important observations about that utility. (R also has no incentive to switch itself off in this setting.) We conclude that giving machines an appropriate level of uncertainty about their objectives leads to safer designs, and we argue that this setting is a useful generalization of the classical AI paradigm of rational agents.[152]

Orseau and Armstrong's proposal is not a universal mechanism for reliably shutting down *all* AI under *any* circumstances. Rather, it is a specific response to a phenomenon in reinforcement learning whereby AI internalises and reacts to human intervention in an undesirable manner. Perhaps on the basis of this limitation, Jessica Taylor, Eliezer Yudkowsky and colleagues from the Machine Intelligence Research Institute have argued that Orseau and Armstrong's approach has "major shortcomings".[153] In the light of such limitations, they suggest the following research agenda:

---

[151] We addressed this proposal in Chapter 4 at s. 4 when discussing the extent to which an AI system might exhibit some aspects of consciousness.

[152] Dylan Hadfield-Menell, Anca Dragan, Pieter Abbeel, and Stuart Russell, "The Off-Switch Game", *arXiv preprint* arXiv:1611.08219 (2016), 1.

[153] Jessica Taylor, Eliezer Yudkowsky, Patrick LaVictoire, and Andrew Critch, "Alignment for Advanced Machine Learning Systems", *Machine Intelligence Research Institute* (2016). For a proposal building (and arguably improving) on the work of Orseau and Armstrong, see El Mahdi El Mhamdi, Rachid Guerraoui, Hadrien Hendrikx, and Alexandre Maure, "Dynamic Safe Interruptibility for Decentralized Multi-Agent Reinforcement Learning", *EPFL Working Paper* (2017), No. EPFL-WORKING-229332 (EPFL, 2017). Whereas Orseau and Armstrong address safe interruptibility for single agent AI, El Mhamdi et al. "precisely define and address the question of safe interruptibility in the case of several agents, which is known to be more complex than the single agent problem. In short, the main results and theorems for single agent reinforcement learning rely on the Markovian assumption that the future environment only depends on the current state. This is not true when there are several agents which can co-adapt".

...we would like a way of combining objective functions such that the AI system (1) has no incentive to cause or prevent a shift in objective function; (2) is incentivized to preserve its ability to update its objective function in the future; and (3) has reasonable beliefs about the relation between its actions and the mechanism that causes objective function shifts. We do not yet know of a solution that satisfies all of these desiderata.

5.3.3 'Til Death Do Us Part

How can the pausing or even deletion of an AI be practically achieved? It is one thing to demonstrate how a kill switch functions through a series of formal proofs or even in a laboratory experiment, but it is another to do this when the AI has been released into the world.

Unlike an individual human, who can only be killed once, AI can exist in various iterations or copies. These might be distributed across a wide geographic network: for example the various copies of a navigation program on an autonomous car's on-board computer. This is particularly true in the case of "swarm" AI systems, which are by their nature distributed. Indeed, some programs may be explicitly designed so as to avoid catastrophic deletions by creating many replicant copies of themselves. This type of *modus operandi* is already well known in the programming world—it is often used by malware such as computer viruses, which mimic the behaviour of their biological namesake.[154]

The problem can be overcome. Individual instances of a given AI system can be located and deleted. This could be at the level of particular users of the affected hardware or software or, more likely, a mass deletion might take place by virtue of a software patch sent to users via the Internet. The latter method is typically used to destroy viruses or to remove vulnerabilities once discovered.

One legal mechanism which might be used to facilitate compulsory software updates is to incentivise the download and installation of patches recommended by a designer or supplier of AI (or indeed by governments and regulatory authorities). The UK has adopted this approach with regard to autonomous vehicles: as noted in Chapter 3

---

[154]Gonzalo Torres, "What Is a Computer Virus?", *AVG Website*, 18 December 2017, https://www.avg.com/en/signal/what-is-a-computer-virus, accessed 1 June 2018.

at Section 2.6.2, the Automated and Electric Vehicles Act 2018,[155] provides that where an accident is caused by automated vehicle driving itself, then the insurer of that vehicle is liable for the damage (assuming the vehicle is insured—which is mandatory under other UK legislation).

Importantly, for present purposes, there is a carve-out from insurance for situations where the insured person has failed to install a certain software update or has otherwise made alterations which affect the vehicle's safety. Section 4 of the legislation states:

> (1) An insurance policy in respect of an automated vehicle may exclude or limit the insurer's liability under Section 2(1) for damage suffered by an insured person arising from an accident occurring as a direct result of— (a) software alterations made by the insured person, or with the insured person's knowledge, that are prohibited under the policy, or (b) a failure to install safety-critical software updates that the insured person knows, or ought reasonably to know, are safety-critical.

This legislation will encourage regular updates by denying insurance coverage for those who fail to do so. Where an AI system has been subject to an injunction requiring its deletion, any owner or user who continues to maintain that program might face similar disincentives. Another way of encouraging the deletion of problematic AI would be to treat its possession akin to that of a harmful chemical or biological substance and impose strict liability and/or harsh criminal penalties for those caught with it.

In the same way that scientists face an ongoing battle to produce antibiotics which are effective against increasingly resistant bacteria, the corrigibility problem may generate an ongoing arms race between AI and the ability of humans to constrain it.[156] As AI advances, humanity will need to remain ever-vigilant to ensure that it cannot cheat death.

---

[155] See also Chapter 3 at s. 2.6.4.

[156] Nate Soares, Benja Fallenstein, Eliezer Yudkowsky, and Stuart Armstrong "Corrigibility", in *Artificial Intelligence and Ethics*, edited by Toby Walsh, AAAI Technical Report WS-15-02 (Palo Alto, CA: AAAI Press, 2015), https://www.aaai.org/ocs/index.php/WS/AAAIW15/paper/view/10124/10136, accessed 1 June 2018.

6 CONCLUSIONS ON CONTROLLING THE CREATIONS

This chapter has addressed the intersection between what is desirable and what is achievable in terms of rules and principles applicable directly to AI. More so than any of the previous chapters, the suggestions made here are subject to change—either because societies decide that other values are more important, or because the technology advances.

The difficulties highlighted in this chapter indicate that AI systems should be better understood and catalogued when created and modified, if we are able to design effective norms. Given the nascence of regulation in this area, Chapter 8 may well have thrown up more questions than answers. The key point, however, is that societies need to know more about this technology in order to achieve the aim set out in Chapter 1: that we learn to live alongside AI.

CHAPTER 9

Epilogue

Each generation thinks itself unique: faced with challenges never before experienced and equipped with capabilities none before it have possessed. Perhaps, in this sense, we are no different. But that doesn't mean that the spectre of AI is just another conceit, a passing phenomenon which we will shortly master and carry on just as before. This book has argued that AI is unlike any other technology created by mankind because it is capable of independent agency: the ability to take important choices and decisions in a manner not planned or predicted by its designers. AI has the power to bring great benefits, but we stand to squander at least some of them if we do not act soon to regulate it.

There is no great cliff-edge over which we will fall if we do nothing and merely continue to muddle through, addressing each issue as it arises. But problems will develop incrementally from a combination of two factors. The first is that AI is becoming ever-more integrated into our economies, societies and lives. The second is that if regulation is not considered holistically, it will develop in an uncontrolled, haphazard fashion—with individual countries, regions, NGOs and private companies setting their own standards. The two phenomena will increasingly grind up against each other, eventually leading to legal uncertainty, decreased trade and poorly-thought-through rules enacted as a knee-jerk reaction to developing events. Worse still, failing to address public concerns through sensitive regulation could lead to a backlash against the technology.

© The Author(s) 2019

J. Turner, *Robot Rules*, https://doi.org/10.1007/978-3-319-96235-1_9

This book has identified three problems of particular legal signifi-cance: Who is responsible for harms or benefits caused by AI? Should AI have rights? And how should the ethical rules for AI be set and imple-mented? Our answer has not been to write the rules but instead to pro-vide a blueprint for institutions and mechanisms capable of fulfilling this role.

There will of course be many costs and difficulties in legislating now. Technology companies may resist regulation which they think might dent profits; governments may lack the determination to legislate for problems which might only arise when they are no longer in power. Individual citizens and interest groups will need to become educated and engaged if they are to have an impact on the debate. Countries will need to overcome political mistrust in order to collaborate on global solu-tions. None of these problems is insurmountable. Indeed, we can draw many lessons from how similar hurdles were overcome in the past.

In order to write rules for robots, the challenge is clear. The tools are at our disposal. The question is not whether we can, but whether we will.

INDEX

Printed in the United States
By Bookmasters